PROTEINASE INHIBITORS

PROTEINASE INHIBITORS

Medical and Biological Aspects

Edited by
Nobuhiko Katunuma, Hamao Umezawa, and Helmut Holzer

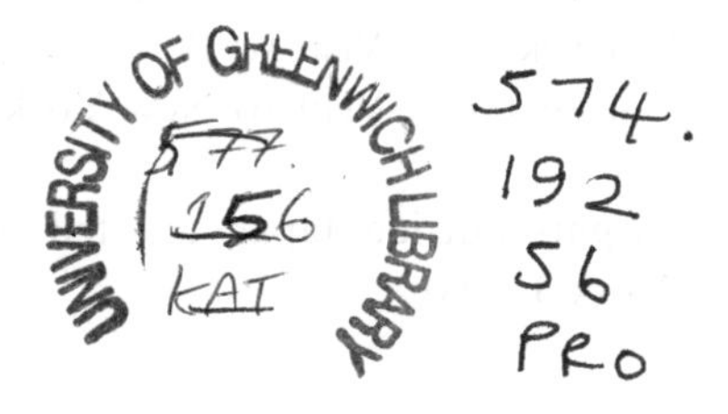

JAPAN SCIENTIFIC SOCIETIES PRESS Tokyo
SPRINGER-VERLAG Berlin Heidelberg New York Tokyo

1983

Published jointly by
JAPAN SCIENTIFIC SOCIETIES PRESS Tokyo Tokyo
ISBN 4-7622-6371-0
and
SPRINGER-VERLAG Berlin Heidelberg New York Tokyo
ISBN 3-540-12770-4 SPRINGER-VERLAG Berlin Heidelberg New York Tokyo
ISBN 0-387-12770-4 SPRINGER-VERLAG New York Heidelberg Berlin Tokyo

Sole distribution rights outside Japan granted to SPRINGER-VERLAG Berlin Heidelberg New York Tokyo

Printed in Japan

Preface

I have been studying antibiotics for the last 35 years, and whenever I report a new antibiotic, I receive many letters requesting a sample. Even more requests came to my office when I reported a new inhibitor of a protease. At present, I receive several letters each week requesting leupeptin, pepstatin, phosphoramidon, *etc.*, which indicated that many biochemists and biologists are involved in the study of proteases.

In the last 20 years, in the study of molecular biology great progress has been made, especially in the structure, function, and biosynthesis of DNA, RNA, and protein; and at the present, the study of proteases is increasing its importance. Moreover, proteases appear to be involved in all kinds of biological phenomena and disease processes. It is said that the number of researchers involved in protease studies is now increasing.

It seems that, in general, corresponding to each protease, there exists an inhibitor. It is certain that the study of proteases and their inhibitors is absolutely necessary for an understanding of biological functions and disease processes.

At present, the structures of these proteins are rapidly determined, and their structures can be displayed on a computer screen, being generated from data-base center. Even at this time, low molecular weight inhibitors are contributing to the identification of proteases and their functions. Therefore, I think it was very fortunate that this book included not only the study of endogenous protease inhibitors but also of low molecular weight inhibitors. Also I am pleased that possible applications of these inhibitors for the treatment of diseases are discussed in this book. I should like to thank all contributors to this book, and I trust it will surely stimulate us to continue the search for new types of enzyme inhibitors.

July 1983

H. Umezawa

Hamao Umezawa

Editorial Note

During the past decade the importance of proteolytic processes in the regulation of intracellular protein metabolism and the post-translational processing of precursor proteins and peptide hormones has been recognized. Various pathological phenomena caused by disorders of these proteolytic processes have also been described. Because of this, increasing attention is being given to the rapid development of biochemical bases and applications of proteinase inhibitors.

Proteinase inhibitors can be classified into those which are synthetic and those occurring naturally, the latter obtained from microbial, plant, animal and other sources. From a biological aspect, these proteinase inhibitors can be separated into endogenous and exogenous. Synthetic and microbial proteinase inhibitors serve as exogenous inhibitors not only for the basic biochemistry and biology of proteolysis but also for various pathological phenomena and clinical application toward certain diseases. Proteinase inhibitors of animal origins play an important role in regulation of the self-defense mechanism, blood coagulation, intracellular protein catabolism and processing of precursor proteins. These inhibitors or their active fragments, if obtained, may find medical application in the future.

This monograph is a compilation of a substantial portion of the recent knowledge gained in the medical and biological aspects of proteinase inhibitors. It is comprised of three sections. The biochemical and chemical bases of a number of microbial, plant and synthetic proteinase inhibitors are described in the first section. Their basic and clinical applications are also discussed. The second section includes topics on endogenous proteinase inhibitors from mammalian tissues and plasma. The role of these inhibitors in the regulation of proteinase activities and physiological functions is described. In the third section the biological and pathological significance of intracellular proteinases is detailed, and the role of these proteinases in intracellular protein degradation, defense mechanism and processing of precursor proteins is discussed. Thus the publication offers information on the most recent advances in this particular area of science in the fields of biochemistry, biology, and medicine.

On behalf of editors, I wish to cordially thank the authors for their cooperation and submission of manuscripts, and I also acknowledge Dr. P.C. Heinrich (Universität Freiburg) and Dr. E. Kominami (Tokushima University) for their assistance in the editing,

July 1983

Nobuhiko Katunuma

Contents

Preface . . . . *v*
Editorial Note . . . . *vii*

Microbial, Plant and Synthetic Inhibitors

A) Enzymological Aspects
Trends in Research of Low Molecular Weight Protease Inhibitors of Microbial Origin . . . . *H. Umezawa and T. Aoyagi* 3
Absorption, Distribution, Metabolism, and Excretion of Leupeptin . . . . *W. Tanaka* 17
Characterization of the Three New Analogs of E-64 and Their Therapeutic Application . . . . . *K. Hanada, M. Tamai, T. Adachi, K. Oguma, K. Kashiwagi, S. Ohmura, E. Kominami, T. Towatari, and N. Katunuma* 25
Synthetic Inhibitors of Proteases in Blood . . . *S. Fujii, Y. Hitomi, Y. Sakai, N. Ikari, M. Hirado, and M. Niinobe* 37
Legume Protease Inhibitors: Inhibition Mechanism of Peanut Protease Inhibitors . . . . *T. Ikenaka and S. Norioka* 45
Relationship between the Amino Acid Sequence and Inhibitory Activity of Protein Inhibitors of Proteinases . . . . *M. Laskowski, Jr., M. Tashiro, M.W. Empie, S.J. Park, I. Kato, W. Ardelt, and M. Wieczorek* 55

B) Medical Aspects
Ca-activated Neutral Protease (CANP) and Its Inhibitors in Pathological States . . . . *H. Sugita, S. Ishiura, and I. Nonaka* 69
Inhibition of Sister Chromatid Exchange and Mitogenesis by Microbial Proteinase Inhibitors . . . . *K. Umezawa* 77
Proteinases and Their Inhibitors in Inflammation: Basic Concepts and Clinical Implication . . . . *M. Jochum, K.-H. Duswald, S. Neumann, J. Witte, H. Fritz, and U. Seemüller* 85
An Investigation of Intracellular Proteinases during Differentiation of Cultured Muscle Cells . . . . *F.J. Roisen, H. Kirschke, R. Colella, L. Wood, A.C.St. John, E. Fekete, Q-S. Li, G. Yorke, and J. W.C. Bird* 97

Endogenous Proteinase Inhibitors

A) Thiol Proteinase Inhibitors
Cysteine Proteinase Inhibitors in Mammalian Plasma . . *J.F. Lenney* 113
Lysosomal Cysteine Proteinases and Their Protein Inhibitors—Structural Studies . . . . *V. Turk, J. Brzin, M. Kopitar, I. Kregar, P. Ločnikar, M. Longer, T. Popović, A. Ritonja, Lj. Vitale, W. Machleidt, T. Giraldi, and G. Sava* 125
Structure, Function, and Regulation of Endogenous Thiol Proteinase Inhibitor . . . . *N. Katunuma N. Wakamatsu, K. Takio, K. Titani, and E. Kominami* 135
Intracellular Proteinases Catalyzing Limited Proteolysis and Their Endogenous Inhibitors
. . . . *S. Pontremoli, E. Melloni, and B.L. Horecker* 147
Calpastatin, an Endogenous Inhibitor Protein Acting Specifically on Calpain . . . . . . *T. Murachi, E. Takano, and K. Tanaka* 165
Calcium Activated Neutral Protease (CANP) and Its Exogenous and Endogenous Inhibitors
. . . . *K. Imahori, K. Suzuki, and S. Kawashima* 173

B) The Other Proteinase Inhibitors
Regulation of Proteinases in *Saccharomyces cerevisiae* . . *H. Holzer* 181
Biochemical Aspects of an Inhibitor Protein from Uterine Myometrium . . . . *E.-G. Afting* 191
Naturally-Occurring Inhibitors of Aspartic Proteinases
. . . . *J. Kay, M.J. Valler, and B.M. Dunn* 201

Human Leukocyte Collagenase and Regulation of Activity
. . . . *H. Tschesche and H.W. Macartney* 211
Intracellular Proteinases and Inhibitors Associated with the Hemolymph Coagulation System of Horseshoe Crabs (*Tachypleus tridentatus* and *Limulus polyphemus*) . . *T. Morita, T. Nakamura, S. Ohtsubo, S. Tanaka, T. Miyata, M. Hiranaga, and S. Iwanaga* 229
Rat Plasma Proteinase Inhibitors: The Biosynthesis of α_1-Proteinase Inhibitor and α_2-Macroglobulin . . . . *T. Andus, V. Gross, W. Northemann, T-Anh T-Thi, and P.C. Heinrich* 243
Renin, Prorenin, and Renin Inhibitor
. . . . *T. Inagami, K.S. Misono, T. Inagaki, and Y. Takii* 251

Regulation of Intracellular Proteinases

Bifunctional Activities and Possible Modes of Regulation of Some Lysosomal Cysteinyl Proteases . . . . *G. Kalnitsky, R. Chatterjee, H. Singh, M. Lones, and A. Paszkowski* 263
Role of Medullasin in the Defense Mechanism against Cancer Development . . . . *Y. Aoki* 275
Intracellular Protein Degradation: Studies on Transplanted Mitochondrial and Microinjected Cytosol Proteins
. . . . . *P.J. Evans, S.M. Russell, F. Doherty, and R.J. Mayer* 283
Accumulation of Autolysosomes in Hepatocytes
. . . . *K. Furuno, T. Ishikawa, and K. Kato* 293
Cotranslational Mode of Protein Translocation across Membranes
. . . . *M. Müller, P. Walter, R. Gilmore, and G. Blobel* 301
Proteolytic Processing of Enzyme Precursors by Liver and Adrenal Cortex Mitochondria
. . . . . . *T. Omura, A. Ito, Y. Okada, Y. Sagara, and H. Ono* 307

Subject Index 313
Author Index 315

MICROBIAL, PLANT AND SYNTHETIC INHIBITORS

Proteinase Inhibitors: Medical and Biological Aspects, eds. N. Katunuma et al., pp. 3–15 (1983).
Japan Sci. Soc. Press, Tokyo/Springer-Verlag, Berlin

Trends in Research of Low Molecular Weight Protease Inhibitors of Microbial Origin

Hamao UMEZAWA and Takaaki AOYAGI

*Institute of Microbial Chemistry**

Thirteen years have passed since we reported the first discovery of a low molecular weight protease inhibitor produced by streptomyces (*3*). In a continuation of such screening studies, we have additionally found inhibitors of various endopeptidases and exopeptidases. In order to be successful in the screening for enzyme inhibitors, it is necessary to use purified enzymes; or if they are not purified enough, then specific substrates must be used. Following the development of new highly purified proteases and/or new specific substrates, new types of protease inhibitors will be discovered in microbial culture filtrates, because, as shown by the study of antibiotics and enzyme inhibitors, microorganisms produce an almost unlimited number of compounds having various structures. In this paper, we will review studies on protease inhibitors, including those which we have recently found, and discuss their biological activities. We will also discuss their applications, especially those of inhibitors which bind to cell surfaces or membranes.

INHIBITORS OF ENDOPEPTIDASES

Leupeptin, inhibiting plasmin, trypsin, papain, cathepsin B, *etc.*; anti-

* 3-14-23 Kamiosaki, Shinagawa-ku, Tokyo 141, Japan.

pain, inhibiting trypsin, papain, cathepsin B, *etc.*; chymostatin, inhibiting chymotrypsins, papain, *etc.*; and elastatinal, inhibiting pancrease elastase have all been found from culture filtrates of streptomyces strains (Fig. 1) (*4, 16, 17, 19, 20*). All these inhibitors have an α-amino aldehyde group in the C-terminal part of their peptide molecules. The terminal aldehyde group is concerned with the specific binding to enzymes and binds to the hydroxyl or thiol group of serine or thiol proteinases.

We have studied the biosynthesis of leupeptin, acetyl-L-leucyl-L-leucyl-L-argininal and were successful in the isolation of all three enzymes which are involved in leupeptin synthesis (*14*). One of these enzymes is leucine acetyltransferase. The second is leupeptin acid synthetase, a multifunctional enzyme on which leupeptin acid (acetyl-L-leucyl-L-leucyl-L-arginine) is synthesized in medium containing acetyl-L-leucine, L-leucine, L-arginine, and ATP. Leupeptin acid is reduced to leupeptin by the third enzyme, leupeptin acid reductase which requires ATP and NADH. This enzyme is located in cell membranes.

An inhibitor recently found by Murao *et al.* (*21*) and designated as β-microbial alkaline protease inhibitor (β-MAPI) is also produced by streptomyces (Fig. 1). It has a very similar structure to antipain, the only difference being that β-MAPI contains an phenylalaninal group

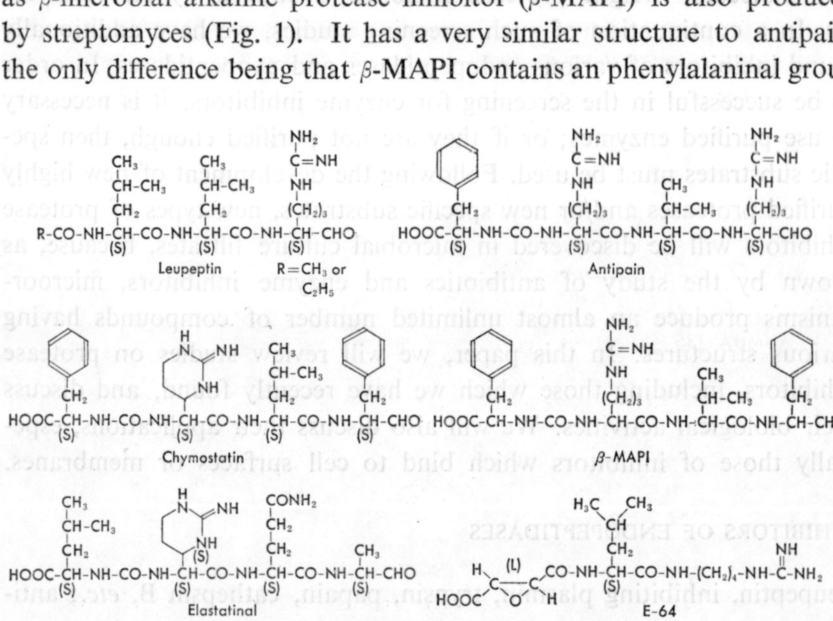

Fig. 1. Inhibitors of serine and thiol proteinases.

instead of the argininal group of antipain. As shown by this structure, it inhibits chymotrypsin. It also inhibits thiol proteinases. An inhibitor of thiol proteinases, E-64, has been found in culture filtrates of *Aspergillus japonicus*, and its structure was elucidated by Hanada *et al.* (Fig. 1) (*10*). The formation of a covalent bond between this inhibitor and a thiol group of thiol proteinases has been confirmed by Katunuma *et al.*

Pepstatin, pepstanone, and hydroxypepstatin have been found by our screening for inhibitors of pepsin (Fig. 2) (*4*, *16*, *17*, *19*, *20*). They inhibit pepsin, cathepsin D, and renin. Pepstatin, which has been used widely in biochemical studies, contains the isovaleryl group. Containing the lactoyl or benzoyl group instead of the isovaleryl group, pepstatin is more water-soluble than that containing the isovaleryl group and

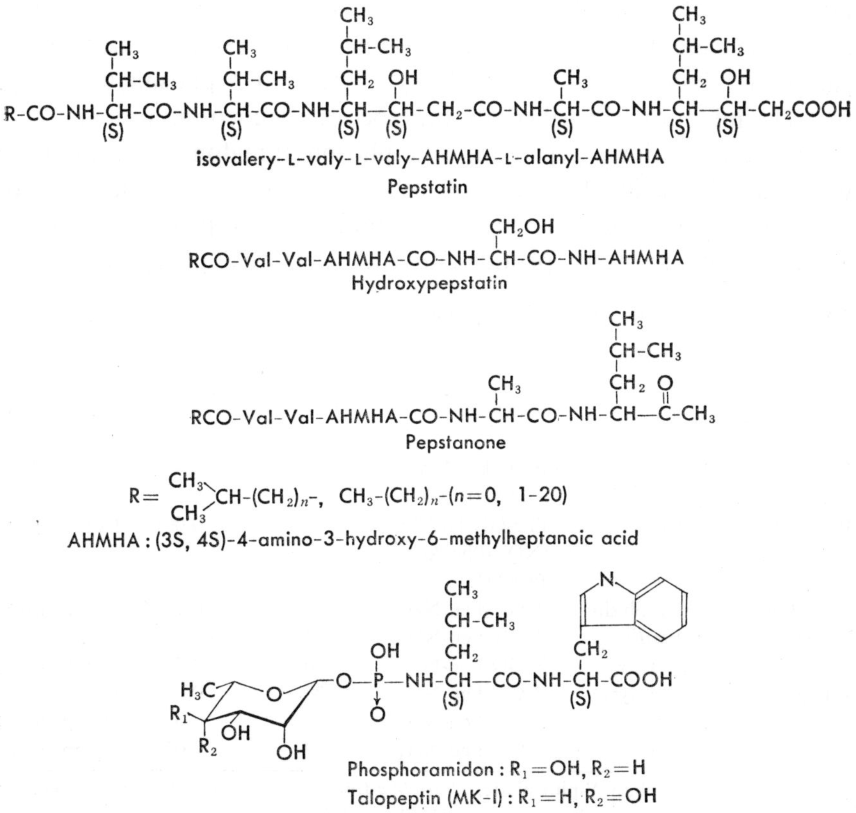

R-CO-NH-CH(CH-CH$_3$, CH$_3$)-CO-NH-CH(CH-CH$_3$, CH$_3$)-CO-NH-CH(CH$_2$-CH(CH$_3$)-CH$_3$)—CH(OH)-CH$_2$-CO-NH-CH(CH$_3$)-CO-NH-CH(CH$_2$-CH(CH$_3$)-CH$_3$)—CH(OH)-CH$_2$COOH
(S) (S) (S) (S) (S) (S) (S)

isovalery-L-valy-L-valy-AHMHA-L-alanyl-AHMHA

Pepstatin

RCO-Val-Val-AHMHA-CO-NH-CH(CH_2OH)-CO-NH-AHMHA

Hydroxypepstatin

RCO-Val-Val-AHMHA-CO-NH-CH(CH_3)-CO-NH-CH(CH_2-CH(CH_3)-CH_3)—C(=O)-CH_3

Pepstanone

R= (CH_3)(CH_3)>CH-(CH_2)$_n$-, CH_3-(CH_2)$_n$-(n=0, 1-20)

AHMHA : (3S, 4S)-4-amino-3-hydroxy-6-methylheptanoic acid

[sugar ring: H_3C, R_1, R_2, OH, OH, O]-O—P(OH)(→O)—NH-CH(CH_2-CH(CH_3)-CH_3)—CO-NH-CH(CH_2-[indolyl, N])-COOH
(S) (S)

Phosphoramidon : R_1=OH, R_2=H

Talopeptin (MK-I) : R_1=H, R_2=OH

Fig. 2. Inhibitors of carboxyl and metallo proteinases.

has the same degree of activity in inhibiting pepsin and cathepsin D. However, the former displays lower activity than the latter against renin. Corvol *et al.* have reported that the addition of aspartic acid or arginine to pepstatin increases its water solubility but does not decrease its renin-inhibiting activity.

Phosphoramidon is also produced by streptomyces and inhibits thermolysin (Fig. 2) (*4*, *16*, *17*, *19*, *20*). Hydrolysis of phosphoramidon gives leucyltryptophan *N*-phosphate. This hydrolysis product has a stronger activity than phosphoramidon itself. Knowledge of this inhibitor structure has stimulated the chemical synthesis of various analogs.

The structure of an inhibitor of metalloproteinases found by Murao *et al.* was recently elucidated, and this inhibitor was named "talopeptin" (Fig. 2). Structurally it is very similar to phosphoramidon. It contains 6-L-deoxytalose instead of the rhamnose moiety of phosphoramidon. As is the case for phosphoramidon, the structural part involved in enzyme inhibition should be the leucyltryptophan *N*-phosphate moiety.

The inhibitors described above, especially leupeptin, antipain, chymostatin, elastatinal, pepstatin, and phosphoramidon, have all been widely used for identification of proteases, analysis of the roles of proteases in biological functions, for inhibition of proteolysis during extraction of important peptides, and as functional groups in affinity

TABLE I. Kinetic Constants of Endopeptidase Inhibitors

Inhibitor	Enzyme	Substrate	K_m ($\times 10^{-4}$ M)	K_i ($\times 10^{-7}$ M)	Type of inhibition
Leupeptin	Trypsin	TAME[b]	7.1	3.4	Competitive
		BAEE[c]	4.8	1.3	Competitive
Elastatinal	Elastase	Ac-$(Ala)_3$-NA[d]	3.2	2.4	Competitive
		Ac-$(Ala)_3$-NA	4.3	2.1	Competitive
Pepstatin	Pepsin	Phe-Gly-His-Phe (NO_2)-Phe-Ala Phe-ME	0.4	0.001	Competitive
Phosphoramidon	Thermolysin	Z-Gly-Leu-NH_2[e]	20	0.28	Competitive
	B. subtilis	Z-Gly-Leu-NH_2	33	65.0	Competitive
	B. griseus	Z-Gly-Leu-NH_2	40	0.2	Competitive
P-Leu-Trp[a]	Thermolysin	Z-Gly-Leu-NH_2	20	0.02	Competitive
	B. subtilis	Z-Gly-Leu-NH_2	33	3.0	Competitive
	B. griseus	Z-Gly-Leu-NH_2	40	0.01	Competitive

[a] L-Leucyl-L-tryptophan *N*-phosphate. [b] *α-N*-(*p*-toluene sulfonyl)-L-arginine methyl ester HCl. [c] *α-N*-benzoyl-L-arginine ethyl ester HCl. [d] Acetyl-L-alanyl-L-alanyl-L-alanine *p*-nitroanilide. [e] Carbobenzoyl-glycyl-L-leucineamide.

chromatography. The K_i values of these inhibitors are shown in Table I. Renin was for the first time purified by pepstatin-affinity chromatography. Leupeptin has exhibited a protective effect against mouse muscular dystrophy (*13*), and its possible usefulness in the treatment of muscular dystrophy in man is being studied.

INHIBITORS OF EXOPEPTIDASES

We have also discovered inhibitors of exopeptidases in culture filtrates of streptomyces. As described later, inhibitors of aminopeptidases enhanced immune responses, and this effect stimulated interest in studies of inhibitors of these enzymes.

Bestatin inhibits aminopeptidase B, leucine aminopeptidase, alanine aminopeptidase, as well as tripeptidyl and tetrapeptidyl aminopeptidases. Amastatin inhibits aminopeptidase A, leucine aminopeptidase, tyrosine aminopeptidase, and tripeptidyl and tetrapeptidyl aminopeptidases. Their structures and K_i values are shown in Fig. 3 and Table II (*5*, *6*, *19*).

As reported for the first time in this paper, we have discovered two inhibitors which are completely specific for aminopeptidase B. We elucidated their structures and named them arphamenines A and B (Fig. 3). These were isolated from culture filtrates of a bacterial strain classified as *Chromobacterium violaceum*.

Arphamenines A and B inhibit aminopeptidase B very strongly, but not aminopeptidase A and leucine aminopeptidase. An interesting new structure for specific inhibition of aminopeptidase B was shown by the discovery of arphamenine. In the screening for aminopeptidase B inhibitors, we also found L-isoleucyl-L-arginine, L-leucyl-L-arginine, and L-valyl-L-arginine (Fig. 3).

Ebelactones A and B, which were found by screening for inhibitors of esterase, inhibited not only esterase but also formylmethionine aminopeptidase (Fig. 3) (*6*, *19*).

In the screening for inhibitors of dipeptidyl aminopeptidase IV, diprotins A and B were discovered (Fig. 3).

An inhibitor of carboxypeptidase A which we found in culture filtrates of a strain of actinomycetes and which we report for the first time in this paper has been identified as (S)-α-benzylmalic acid (Fig.

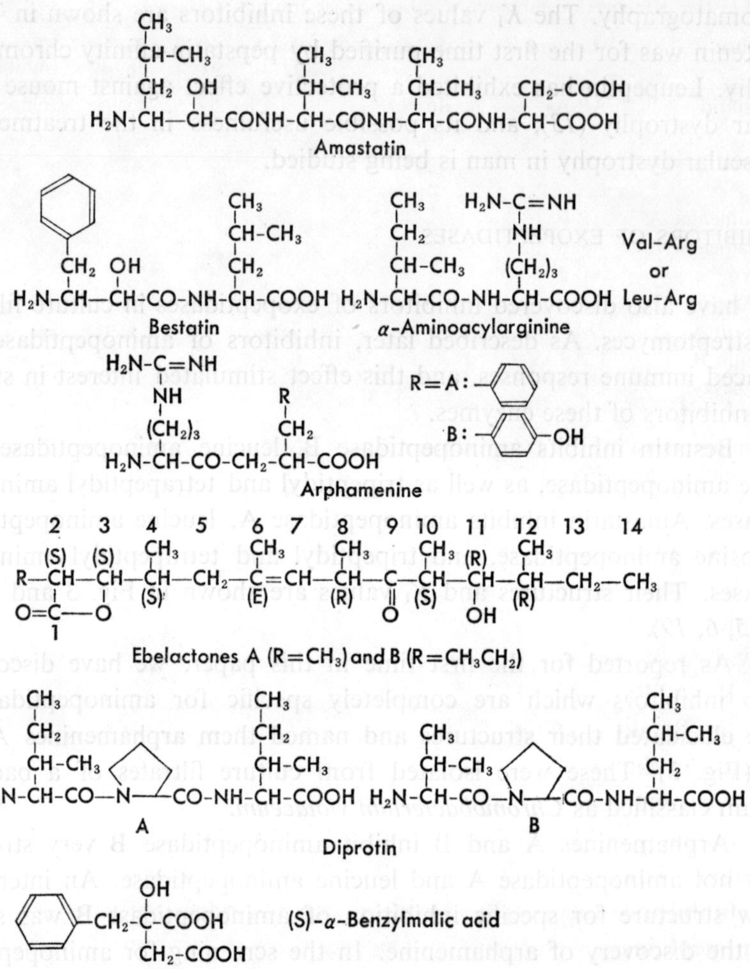

Amastatin

Bestatin

α-Aminoacylarginine (Val-Arg or Leu-Arg)

Arphamenine

Ebelactones A ($R=CH_3$) and B ($R=CH_3CH_2$)

Diprotin (A, B)

(S)-α-Benzylmalic acid

Fig. 3. Structures of exopeptidase inhibitors.

3). It is a specific inhibitor of carboxypeptidase A, and it also weakly inhibits carboxypeptidase B.

The K_i values of the above mentioned inhibitors are shown in Table II.

IMMUNITY-ENHANCING EFFECTS OF AMINOPEPTIDASE INHIBITORS

In 1972, we found that the administration of very small doses of cori-

TABLE II. Kinetic Constants of Aminopeptidase Inhibitors

Inhibitor	Enzyme	Substrate	K_m ($\times 10^{-4}$ M)	K_i ($\times 10^{-8}$ M)	Type of inhibitor
Amastatin	AP-A[a]	Glu-NA[d]	1	15	Competitive
	Leu-AP	Leu-NA	37	160	Competitive
Bestatin	AP-B	Arg-NA	1	6	Competitive
	Leu-AP	Leu-NA	5.8	2	Competitive
Arphamenine A	AP-B	Arg-NA	1	0.25	Competitive
Arphamenine B	AP-B	Arg-NA	1	0.08	Competitive
Ile-Arg	AP-B	Arg-NA	1	50	Competitive
Val-Arg	AP-B	Arg-NA	1	210	Competitive
Diprotin A	DAP-IV[b]	Gly-Pro-NA	2.3	350	Competitive
Diprotin B	DAP-IV	Gly-Pro-NA	2.3	3,000	Competitive
(S)-α-beyzyl-malic acid	CP-A[c]	Hip-Phe[e]	10	67	Competitive

[a] Aminopeptidase A. [b] Dipeptidylaminopeptidase IV. [c] Carboxypeptidase A. [d] L-Glutamic acid β-naphthylamide. [e] Hippuryl-L-phenylalanine.

olins A and C and diketocoriolin B increased the number of antibody-forming cells in mouse spleen (*11*). These substances are antitumor antibiotics produced by *Coriolus consor*, and diketocoriolin B is a derivative of coriolin B. We also found that diketocoriolin B inhibits Na^+-K^+-ATPase (*19*). Therefore, we thought that the binding of diketo-coriolin B and coriolins to ATPase in cell membranes resulted in an increase in the number of antibody-forming cells.

We assumed that the screening of compounds binding to cell membranes or cell surfaces would result in the finding of immuno-modifiers. Thus, we studied enzyme activities of intact cells and searched for inhibitors of enzymes located on cell surfaces. In this study, all aminopeptidases were found to be not only located in cells but also on their surfaces without being released extracellularly. Alkaline phos-phatase and esterase were also found to be located on cell surfaces (*1*, *2*). As described in a previous paragraph, we found inhibitors of aminopeptidases. We also found inhibitors of alkaline phosphatase and esterase. All these inhibitors enhanced immune responses except for esterastin, an esterase inhibitor which suppresses immune responses (*1*, *2*, *19*, *20*). Amastatin increased the number of antibody-forming cells in mouse spleen. Bestatin, arphamenines A and B, α-aminoacyl arginine, ebelactones A and B, and diprotins A and B enhanced delayed-type hypersensitivity as judged by the footpad test in mice. Forphenicine, L-(4-formyl-3-hydroxyphenyl)glycine, which inhibits alkaline phosphatase, enhanced both delayed-type hypersensitivity and antibody formation (*18*).

Arphamenines A and B and bestatin, all of which enhance delayed-type hypersensitivity, are different in their effects on lymphocyte blastogenesis induced by concanavalin A (Con A). Arphamenine A suppresses the blastogenesis induced by Con A, whereas bestatin does not.

Among inhibitors which enhance immune responses, bestatin has been studied in most detail. Experimental results from testing the effects of bestatin on the mouse immune system are summarized as follows:

1) Low doses of bestatin enhance delayed-type hypersensitivity and increase thymidine incorporation into spleen cells.

2) High doses of bestatin increase the number of mouse spleen cells producing antibody.

3) Bestatin has antitumor effects, enhances the effect of antitumor antibiotics, and promotes the production of interleukins 1 and 2.

4) Bestatin restores the decreased resistance against bacterial infections.

As reported by Müller (*12*), the administration of bestatin to mice induces an increase of ^{3}H-thymidine incorporation into the DNA of T-cells but not of B-cells. An increased activity of DNA polymerase α is observed in T-cells but not in B-cells, whereas DNA polymerase β activity is not influenced by bestatin administration. Terminal deoxynucleotidyl transferase activity in bone marrow is also increased by bestatin administration.

In clinical studies, the oral administration of 30 or 60 mg of bestatin daily has been confirmed to increase the reduced T-cell percentage. It increased the reduced NK-cell activity in cancer patients and improved the bone marrow picture of cancer patients (*18*). The suppressor activity of cancer patient serum in inhibiting IgG formation is nullified by bestatin both *in vitro* and *in vivo*.

All the effects of bestatin on the immune system seem to be due not to the inhibition of aminopeptidases but to its binding to cells involved in immune responses.

THE ACTION OF LOW MOLECULAR WEIGHT ENZYME INHIBITORS ON MURINE MUSCULAR DYSTROPHY

There have been many studies reporting an increase of endopeptidase activities in muscles of muscular dystrophy patients or animals. Leupeptin has been reported by Stracher *et al.* (*13*) to exhibit a preventive effect against genetic muscular dystrophy in mice. We observed that not only endopeptidases but also aminopeptidases increased in dystrophy muscles (*7–9*). Therefore, we asked Prof. Matsushita of the Wakayama Medical School to test the effects of bestatin and forphenicinol on murine muscular dystrophy. Both bestatin (200 μg/mouse intraperitoneally, daily starting 2 weeks after the birth) and forphenicinol (500 μg/mouse) showed a marked therapeutic effect. Forphenicinol, L-(3-hydroxy-4-hydroxymethylphenyl)glycine, is a derivative of forphenicine. It does not inhibit alkaline phosphatase, but binds to cells and enhances delayed-type hypersensitivity. Therefore, the effect of these compounds in preventing muscular dystrophy was suggested

TABLE III. Low Molecular Weight Inhibitor-induced Changes in Various Enzymatic Activities in Forelimb Muscles of Dystrophic Mice

Enzyme	Specific activity±S.D.			
	None	Leupeptin	Bestatin	Forphenicinol
AP-A	1.29± 0.22	0.95± 0.60	1.26± 0.22	1.28± 0.06
AP-B	10.28± 1.61	8.39± 2.29	9.22± 1.14	9.08± 1.42
Pro-AP	2.97± 1.75	2.42± 0.49	1.45± 0.43	2.54± 0.85
Leu-AP	8.13± 0.90	5.68± 3.45	7.67± 0.74	7.58± 0.12
fMet-AP	5.33± 0.40	6.34± 1.41	5.17± 0.67	4.60± 0.07
Phe-AP	13.17± 0.91	10.87± 5.32	10.93± 2.31	13.59± 0.58
Gly-Pro-Leu-AP	1.00± 0.02	0.94± 0.48	0.97± 0.18	1.03± 0.15
Trypsin-like	86.62± 5.20	102.07±47.93	54.42±21.40	55.11±13.43
Chy-Try-like[a]	100.97±17.47	134.26±25.32	63.53±23.28	90.06± 5.96
Elastase-like	56.35± 7.78	48.74±18.50	45.76±18.82	54.11± 7.22
Cathepsin C	1.19± 1.01	2.37± 0.49	0.99± 0.19	1.38± 0.11
α-D-Glucosidase	1.04± 0.42	0.74± 0.16	0.40± 0.04	0.43± 0.00
α-D-Mannosidase	0.16± 0.28	0.31± 0.28	0.14± 0.03	0.19± 0.05
Glc-NH_2ase[b]	2.67± 0.91	2.22± 1.07	1.22± 0.25	1.13± 0.14
CK[c]	22.51± 4.76	19.75±10.43	16.05± 2.74	19.27± 3.48
Phosphatase	6.27± 1.58	1.67± 1.23	1.46± 0.20	1.42± 0.19
Esterase	206.20±79.30	67.62±16.80	62.57± 2.85	57.31± 4.81

[a] Chymotrypsin-like. [b] *N*-acetyl-β-D-glucosaminidase. [c] Creatine kinase.

to be due to their indirect effects on the biosynthesis or degradation of various hydrolytic enzymes. Consequently, we tested the activities of hydrolytic enzymes after the administration of leupeptin (500 μg/mouse, 7 days), bestatin (200 μg/mouse, 7 days), or forphenicinol (500 μg/mouse, 7 days). As shown in Table III, all these compounds induced changes in the activities of proteases and other hydrolytic enzymes in forelimb muscles; that is, in general, the activities were lowered. Especially, phosphatase and esterase activities were markedly decreased. It seems that the therapeutic effect of these compounds, particularly of bestatin and forphenicinol, on murine muscular dystrophy is not only due to their direct action on enzymes sensitive to these compounds but also to the sequential changes induced by inhibition of enzymes or their binding to cell membranes (*7*).

CHANGES IN ENZYME PATTERNS INDUCED BY INHIBITORS OF SERINE AND THIOL PROTEINASES

Enzyme inhibitors are known to have a wide variety of biological actions *in vivo*. The effects on immunity and muscular dystrophy described above are examples. We analyzed the movements of enzyme networks triggered by the administration of leupeptin, pyroglutamyl-L-leucyl-L-argininal (Pyr-Leu-Argal), and dansyl-L-leucyl-L-argininal (Dan-Leu-Argal) which inhibit trypsin, plasmin, papain, and kallikrein. Pyr-Leu-Argal and Dan-Leu-Argal also inhibit urokinase or thrombin, respectively. The effects of other thiol proteinase inhibitors, E-64-C and its analogue, Ep-459 (supplied by Hanada *et al.*) (*15*), were also studied.

Activities of 17 enzymes in 6 organs (forelimb muscle, hindlimb muscle, heart, spleen, liver, and kidney) of mice were examined after injections of these inhibitors (500 μg/mouse, i.p., daily for 8 days). In forelimb muscles the activity of trypsin-like enzymes was significantly decreased by leupeptin but increased by Pyr-Leu-Argal and Dan-Leu-Argal. E-64-C and Ep-459 did not cause such changes. On the other hand, the activity of these enzymes in liver was significantly increased by leupeptin and E-64-C, whereas the others did not affect them significantly.

Changes in activities of enzymes which are not inhibited by the inhibitors *in vitro* also occurred after injection of these inhibitors, and the changes in these enzyme patterns were extensive. Activities of aminopeptidases, endopeptidases, glycosidases, phosphatase, and esterase changed variously depending on the organs. We also studied the time course of enzymatic activity changes in each organ, and the changes observed for 16 enzyme activities were subjected to multivariate analysis. Then the effects of leupeptin on the activity profiles of various enzymes were clearly separated from those of Pyr-Leu-Argal and Dan-Leu-Argal. The effects of the latter two were not separated from each other. E-64-C and Ep-459 showed similar effects which were clearly separated from those of leupeptin.

CONCLUSION

An unlimited number of genes involved in the biosynthesis of secondary metabolites are distributed among microorganisms in nature, and screening studies have resulted in the discovery of many inhibitors of various enzymes. It has been confirmed that the enzyme for the last step of leupeptin biosynthesis is inhibited strongly by leupeptin, and the biosynthesized leupeptin is rapidly released extracellularly. Low molecular weight protease inhibitors seem to play no role in the growth processes of the cells producing the inhibitors.

Low molecular weight protease inhibitors which we discovered have been widely used in the identification of proteases, the analysis of the roles of proteases in biological functions and disease processes, as functional groups for affinity chromatography, and for the protection of important peptides and proteins against hydrolysis during their extraction. We have also found inhibitors of exopeptidases such as bestatin and amastatin. In this paper, we reported arphamenines A and B inhibiting aminopeptidase B, and diprotins A and B inhibiting dipeptidylaminopeptidase IV. We also found (S)-α-benzylmalic acid inhibiting carboxypeptidase A.

Aminopeptidases are not only in cells but are also located on cell surfaces. They are not released extracellularly; thus, their inhibitors bind to the cells. Probably due to this binding, these inhibitors can modify immune responses. Moreover, both bestatin and forphenicinol showed a therapeutic effect against muscular dystrophy and induced changes in the activities of hydrolytic enzymes in the muscles of dystrophic mice.

We also studied the *in vivo* effects of leupeptin, Pyr-Leu-Argal, Dan-Leu-Argal, E-64-C, and Ep-459 on enzyme networks in various organs. They induced changes in activities of enzymes which were not susceptible to them *in vitro*. Multivariate analysis of the results indicated that the enzyme networks where the changes are induced by the inhibitors injected are not the same among all organs and differ among leupeptin, Pyr-Leu-Argal (and Dan-Leu-Argal), and E-64-C (and Ep-459).

REFERENCES

1. Aoyagi, T., Nagai, M., Iwabuchi, M., Liaw, W. S., Andoh, T., and Umezawa, H. *Cancer Res.*, **38**, 3505 (1978).
2. Aoyagi, T., Suda, H., Nagai, M., Ogawa, K., Suzuki, J., Takeuchi, T., and Umezawa, H. *Biochim. Biophys. Acta*, **452**, 131 (1976).
3. Aoyagi, T., Takeuchi, T., Matsuzaki, A., Kawamura, K., Kondo, S., Hamada, M., Maeda, K., and Umezawa, H. *J. Antibiot.*, **22**, 283 (1969).
4. Aoyagi, T. and Umezawa, H. *In* "Proteases and Biological Control," eds. E. Reich, D. B. Riffkin, and E. Shaw, p. 429 (1975). Cold Spring Harbor Laboratory, New York.
5. Aoyagi, T. and Umezawa, H. *In* "Industrial and Clinical Enzymology," eds. Lj Vitale and V. Simeon, p. 89 (1980). Pergamon Press, Oxford.
6. Aoyagi, T. and Umezawa, H. *Acta Biol. Med. Germ.*, **40**, 1523 (1981).
7. Aoyagi, T., Wada, T., Iwabuchi, M., Kojima, F., Nagai, M., and Umezawa, H. *Biochem. Int.*, **5**, 97 (1982).
8. Aoyagi, T., Wada, T., Kojima, F., Nagai, M., and Umezawa, H. *J. Clin. Invest.*, **67**, 51 (1981).
9. Aoyagi, T., Wada, T., and Umezawa, H. *In* "Muscular Dystrophy," ed. S. Ebashi, p. 239 (1980). Univ. Tokyo Press, Tokyo.
10. Hanada, K., Tamai, M., Ohmura, S., Sawada, J., Seki, T., and Tanaka, I. *Agric. Biol. Chem.*, **42**, 529 (1978).
11. Ishizuka, M., Iinuma, H., Takeuchi, T., and Umezawa, H. *J. Antibiot.*, **25**, 320 (1972).
12. Müller, W. E. G. *In* "Small Molecular Immunomodifiers of Microbial Origin," ed. H. Umezawa, p. 39 (1980). Japan Sci. Soc. Press, Tokyo/Pergamon Press, Cxford.
13. Stracher, A., McGowan, E. G., and Shafiq, S. A. *Science*, **200**, 50 (1979).
14. Suzukake, K., Hori, M., Hayashi, M., and Umezawa, H. *In* "Peptide Antibiotics-Biosynthesis and Functions," eds. H. Kleinkauf and H. V. Döhren, p. 325 (1982). Walter de Gruyter & Co., Berlin.
15. Tamai, M., Hanada, K., Adachi, T., Ogawa, K., Kashiwagi, K., Omura, S., and Ohzeki, M. *J. Biochem.*, **90**, 255 (1981).

16. Umezawa, H. *In* "Enzyme Inhibitors of Microbial Origin," p. 1 (1972). Univ. Tokyo Press, Tokyo.
17. Umezawa, H. *Methods Enzymol.*, **45**, 678 (1976).
18. Umezawa, H. *In* "Small Molecular Immunomodifiers of Microbial Origin," ed. H. Umezawa, p. 1 (1980). Japan Sci. Soc. Press, Tokyo/Pergamon Press, Oxford.
19. Umezawa, H. *Annu. Rev. Microbiol.*, **36**, 75 (1982).
20. Umezawa, H. and Aoyagi, T. *In* "Proteinases in Mammalian Cells and Tissues," ed. A. J. Barrett, p. 637 (1977). North-Holland, Amsterdam.
21. Watanabe, T., Fukuhara, K., and Murao, S. *Tetrahedron*, **38**, 1775 (1982).

Proteinase Inhibitors: Medical and Biological Aspects, eds. N. Katunuma et al., pp. 17–24 (1983).
Japan Sci. Soc. Press, Tokyo/Springer-Verlag, Berlin

Absorption, Distribution, Metabolism, and Excretion of Leupeptin

Wataru TANAKA

*Pharmaceuticals Division, Nippon Kayaku Co.**

The decrease of myofibrillar protein in dystrophic muscle was found to be attributable to the increase of protease (cathepsins (*1*), Ca-activated neutral protease (CANP) (*2*), and serine proteases (*3*)) activity in the muscle. Actually, leupeptin (*8*), which inhibits cathepsin B, CANP, and serine proteases, is effective in reducing the protein degradation in denerved rat and chicken muscle (*4, 7*), dystrophic mouse muscle (*4, 5*), and dystrophic chicken muscle (*6*). These results are encouraging for clinical study, but some of them are results of *in-vitro* study (*4*) and others are those of intramuscular application (*5–7*).

The problem is how to bring enough leupeptin to the target organ, the muscle. This is not easily resolved because leupeptin is chemically reactive and biodegradable.

LEUPEPTIN AND ITS METABOLITES IN BLOOD AND EXCRETA

[^{14}C]-leupeptin (acetyl-L-[^{14}C]-L-Leu-L-Argininal) was given to rats.

When intravenously administered urine was the main route of excretion. The radioactivity in the urine was found to be 90.4% as the

* 15-5 Ichibancho, Chiyoda-ku, Tokyo 102, Japan.

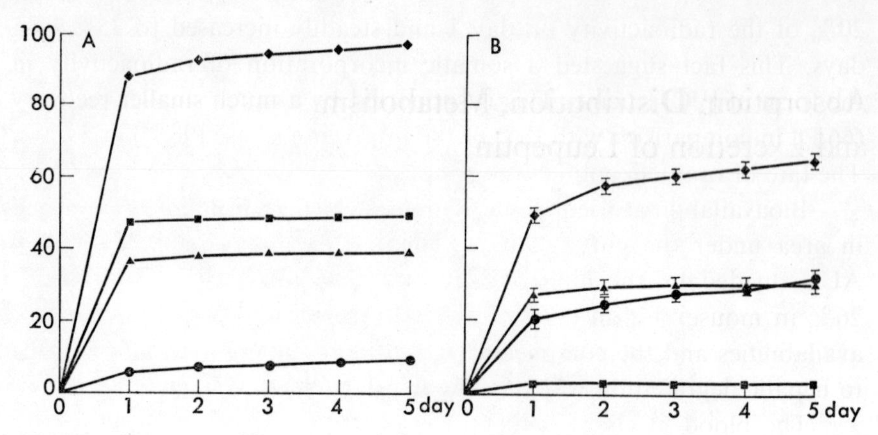

Fig. 1. Excretion of radioactivity in urine, feces, and expired air after intravenous (A) and oral (B) administration of [^{14}C]-leupeptin to rats.
◆ total; ■; urine; ▲ feces; ● expired air.

intact form, 3.5% as Ac-L-Leu-L-Leu, 1.4% as L-Leu, and 4.7% as unknown metabolites. The next excretion route was feces, which amounted to 37% on day 1. Biliary excretion of radioactivity of bile duct-cannulated rat was 21.2% at 0–3 hr, 2.5% at 3–6 hr, and 2.0% at 6–24 hr, following administration.

In oral administration of [^{14}C]-leupeptin, urine was the minor route of excretion while feces and expired air were the major ones. The radioactivity in the urine was only 27.6% as the intact form plus leupeptinic acid, 16.9% as Ac-L-Leu-L-Leu, 40.6% as L-Leu, and 15.0% as unknown metabolites. The radioactivity in the feces, a major excreta, reached 27% on day 1, and increased to 31% in 5 days. But the biliary excretion of bile duct-cannulated rat was only 0.74% at 0–3 hr, 0.14% at 3–6 hr, and 0.43% at 6–24 hr. So, the major part of the excreted radioactivity on day 1 might be an indication of the total which was intestinally unabsorbed. The urinary metabolite pattern of intravenous administration, although more than 40% is fed into the intestine as bile, was simple as mentioned above. The intestinal metabolism is thus assumed to produce little readily absorbable leupeptin fraction such as leucine. The intestinal absorption of leupeptin is, consequently, thought to be a little less than 73% (=100−27%), because absorption after intestinal metabolism is supposed to be small. The expired air shared

20% of the radioactivity on day 1 and steadily increased to 35% in 5 days. This fact suggested a somatic incorporation of radioactivity in the form of [^{14}C]-leucine. This is supported by a much smaller recovery (66%) in comparison with that of the intravenous case (98%) in 5 days. The fate of oral leupeptin is not simple.

Bioavailability of leupeptin in mouse, chicken or dog was expressed in area under the curve (AUC). The relative bioavailability, or oral AUC divided by the highest AUC, was low for oral administration: 26% in mouse, 4% in chicken, and 19% in dog. These low oral bioavailabilities and the complicated fate of oral leupeptin are attributable to hepatic degradation which is the first pass effect of leupeptin.

The blood levels of leupeptin orally administered to healthy volunteers and dystrophic patients in two hospitals were measured by a competitive binding assay method using [^{3}H]-leupeptin and trypsin.

The blood levels in the patients were much higher than those in the volunteers. In volunteers (adult, 100 mg/man or 1.6 mg/kg, n=5), patient group 1 (child, 100 mg equivalence by Ausberger formula, n=8), and patient group 2 (adult, 100 mg, n=7), AUC's were 0.145,

TABLE I. Bioavailability of Leupeptin in Mouse, Chicken, and Dog

Animal	Route of administration	Dose (mg/kg)	$t_{1/2}$(hr)	AUC (μg·hr/ml)	Maximum conc. of leupeptin (μg/ml (hr))	
Mouse	p.o.	30	1.6	4.3 (0.26)	1.6	(0.5)
	s.c.	30	0.3 (α) 2.6 (β)	16.8 (1.00)	1.47	(0.25)
	i.v.	30	0.2 (α) 1.7 (β)	10.9 (0.65)	12.8	(0.5)
Chicken	p.o.	30	1.2	0.6 (0.04)	0.23	(2.0)
	s.c.	10	2.0	6.7 (1.00)	3.13	(0.6)
	i.p.	10	0.2 (α) 2.3 (β)	3.4 (0.51)	2.44	(0.5)
	i.m.	10	0.7	1.4 (0.21)	1.05	(0.6)
Dog	i.v.	10	0.25 (α) 1.37 (β)	15.31(1.00)	21.05	(0.2)
	p.o.	10	1.38	2.76(0.18)	1.15	(0.5)
	Sublingual	10	1.29	3.09(0.20)	0.82	(2.0)
	Intra-rectal	10	1.40	3.96(0.26)	1.18	(0.5)
	Percutaneous	10	—	0.56(0.04)	0.03	(8.0)

[a] Leupeptin was measured by competitive binding assay.
p.o., *per os*; s.c., subcutaneous; i.v., intravenous, i.m., intramuscular, i.p., intraperitoneal.

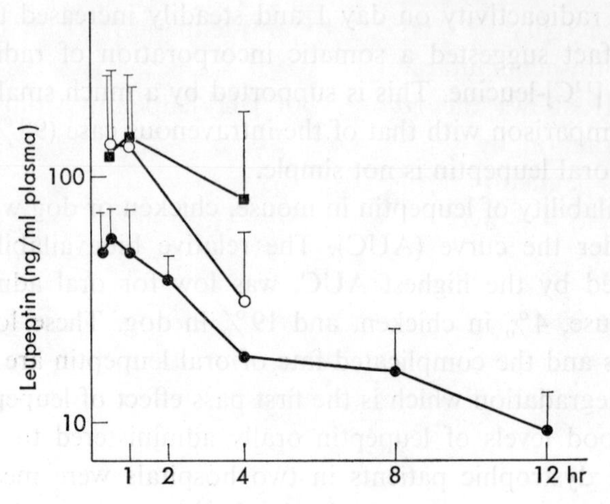

Fig. 2. Serum levels of leupeptin after oral administration.
● 100 mg/volunteer (healthy adult); ■ 100 mg/patient with muscular dystrophy (adult); ○ 28–44 mg/patient with muscular dystrophy (child).

0.350, and 0.439 μg•hr/ml, respectively. The values were comparable to AUC's in Table I, except that of chicken, which was exceptionally low. The maximum blood levels (456 ng/ml plasma in patient group 1 and 403 ng/ml plasma in patient group 2) were one order lower than the level, where the blood coagulation system begins to be hindered by the anti-thrombin effect of leupeptin (3–6 μg/ml plasma).

LEUPEPTIN IN THE TARGET ORGAN, THE MUSCLE

Leupeptin thus detected in the bloodstream is not at work. The leupeptin which exists in the muscle and binds to the target enzyme is that in the reserve and at work. The target enzyme is assumed to be some protease, such as cathepsins, CANP and/or serine proteases. For the purpose of measuring the bound leupeptin, a new method was developed. Its principle is: to purge the leupeptin from the bound protease with an excess of second inhibitor (*i.e.*, leupeptin analogue), and the released leupeptin is measured by RIA which is insensitive to the second inhibitor.

n-C_8H_{17}CO-L-Leu-L-argininal was selected as the second inhibitor, because of a low IC_{50} to papain (2.5×10^{-7} M), trypsin (1.8×10^{-6} M), kallikrein (1.3×10^{-5} M), and plasmin (2.7×10^{-6} M), and because of a low cross-reactivity (<0.005%) to anti-leupeptin serum (hapten: dihydroleupeptin-hemiglutarate).

Table II shows the recovery of [^{3}H]-leupeptin from a rat liver lysosomal fraction by several kinds of second inhibitors. Even 1,000 times excess leupeptin itself failed to achieve 100% recovery, while iodoacetic acid was avoided because it does not bind with serine proteases.

TABLE II. Effect of Protease Inhibitors on Binding of [^{3}H]-leupeptin to Liver Lysosomal Protease

Inhibitor	Added amount (μg)	Recovery (%)
None	0	41.0
Leupeptin	10	92.5
	100	92.0
	1,000	93.1
n-C_8H_{17}CO-L-Leu-L-Argal	10	90.8
	100	92.4
	1,000	92.7
Iodoacetic acid	10	78.2
	100	96.6
p-Aminobenzamidine	100	61.8

TABLE III. Total and Free Leupeptin in Normal and Denerved Muscle of Rat after Continuous Administration of Leupeptin by Osmotic Mini-pump at a Dose of 30 mg/kg/day

Rat No.		Leupeptin (ng/mg or ng/g)			(T)/(F)	D/N in (T)
		Plasma	Muscle			
		(T)	(F)	(T)		
351	Normal (N)	ND	ND	ND	—	>2.3
	Denerved (D)		ND	46.8	4.7	
352	Normal	642	39.6	42.9	1.1	5.02
	Denerved		41.9	215.5	5.1	
353	Normal	312	55.4	100.5	1.8	2.15
	Denerved		85.7	215.9	2.5	
354	Normal	369	38.7	282.6	7.3	0.67
	Denerved		38.6	189.9	4.9	
355	Normal	233	38.7	53.3	1.3	8.90
	Denerved		194.6	474.4	2.4	
356	Normal	255	20.3	55.9	2.8	(534.9)
	Denerved		(30.1 μg)	(29.9 μg)	(1.0)	

(F), free leupeptin; (T), total leupeptin; ND, plasma <15 ng, muscle <20 ng.

Akio Takagi of the National Center for Nervous, Mental and Muscular Disorders (Tokyo) used an osmotic mini-pump (Alzet model 2002) subcutaneously with leupeptin of 30 mg/kg/day for 7 or 12 days to sciatic-denerved rat. The decrease of hindlimb muscle weight at the denerved side was significantly reduced by leupeptin. Free and total leupeptin were measured as described above. Leupeptin in the muscle is higher at the denerved side than at the normal side. The leupeptin levels in the denerved side muscle (200 ng/g) gave a measure of the effective level.

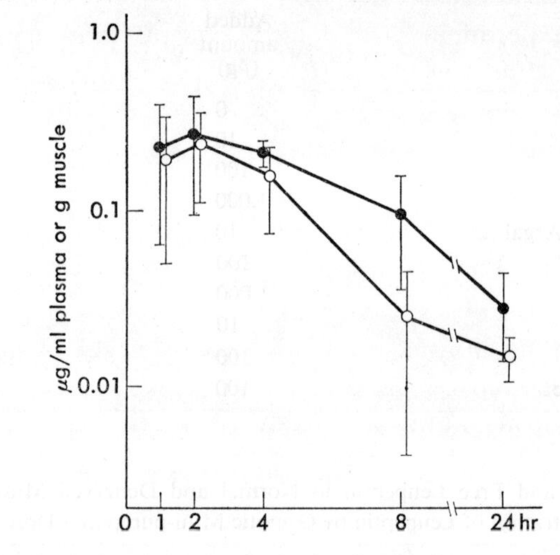

Fig. 3. Leupeptin levels in plasma and muscle after percutaneous administration of leupeptin ointment at a single dose of 30 mg/kg.

TABLE IV. Plasma and Muscle Concentration of Leupeptin 1 hr after Percutaneous Administration (30 mg/kg) to Three Muscular Dystrophic Mice (n=2–4)

Mouse	Age	Plasma	Muscle
		(μg/ml or μg/g)	
S	13w	8.3	10.7
R		2.3	3.3
S	12w	3.8	4.4
R		6.4	4.3
S	11w	9.3	2.5
R		2.1	2.7
S	10w	1.9	2.0
R		1.4	2.2

S, single administration; R, repeated administration for 1 month.

The next problem is how to attain a high enough leupeptin level in the muscle. A whole body adsorption auto-radiogram of laterally sliced mouse showed a marked accumulation of [^{14}C]-leupeptin in brain, muscle, intestine, cartilage and heart>spleen, liver and kidney>lung and fat. This fact suggested the affinity of leupeptin to the muscle. One trial is dermal application of leupeptin ointment. Rat hair was removed with Ebacream, the cream was washed off, and 18 hr later 20% leupeptin in vaseline-ointment was applied to 8 cm^2 of the hindleg at a dose of 30 mg/kg. The muscle levels, shown in Fig. 3, were high enough (300 ng/g) but did not last long enough.

A similar experiment was done with muscular dystrophic mice, C57BL/6Jdy(—), 10–13 weeks of age with symptoms already apparent. The hair was not removed in this experiment. The plasma and muscle levels were extremely high, and would not ascend in spite of repeated administration for 1 month. This was an unexpected result. The blood levels of human patients show a decreasing tendency by monthly repeated oral dosage. The symptoms of dystrophic mice were not reduced according to general observation and with a revolving cage. This result is not disappointing, however, because 10–13 weeks of age is too late for the cure of this type of mouse.

From a pharmacokinetical point of view, bringing a sufficient amount of leupeptin or a leupeptin-analogue to the muscle is thought to be of primary necessity. For this purpose, the first means is a topical application such as an osmotic mini-pump or ointment. The second means is an intravenous liposome of affinity to the muscle. And the third means is a leupeptin-analogue, which is indifferent to hepatic metabolism and is endowed with affinity to the muscle. These three means are all worth trying in the development of leupeptin as a new drug for muscular dystrophy.

APPENDIX: TOXICOLOGICAL STUDIES

Toxicological studies of leupeptin are briefly reviewed in Table V.

The cause of acute death was respiratory paralysis. A sympathetic ganglion blocking effect was observed in a general pharmacological study of leupeptin. Gallamine-immobilized cats were intravenously infused with 1, 3, and 10 mg/min/kg leupeptin for 10 min. At the highest

TABLE V. LD_{50} and Teratological Thresholds in mg/kg

		Mouse	Rat	Dog	Rabbit
Acute	p.o.	870	730	350	
	s.c.	570	1,130		
	i.v.	76	84	120	
5 weeks	p.o.		400	250	
	i.v.			>20	
6 month	p.o.		400		
Teratological study	p.o.		100–200		30–100

dose, the intervals of RR, QT and PR of ECG in lead II increased while blood pressure and heart rate decreased. Leupeptin blood levels, immediately after the end of infusion were, in these three doses, 32, 140, and 397 μg/ml, respectively. The cause of chronic death was intestinal ulcer followed by peritoneal perforation or severe hemorrhage. Other target organs were the liver and kidney. The function of the pancreas was enhanced and its weight increased. All the toxic damages were recoverable with cessation of the drug. Antigenicity (active and passive anaphylaxis tests, Schultz-Dale reaction, passive cutaneous anaphylaxis (PCA) reaction, Ouchterony's tanned red cell hemagglutination test, FDA method), irritative effect on eye mucous membrane and mutagenicity test (6 strains) were negative.

Acknowledgments

This paper is a review of work in our laboratory done by the New Drug Development Project Team led by Prof. H. Umezawa and sponsored by the Japanese Ministry of Health and Welfare.

REFERENCES

1. Iodice, A. A., Chin, J., Perker, S., and Weinstock, I. M. *Arch. Biochem. Biophys.*, **152**, 166 (1972).
2. Ishiura, S., Sugita, H., Nonaka, I., and Imahori, K. *J. Biochem.*, **87**, 343 (1980).
3. Katunuma, N., Yasogawa, N., Kito, K., Sanada, Y., Kawai, H., and Miyoshi, K. *J. Biochem.*, **83**, 625 (1978).
4. Libby, P. and Goldberg, A. L. *Science*, **199**, 534 (1978).
5. Sher, J. H., Stracher, A., Shafiq, S. A., and Hardy-Stashin, J. *Proc. Natl. Acad. Sci. U.S.*, **78**, 7742 (1981).
6. Stracher, A., McGowan, E. B., and Shafiq, S. A. *Science*, **200**, 50 (1978).
7. Stracher, A., McGowan, E. B., Hedrych, A., and Shafiq, S. A. *Exp. Neurol.*, **66**, 611 (1979).
8. Umezawa, H. *In* "Enzyme Inhibitors of Microbial Origin," p. 15 (1972). Univ. Tokyo Press, Tokyo.

Proteinase Inhibitors: Medical and Biological Aspects, eds. N. Katunuma et al., pp. 25–36 (1983).
Japan Sci. Soc. Press, Tokyo/Springer-Verlag, Berlin

Characterization of the Three New Analogs of E-64 and Their Therapeutic Application

Kazunori HANADA,[*1] Masaharu TAMAI,[*1] Takashi ADACHI,[*1] Kiyoshi OGUMA,[*1] Keiko KASHIWAGI,[*1] Sadafumi OHMURA,[*1] Eiki KOMINAMI,[*2] Takae TOWATARI,[*2] and Nobuhiko KATUNUMA[*2]

Research Center, Taisho Pharmaceutical Co. Ltd.[*1] *and Department of Enzyme Chemistry, Institute for Enzyme Research, School of Medicine, The University of Tokushima*[*2]

The direct role of a specific protease in protein breakdown in cell necrosis including muscular dystrophy, myocardial infarction and some kinds of myopathy, has not been well established. However, there seems to be little doubt that proteases play some key roles in the diseases (*3*, *15*). Studies from several laboratories have indicated that some cysteine proteases such as cathepsins B, L, and H and Ca^{2+}-activated neutral protease (CANP) are definitely suspect (*4*, *10*).

Our approach has been to search for specific inhibitors to suppress the increased activity of these enzymes in the pathological state and to ascertain the possibility of their practical use as therapeutic agents using model animal diseases. Leupeptin and E-64 (*5*), are both of microbial origin and are known as potent inhibitors of thiol proteases with low toxicity. Leupeptin, a peptide aldehyde derivative produced by Actinomycetes, binds tightly to the active site of various serine and thiol proteases (*24*). In contrast, E-64 has a strong affinity only for thiol proteases and is found to react irreversibly with a thiol group at the active site of papain or cathepsin B to form a thioether linkage (*6*, *8*). In this connection, we tried to synthesize a variety of E-64 de-

*1 1-403 Yoshino-cho, Ohmiya, Saitama 330, Japan.
*2 3-18-15 Kuramoto-cho, Tokushima 770, Japan.

TABLE I. Structure of E-64 and Its Analogs

General structure: epoxide ring C—C (bridged by O) bearing H on each carbon; one carbon bears R_1-OC, the other bears CO-NH-CH-CH_2-CH-CH_3, with CO-R_2 on the first CH and CH_3 on the second CH.

	R_1	R_2
E-64	HO-	-NH-$(CH_2)_4$-NH-C(=NH)-NH_2
Ep-459	HO-	-NH-$(CH_2)_4$-NH_2
Ep-475	HO-	-NH-CH_2-CH_2-CH(CH_3)-CH_3
Ep-453	EtO-	-NH-CH_2-CH_2-CH(CH_3)-CH_3

rivatives originally produced by *Aspergillus japonicus* (*5*), evaluated them by *in vivo* inhibitory activity, toxicity, bioavailability, distribution, and so on, and finally picked up three promising compounds, Ep-459, Ep-475, and Ep-453.

During the present study, we obtained new synthetic substrates specific for cathepsins B and L which were found to be very helpful in determining the activity of these cathepsins in tissue homogenate. The aim of this paper is to summarize our recent studies on E-64 analogs focused on their application for clinical use.

PREPARATION OF E-64 ANALOGS

During fundamental studies on the structure-activity relationships for the inhibition of papain (*6*, *22*, *23*), cathepsins (*2*, *8*), and CANP (*19*), the following analogs of E-64 were chosen as strong candidates and subjected to a further series of experiments. All were synthesized by the method previously reported (*23*).

IN VIVO INHIBITION OF CATHEPSINS B AND L

L-*trans*-Epoxysuccinic acid is the essential active component of E-64 (*7*). On the basis of this finding, we synthesized a variety of derivatives of the acid and studied the inhibition of cathepsins *in vitro* and *in vivo* (*2*, *8*). As with E-64 clarified by Kominami *et al.* (*14*), some analogs including Ep-459 and Ep-475 could be incorporated into lysosome of the liver cells and retarded the turnover rates of intracellular proteins in rats injected subcutaneously (*8*).

Suzuki and his coworkers reported in detail on the reactivity with CANP (*20*, *21*). Studies from several laboratories have indicated that lysosomal thiol proteases are involved in protein degradation in diseased muscles (*13*, *17*). Therefore, inhibition of such specific proteases affecting muscle breakdown would contribute to improvement of the pathological condition. This has been strongly supported by the recent studies by Goldberg and his coworkers in which Ep-475 reduced protein degradation significantly and improved net protein balance in the isolated rat diaphragm and soleus by inhibiting cathepsin B (*17*). They also found that denervation atrophy in rats was significantly suppressed by subcutaneous injection of Ep-475 at a dose of 20 mg/kg in the weight loss of plantaris, gastrocnemius and soleus muscle, accompanied by complete inhibition of cathepsin B (personal communication). However, the contribution of the other thiol proteases such as cathepsins L and H to the degradation system in muscle remains to be clarified. For an approach to the problem, a specific and simple method is required for assay of the activity of these enzymes

TABLE II. Effect of Cathepsins B, L, and H from Rat Liver on Various Amino-terminus Blocked Peptide 2-Naphtylamide Substrates

Substrate	m.p.(°C)	$[\alpha]_D$	Activity (μmol × min^{-1} × mg^{-1})		
			Cathepsin L	Cathepsin B	Cathepsin H
BANA			0.117	0.669	1.884
			(1)	(1)	(1)
Z-L-Arg-L-Arg-NA			0.495	1.500	0.010
			(4.23)	(2.24)	(0.00)
Suc-L-Tyr-L-Met-NA	220–221	+16.0	5.107	1.925	0.068
		(*c*=1,DMSO)	(43.65)	(2.88)	(0.04)
β-Ala-L-Tyr-L-Met-NA	188–190	+14.8	7.582	1.901	0.026
HCOOH		(*c*=1,DMSO)	(64.80)	(2.84)	(0.01)
D-Leu-L-Tyr-L-Met-NA	223–225	−30.4	7.583	1.246	0.020
HCOOH		(*c*=1,MeOH)	(64.81)	(1.86)	(0.01)

Activity was measured in 0.1 M potassium phosphate buffer, pH 6.0, according to method of Barrett[2a] Z-Arg-Arg-NA was added at 0.2 mM, Bz-D,L-Arg-NA (BANA) at 5 mM, and other substrates at 2 mM. Activity was calculated as mol of substrate hydrolyzed (× min^{-1} × mg protein^{-1}). Rates relative to that with BANA are shown in parentheses.

in various tissue extracts. In particular, a specific substrate for cathepsin L is required since this enzyme has very little activity on usual synthetic substrates. We studied the selective cleavage of the peptide bond of the peptide 2-naphthylamide substrates. As the synthetic peptide 2-naphthylamide substrates used in this study were synthesized after selective cleavage of the hexapeptide Leu-Trp-Met-Arg-Phe-Ala, all had methionine as a penultimate residue (*11*). Results showed that the 2-naphthylamide compounds with a blocked NH_2-terminus are specific and sensitive substrates for cathepsins L and B; they are not specific only for cathepsin L because they are also hydrolyzed by cathepsin B (Table II) (*12*).

However, these new substrates could offer an advantageous method of assay for total activity of cathepsins B and L (shown as cathepsin B & L hereafter) in various tissues homogenates because of their high sensitivity and specificity for both enzymes. Of these, Suc-Tyr-Met-NA was used for further experiments; the activity cleaving this substrate

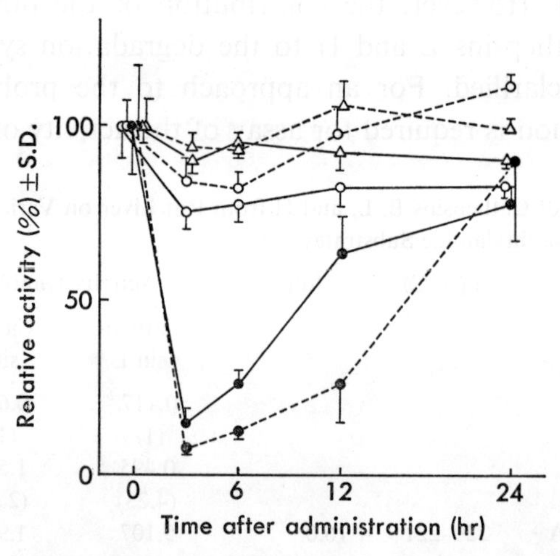

Fig. 1. Changes in lysosomal enzyme activity in rats after intraperitoneal administration of Ep-459.
Four male Wistar rats at the age of 7 weeks were used for each experiment. Animals were injected with 100 mg/kg of Ep-459 and killed at the times indicated and then their hearts and quadriceps femoris were removed. Cathepsin B & L (●), cathepsin D (○), and acid phosphatase (△) of these homogenates were determined at pH 5.0, 3.8, and 5.5 using Suc-L-Tyr-L-Met-NA, ^{14}C-hemoglobin and *p*-nitrophenyl phosphate as substrates, respectively. Each point shows relative activity to that of untreated control. —— hearts; ---- quadriceps femoris (mean ± S.D., *n*=4).

is the cathepsin B & L, as mentioned above. Inhibition of cathepsin B & L in heart and hindlimbs of rats by E-64 analogs *in vivo* was investigated with their homogenates 3 hr after intraperitoneal administration of 100 mg/kg of each. With single administrations of Ep-459, Ep-475, and Ep-453, the cathepsin B & L activity (% activity for untreated control expressed as mean$\pm$S.D. for 4 male Wistar rats per point) decreased to 8.2$\pm$1.1, 17.0$\pm$10.2, and 8.2$\pm$2.7 in hindlimb muscles and 14.8$\pm$1.1, 67.1$\pm$16.1, and 46.1$\pm$14.1% in heart, respectively.

Time courses of the change in cathepsin B & L activity in various rat tissues were followed after intraperitoneal administration of Ep-459 (Fig. 1). The activity fell steeply with its minimum at 3 hr after administration and then increased gradually. Activity returned to approximately 75% of the initial figure. It seems likely that this restoration results from the regeneration of the enzymes, since Ep-459 reacts irreversibly with both cathepsins B and L. This result shows a good coincidence with that reported in rat livers with E-64 by Hashida *et al.* (*9*).

TOXICITY

It is essential to deal with the problem of toxicity when considering medical application. Judging from the fact that the LD_{50} s of the above three compounds in rats and mice are over 2,000 mg/kg of body weight in the intraperitoneal, subcutaneous or peroral administration route, their acute toxicities seem extremely low. In addition, no serious finding was observed in any item such as behavior, body and tissue weight changes, hemolytical analysis of blood and urine, or histopathological observation of tissues routinely conducted for toxic effects even when 200 mg or 400 mg/kg of each compound was given successively for a month to rats and dogs.

ABSORPTION, DISTRIBUTION, AND METABOLISM

The time courses of blood levels of E-64, Ep-459, and Ep-475 in rabbits after subcutaneous injection were investigated (Fig. 2A), followed by further comparative studies on the administration route and species difference for Ep-475 (Fig. 2B, 2C). As shown in Fig. 2, it is note-

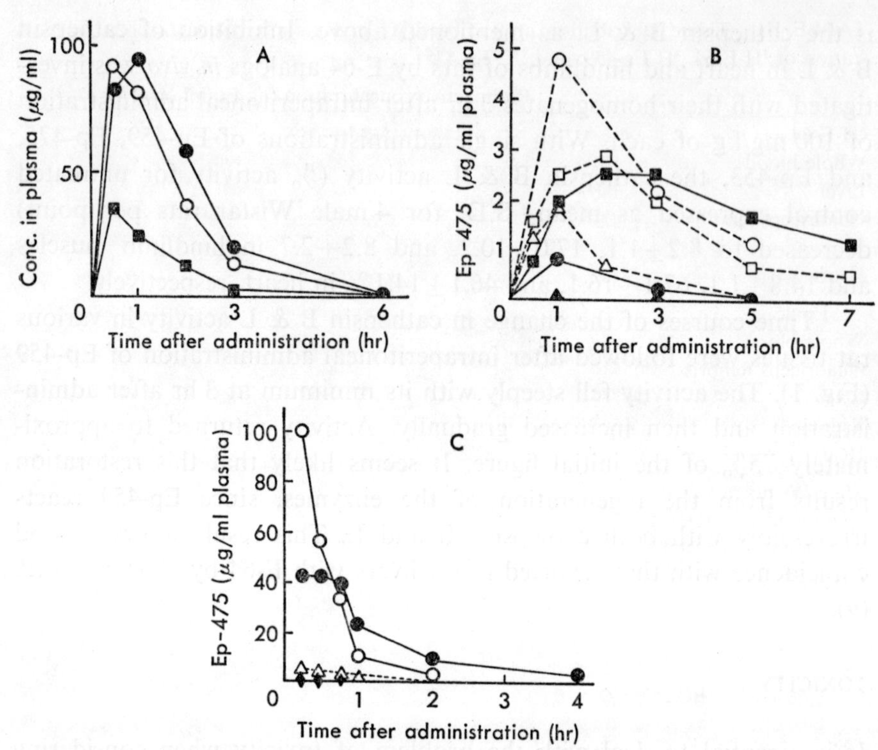

Fig. 2. Time-dependent profiles of the changes in plasma levels of E-64 and its analogs. A: comparison of plasma levels in rabbits after subcutaneous administration of E-64, Ep-459, and Ep-475 at the dose of 50 mg/kg for each. Each point shows the mean value for 2 animals. ● E-64; ○ Ep-459; ■ Ep-475. B: plasma levels of Ep-475 in rats, hamsters, and dogs after oral administration of Ep-475 and Ep-453 at a dose of 50 mg/kg/body weight for each. Both were given as a suspension with 5% gum arabic for rats and hamsters, and powder in capsule for dogs. Ep-475: ■ dogs (n=2); ▲ rats (n=4); ● hamsters (n=4). Ep-453: □ dogs (n=4); △ rats (n=3); ○ hamsters (n=3). C: comparison of Ep-475 levels in plasma of rabbits treated by various routes at a dose of 50 mg/kg body weight. ○ intravenous; ● subcutaneous; △ rectal; ◆ oral administration. In each experiment, determination was carried out by means of high pressure liquid chromatography (HPLC) methods.

worthy that there exist remarkable differences among animals in the blood level of Ep-475 when given orally, that is, there is an appreciable amount in dogs and hamsters, but little in rats.

Follow-up studies were done using ^{3}H-labelled compounds prepared by substitution with ^{3}H-gas. The time dependent radioactivities in the various tissues in rats after subcutaneous administration were

TABLE III. Tissue Distribution of Radioactivity in Rats after Subcutaneous Administration of ^{3}H-E-64, ^{3}H-Ep-459, and ^{14}C-Ep-475

	Radioactive concentration (ng Eq/ml or g)						
^{3}H-E-64	1/2	3	24 hr	^{3}H-Ep-459	1/2	3	24 hr
Whole blood	3,101	39	25	Whole blood	3,015	122	71
Liver	4,388	3,326	990	Liver	4,862	1,092	799
Quadriceps femoris	342	37	51	Quadriceps femoris	829	204	202
Heart	828	18	0				
^{14}C-Ep-45	1/6	3	24 hr	^{14}C-Ep-475	1/6	3	24 hr
Whole blood	1,937	110	163	Lung	1,679	171	170
Quadriceps femoris	610	91	88	Brain	116	58	60
Liver	16,785	2,220	1,435	Thymus	1,678	110	151
Kidney	19,368	9,358	7,463	Testis	1,065	85	94
Heart	1,653	78	99	Skin	1,188	186	230

Male Wistar rats at the age of 7 weeks were injected with 5 mg/kg body weight of labelled compounds and their tissues were removed at the times indicated. Samples were solubilized with Soluen-350 (Packard). Blood was decolored with 30% hydroperoxide to prevent interference before measurement. Values are averages from four animals.

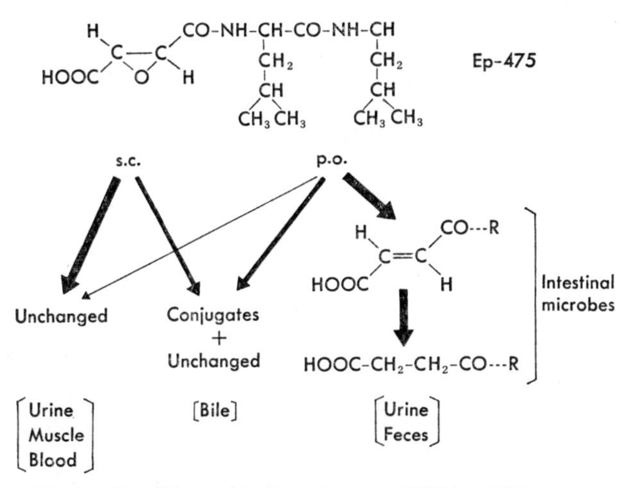

Fig. 3. Possible metabolic pathways of ^{14}C-Ep-475 in rats.

followed. More details were obtained with ^{14}C-Ep-475 labelled on its leucine moiety. There results were summarized in Table III. For the reason given above, we carefully focused attention on skeletal muscle. Although not shown here, microautoradiograms of sliced quadriceps femoris in rats indicated the presence of black grains within the cells, most of which were found to be derived from intact ^{14}C-Ep-475. In other words, the results show that Ep-475 can be delivered into skeletal muscle cells.

Figure 3 is a schematic diagram of the metabolic pathway of ^{14}C-Ep-475 in rats. About 50% of the ^{14}C-Ep-475 given subcutaneously was excreted into urine in an unchanged form. In contrast, unchanged Ep-475 found in urine and bile when given orally, was only 3.5% and 12.6%, respectively, showing a lower utilization in oral than in subcutaneous administration in rats. This seems to be partially due not only to its own properties but also to the cleavage of the oxirane ring of *trans*-epoxysuccinic acid by intestinal microbes, as shown above. To get a higher blood level of Ep-475 by *per os* administration, Ep-453, the ethylester of Ep-475, was subjected to investigation, since the ester form was lipophilic and expected to be readily absorbed through intestinal membrane. As expected, Ep-453 was absorbed well following oral administration and soon converted to Ep-475 by the action of esterases in the blood and liver (Fig. 2B). From the point of view of practical use, oral administration would be very advantageous. Therefore, Ep-453 is sure to be a strong candidate as a prodrug of Ep-475.

APPLICATION TO THE TREATMENT OF MUSCLE NECROSIS

One of our goals is to investigate the clinical effects of our inhibitors on certain muscle diseases. Of particular interest is the data by Jasmin and Proschek showing that both Ep-459 and Ep-475 administered by osmotic minipumps significantly impeded the development of the necrotic changes accompanied by steep decreases in myocardial Ca^{2+} content in cardiomyopathic hamsters UM-X 7.1. By treating with Ep-459 (3–6 mg/kg body weight) during the period from 28 through 58 days of age, the incidence of necrotic changes and Ca^{2+} content in the heart decreased from 100% to 33% and from 228 ± 63 to 44 ± 6 nM·Ca mg^{-1} protein, respectively (mean±S.E. for 45 untreated and 12 treated animals) (G. Jasmin and L. Proschek, unpublished data). This strongly suggests that some thiol proteases such as CANP and cathepsins are involved in the development of early myolytic heart changes and that these inhibitors are of interest in investigation of muscle wasting diseases.

TABLE IV. Effect of Ep-453 on Dystrophic Hamster BIO 14.6

	Normal (n=10)	Dystrophy	
		Control (n=10)	Ep-453 (n=8)
Plasma enzymes[a]			
GOT (K units)	31 ± 2.8	222 ±20.6	162 ±13.8
GPT (K units)	21 ± 3.5	93 ± 9.2	52 ± 7.1
LDH (mIU/ml)	201 ±24.0	441 ±40.2	366 ±20.2
CPK (IU/ml)	1.08± 0.269	14.9 ± 3.64	9.61± 1.48
PK (IU/ml)	0.59± 0.054	7.66± 4.55	15.8 ± 3.16
Tissue proteases			
Cathepsin B & L (units/g tissue)			
Quadriceps femoris muscles	4.43± 0.639	27.6 ± 3.38	4.15± 1.12
Heart muscles	9.48 ± 3.16	116 ±12.2	13.6 ± 3.46
Cathepsin D (10^3 units/g tissue)			
Quadriceps femoris muscles	14.5 ± 0.569	42.8 ± 3.07	27.4 ± 1.66
Heart muscles	89.1 ± 1.45	102 ± 1.49	112 ± 1.06
CANP (10^3 units/g tissue)			
Quadriceps femoris muscles	32.2 ± 1.45	36.6 ± 4.33	30.6 ± 3.35
Necrotic change (grade 0–4)			
Heart	0	2.2 ± 0.4	1.0 ± 0.2
Tibialis anterior muscles	0	1.9 ± 0.5	0.7 ± 0.3
Calcium deposition (grade 0–4)			
Heart	0	2.2 ± 0.4	1.4 ± 0.3

Dystrophic hamsters BIO 14.6 at the age of 8 weeks began receiving intraperitoneal injections of 75 mg/kg/day of Ep-453 and these were continued to the age of 15 weeks. Age-matched control and normal groups were injected with the vehicle alone. Twenty four hr after the final injection, hamsters were killed and their blood, heart, and quadriceps femoris were removed. Plasma enzymes were assayed with an HITACHI 712 autoanalyzer. Cathepsin B & L, CANP, and cathepsin D were measured at pH 5.0, 7.5, and 3.8 using Suc-L-Tyr-L-Met-NA, [^{14}C]methyl casein, and [^{14}C]hemoglobin as substrate, respectively. Sections from muscles and hearts were stained by glyoxal-bis(2-hydroxyanil) (GBHA) and hematoxylin-eosin for the observation of calcium deposition and necrotic changes, respectively.

[a] GOT, glutamic oxalo-acetic transaminase; GPT, glutamic pyruvic transaminase; LDH, lactate dehydrogenase; CPK, creatine phosphokinase; PK, pyruvate kinase.

We have reported that a marked increase in the activity of lysosomal thiol proteases and a considerable, but less marked, rise in the level of their endogeneous inhibitors was observed in dystrophic animals. In contrast, it is noticeable that there is no increase in other lysosomal hydrolases than proteases. This seems to imply a peculiar and characteristic feature in muscle damage (*13*). In this respect, we tried to apply Ep-453 for dystrophic hamster BIO 14.6. The findings show that Ep-453 administered intraperitoneally impeded the development of the necrotic changes in heart and tibialis anterior and the calcium deposition in heart, being consistent with the decrease in the corresponding enzyme activities in plasma and tissue (Table IV). On the other hand, *per os* administration would be a more suitable method for a chronic disease such as muscular dystrophy. We are now therefore trying to give Ep-453 mixed with feed to dystrophic hamsters.

An attempt to treat myocardial infarction is also in progress in collaboration with Shibata and Akagami. Myocardial infarction was produced in mongrel dogs weighing 8–12 kg by the method of Akagami *et al.* (*1*) and the hearts were transversely cut into five round slices of equal thickness 24 hr after ligation and subjected to the determination of the necrotic areas distinguished by phosphorylase staining. The necrotic percentage, which was calculated from total necrotic area/total area of five slices for each subject, was 36.9 ± 4.3 in the control group (mean$\pm$S.D. for six dogs). In the group treated with Ep-459, 10 mg/kg intraarterially just before ligation and 100 mg/kg intravenously after 15 hr, the corresponding figure was 28.0 ± 2.4 for five dogs. This is a 25.1% reduction in myocardial infarction in the animals treated. Student's *t*-test revealed that this difference was statistically significant ($p<0.05$). Not shown here, the degradation of myofiblar proteins checked by SDS-polyacrylamide gel electrophoresis and the leakage of creatine phosphokinase (CPK) from tissues decreased distinctly. These findings strongly suggest that thiol proteases play an important role in ischemic heart injury.

DISCUSSION

Because of recent progress in the clarification of biochemical features of intracellular mechanism for protein degradation, there has been growing interest in the possible use of protease inhibitors in the treatment of muscle wasting diseases. In fact, reports of pioneering studies in this area from some laboratories clearly demonstrate the beneficial effects of leupeptin, pepstatin, and Ep-475 in various types of muscle wasting (*16*, *18*). However, further studies based on pharmacology, biopharmacology, and toxicology could possibly lead to improved therapeutic procedures. The inhibitor for practical use in such treatment should (1) inhibit selectively specific proteases, (2) readily reach the targets, (3) be without toxic effects, and (4) be highly productive. In keeping with this approach, a variety of synthetic analogs of E-64 were investigated and Ep-459, Ep-453, and Ep-475 have recently drawn much attention. Our preliminary observations in the treatment of muscular dystrophy in hamsters and of myocardial infarction in dogs seem to indicate that thiol proteases play an important role in the necrotic process, and that reduction by inhibitors such as Ep-459 and Ep-453 was due to direct inhibition of the target enzyme involved. Although lysosomal cathepsins B and L and Ca^{2+}-protease are inevitably under suspiction, further detailed studies are needed to show which enzyme actually contributes to protein breakdown in a serial pathological process.

Another interesting subject is targeting, the selective delivery of the drugs to the target cells. The inhibitors discussed above do not necessarily fulfill this requirement. Therefore, to target a specific tissue, further study will be required. A specific liposome might be a useful tool and such an approach merits further study.

REFERENCES

1. Akagami, H., Yamagami, T., Shibata, N., and Toyama, S. *In* "Recent Advances in Studies on Cardiac Structure and Metabolism," Vol. 12, eds. T. Kobayashi, Y. Ito, and G. Roma, p. 445 (1978). Univ. Park Press, Baltimore.
2. Barrett, A. J., Kembhavi, A. A., Brown, M. A., Kirschke, H., Knight, C. G., Tamai, M., and Hanada, K. *Biochem. J.*, **201**, 189 (1982).

2a. Barret, A. J. *Annal. Biochem.*, **47**, 280 (1972).

3. Bird, J. W. C., Spanier, A. M., and Schwarz, W. N. *In* "Protein Turnover and Lysosomal Function," eds. H. L. Segal and D. J. Doyle, p. 589 (1978). Academic Press, New York.
4. Busch, W. A., Stromer, M. H., Goll, D. E., and Suzuki, A. *J. Cell Biol.*, **52**, 367 (1972).
5. Hanada, K., Tamai, M., Yamagishi, M., Ohmura, S., Sawada, J., and Tanaka, I. *Agric. Biol. Chem.*, **42**, 523 (1978).
6. Hanada, K., Tamai, M., Morimoto, S., Adachi, T., Oguma, K., Ohmura, S., and Ohzeki, M. *In* "Peptide Chemistry," ed. H. Yonehara, p. 31 (1980). Protein Research Foundation, Osaka.
7. Hanada, K., Tamai, M., Morimoto, S., Adachi, T., Ohmura, S., Sawada, J., and Tanaka, I. *Agric. Biol. Chem.*, **42**, 537 (1978).
8. Hashida, S., Towatari, T., Kominami, E., and Katunuma, N. *J. Biochem.*, **88**, 1805 (1980).
9. Hashida, S., Kominami, E., and Katunuma, N. *J. Biochem.*, **91**, 1373 (1982).
10. Katunuma, N. and Noda, T. *In* "Muscular Dystrophy," ed. S. Ebashi, p. 225 (1980). Univ. Tokyo Press, Tokyo.
11. Katunuma, N., Towatari, T., and Hashida, S. *In* "Proteinase and Their Inhibitors,"

eds. V. Turk and Lj. Vitale, p. 83 (1981). Mladinska Knjiga-Pergamon Press, Ljubljana.
12. Katunuma, N., Towatari, T., Tamai, T., and Hanada, K. *J. Biochem.*, in press.
13. Katunuma, N., Kominami, E., Noda, T., and Isogai, K. *In* "Muscular Dystrophy—Biomedical Aspects," eds. S. Ebashi and E. Ozawa, p. 237 (1983). Japan Sci. Soc. Press, Tokyo/Springer-Verlag, Berlin.
14. Kominami, E., Hashida, S., and Katunuma, N. *Biochem. Biophys. Res. Commun.*, **93**, 713 (1980).
15. Libby, P. and Goldberg, A. L. *In* "Degradative Processes in Heart and Skeletal Muscle," ed. K. Wildenthal, p. 201 (1980). Elsevier, Amsterdam.
16. Libby, P. and Goldberg, A. L. *Science*, **199**, 534 (1978).
17. Rodemann, H.P., Waxman, L., and Goldberg, A. L. *J. Biol. Chem.*, **257**, 8716 (1982).
18. Stracher, A., McGowan, E. B., and Shatiq, S. A. *Science*, **200**, 50 (1978).
19. Sugita, H., Ishiura, S., Suzuki, K., and Imahori, K. *J. Biochem.*, **87**, 339 (1980).
20. Sugita, H., Ishiura, S., Suzuki, K., and Imahori, K. *Muscle and Nerve*, **3**, 335 (1980).
21. Suzuki, K., Tsuji, S., and Ishiura, S. *FEBS Lett.*, **136**, 119 (1981).
22. Tamai, M., Adachi, T., Oguma, K., Morimoto, S., Hanada, K., Ohmura, S., and Ohzeki, M. *Agric. Biol. Chem.*, **45**, 675 (1981).
23. Tamai, M., Hanada, K., Adachi, T., Oguma, K., Kashiwagi, K., Ohmura, S., and Ohzeki, M. *J. Biochem.*, **90**, 255 (1981).
24. Umezawa, H. *Methods Enzymol.*, **45**, 678 (1976).

Proteinase Inhibitors: Medical and Biological Aspects, eds. N. Katunuma et al., pp. 37–44 (1983).
Japan Sci. Soc. Press, Tokyo/Springer-Verlag, Berlin

Synthetic Inhibitors of Proteases in Blood

Setsuro FUJII, Yuji HITOMI, Yuki SAKAI, Nobuhiko IKARI, Masayuki HIRADO, and Michio NIINOBE

*Division of Regulation of Macromolecular Function, Institute for Protein Research, Osaka University**

Several activation systems have been recognised in the blood which are essential for the survival of an animal. These include blood coagulation, fibrinolysis, kallikrein-kinin, and complement systems; all are dependent, at least in part, on the sequential activation of a series of proteolytic zymogens.

The complement system is intimately involved not only in host defence against infectious agents, but also in the pathogenesis of atuoimmune disease. Activation of the complement system can take place by two distinct pathways, the classical and the alternative. Both pathways have the same terminal components, C_5 to C_9, which are directly responsible for membrane damage. The difference between them lies in their early-acting components and thus in the manner by which they are activated by immune complexes or other factors, such as microbial cell wall or lipopolysaccharide.

There are many reports describing inhibitors of the complement systems; Table I lists many of these. In 1976 Doll and Baker (*2*) and in 1977 Hauptmann and Markwardt (*4*) reported that various benza-

* Yamada Oka 3-2, Suita, Osaka 565, Japan

TABLE I. Various Inhibitors on Complement-mediated Hemolysis

Reports		Compounds	Origin	ID_{50} (M) for hemolytic Act.		Site of inhibition
				Classical	Alternative	
Baker	1976	Benzamidine deriv. (Am–⟨C₆H₄⟩–R)	Synthetic	10^{-6}	ND	$C_{\bar{1r}}$, $C_{\bar{1s}}$
Markwardt	1977	Benzamidine deriv. (Am–⟨C₆H₄⟩–R)	Synthetic	7×10^{-5}	ND	$C_{\bar{1r}}$, $C_{\bar{1s}}$
Umezawa	1969	Leupeptin	Microbial	1.4×10^{-5}	1.0×10^{-3}	$C_{\bar{1r}}$, $C_{\bar{1s}}$, $\bar{D}$, $\bar{B}$
Inoue	1979	K-76	Microbial	2.3×10^{-4}	ND	C_5
Kaneko	1980	Complestatin	Microbial	ND	7.5×10^{-6}	B
Lukas	1981	Val-Gln-Val-His-Asn-Ala-Lys-Thr-Lys-Pro-Arg	Synthetic	7.7×10^{-5}	ND	Ag-Ab—C_{1q} (↓)
Lesavre	1982	Gln-Lys-Arg-Lys-Ile-Val	Synthetic	ND	1.0×10^{-4}	$\bar{D}$

ND, not determined.

midine derivatives, and then leupeptin (*1*), K-76 (*7*), and complestatin (*9*) were isolated from microbes. Recently, peptides (*10*, *11*) were reported as inhibitors of complement systems. However, the ID_{50} (concentration for 50% inhibition) value of these inhibitors for complement-mediated hemolysis was from 10^{-4} to 10^{-6} M. Recently, we reported (*3*, *8*) that 6-amidino-2-naphthyl 4-guanidinobenzoate·dimethane sulfonate (FUT-175) is a strong and reversible new synthetic inhibitor of proteases in complement systems and strongly inhibits complement-mediated hemolysis. Further, various immunological reactions *in vivo* were strongly suppressed by treatment with FUT-175 (*6*).

This paper presents a summary of these results.

EVALUATION PROCEDURES OF SYNTHETIC PROTEASE INHIBITOR

FUT-175 prepared in the Research Laboratories of Torii & Co., Ltd. Tokyo, Japan. Its structural formula is shown in Fig. 1. Purified human $C_{1\bar{r}}$, C_1 esterase, factor B, factor $\bar{D}$, C_3, and cobra venom factor (CVF) were prepared in our laboratory, and these components showed a single band of SDS-polyacrylamide gel electrophoresis (SDS-PAGE). Human plasmin and thrombin were purchased from the Green Cross Co., Japan. Porcine plasma and pancreatic kallikrein were obtained from the Research Laboratory of Ono Pharmaceutical Co., Japan. Bovine trypsin and α-chymotrypsin were purchased from Sigma Chemical Co., U.S.A. Cathepsins A, B, C, D, H, and L were partially purified from rat liver lysosome in our laboratory. Cathepsin G was partially purified from rat peritoneal polymorphonuclear leucocytes in our laboratory.

1. Experiments on Inhibition of Various Enzymes and Complement-mediated Hemolysis

The rate of hydrolysis of tosyl-arginine-methylester (TAME) by trypsin, plasmin, plasma kallikrein, pancreatic kallikrein, and thrombin, and of

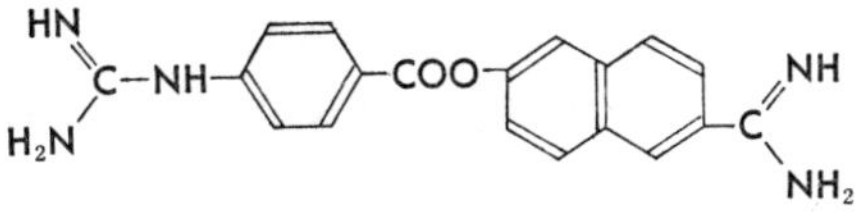

Fig. 1. Structural formula of FUT-175.

acetyl-arginine-methylester (AAME) by $C_{1\bar{r}}$, and of acetyl-tyrosine-ethylester (ATEE) by C_1 esterase and chymotrypsin, were determined as described previously (*3*). The rate of hydrolysis of Leu-Ala-Arg-α-naphthylester by factor B and CVF·Bb were also determined as described previously (*8*). Cathepsins A, B, C, D, G, H, and L were determined as earlier described (*5*), using CBz-Glu-Phe, Bz-Arg-*p*-nitro-anilide, Gly-Tyr-amide, hemoglobin, Bz-Phe-β-naphthylester, Bz-Arg-β-naphthylamide, and azocasein as substrates, respectively. For measurement of inhibitory effects, mixtures of enzyme solution and FUT-175 were preincubated at 37°C for 5 min and then residual enzyme activity was determined. The inhibitory effect of FUT-175 on factor $\bar{D}$ was determined in the following manner. Factor $\bar{D}$ was preincubated with FUT-175, and then factors B, C_3, and Mg^{2+} were added and the mixture was incubated further. After incubation, materials were applied to SDS-PAGE and cleavage of factor B by factor $\bar{D}$ was measured densitometrically. Complement-mediated hemolytic activities were determined as described previously (*3*).

2. Various Immunological Reactions in Animals

Systemic Forssman shock in guinea pigs, passive Arthus reactions in rats, and endotoxin shock in mice were induced as previously described (*6*). The drug or saline used as a control was administered intraperitoneally (i.p.) or *per os* (p.o.) 30 min or 1 hr before eliciting reactions, respectively.

EFFECTS OF FUT-175 ON *IN VITRO* AND *IN VIVO* EXPERIMENTS

1. Inhibition of Various Proteases by FUT-175

The inhibitory effects of FUT-175 on various proteases were examined. The concentrations required for 50% inhibition are shown in Fig. 2. FUT-175 inhibited proteases in the complement system and other trypsin-like proteases in blood and trypsin, but did not inhibit enzyme activities of various cathepsins. It inhibited C_1 esterase more than various proteases in blood. The concentration of FUT-175 causing 50% inhibition of factor B cleaving by factor $\bar{D}$ was 1.2×10^{-4} M.

FUT-175 was not removed from the inhibited enzyme by dialysis, but after dialysis it was completely removed by incubating the inhibited

	Proteases	Concentration for 50% inhibition (M)
Proteases in complement-system	$C_{1\bar{r}}$	2.1×10^{-7}
	$C_{1\bar{s}}$	5.1×10^{-8}
	B	5.8×10^{-5}
	CVF·Bb	4.3×10^{-5}
Other proteases	Thrombin	3.3×10^{-7}
	Plasmin	4.1×10^{-7}
	Plasma kallikrein	3.1×10^{-7}
	Pancreatic kallikrein	8.4×10^{-6}
	Trypsin	1.9×10^{-8}
	Chymotrypsin	5.4×10^{-5}
Cathepsins	A	$>10^{-3}$
	B	$>10^{-3}$
	C	$>10^{-3}$
	D	$>10^{-3}$
	G	7.1×10^{-4}
	H	$>10^{-3}$
	L	$>10^{-3}$

Fig. 2. Concentrations of FUT-175 required for 50% inhibition of various proteases.

enzyme at 37°C. Therefore, FUT-175 was thought to be a reversible inhibitor.

2. *Inhibition of Complement-mediated Hemolysis by FUT-175*

The effects of FUT-175 on complement-mediated hemolysis were examined, and the results are shown in Fig. 3. The concentration of FUT-175 required for 50% inhibition of complement-mediated hemolysis by the classical pathway and the alternative pathway were 6.9×10^{-8} and 5.1×10^{-7} M, respectively. These phenomena indicated that FUT-175 consequently inhibited $C_{1\bar{r}}$ and C_1 esterase in the classical pathway, and factor $\bar{D}$, factor B, and C_3 convertase in the alternative pathway.

FUT-175 is shown to be the strongest inhibitor of all known anti-complement reagents in Table I. These results suggest that FUT-175 treatment should effect various complement-mediated immunological reactions *in vivo*.

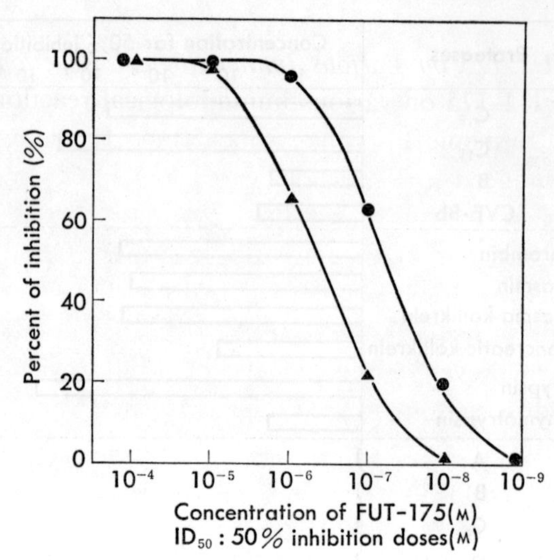

Fig. 3. Concentration of FUT-175 required for 50% inhibition of complement-mediated hemolysis.
● classical pathway ID_{50}: 6.9×10^{-8}; ▲ alternative pathway ID_{50}: 5.1×10^{-7}.

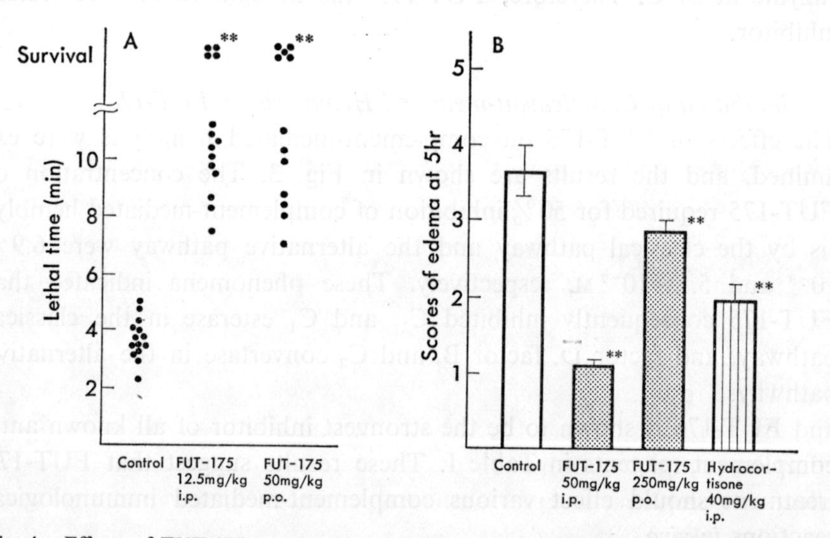

Fig. 4. Effects of FUT-175 on experimental allergy models of types II (Forssman shock, A) and III (Arthus reaction, B).

3. Effect of FUT-175 on Various Immunological Reactions in Animals
The effects of FUT-175 on various immunological reactions in animals were examined. Systemic Forssman shock (type II allergy by Coombs and Gell) was induced in FUT-175-treated and untreated guinea pigs, as shown in Fig. 4. FUT-175 prevented death at a low dose of 2.5 mg/kg i.p. or 50 mg/kg p.o. Passive Arthus reactions (type III allergy by Coombs and Gell) in FUT-175-treated, hydrocortisone-treated and untreated rats, are shown in Fig. 4. FUT-175 was more effective than hydrocortisone in preventing edema at 5 hr. Endotoxin shock in FUT-175-treated, hydrocortisone-treated and untreated mice, is shown in Fig. 5. The substance significantly increased the survival rate at 24 hr. Further, FUT-175 suppressed Masugi nephritis and Adjuvant arthritis in rats.

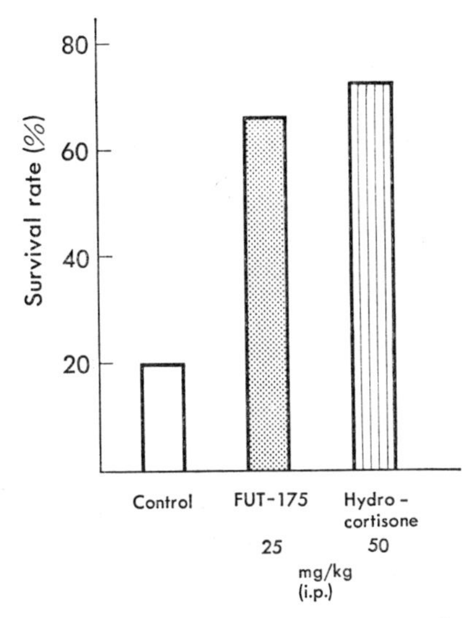

Fig. 5. Effect of FUT-175 on endotoxin shock.

These results concerning the effect of FUT-175 on Forssman shock and Arthus reaction confirm the assumption that the drug's inhibitory effect on complement-mediated hemolysis is paralelled by its inhibitory effect on complement-dependent immunological reactions.

Thus, FUT-175 is a very useful reagent in research on the complement system in various immunological diseases.

REFERENCES

1. Aoyagi, T., Takeuchi, T., Matsuzaki, A., Kawamura, K., Kondo, S., Hamada, M., and Umezawa, H. *J. Antibiot.*, **22**, 283 (1969).
2. Doll, M. H. and Baker, B. R. *J. Med. Chem.*, **19**, 1079 (1976).
3. Fujii, S. and Hitomi, Y. *Biochim. Biophys. Acta*, **661**, 342 (1981).
4. Hauptmann, J. and Markwardt, F. *Biochem. Pharm.*, **26**, 325 (1977).
5. Hirado, M., Iwata, D., Niinobe, M., and Fujii, S. *Biochim. Biophys. Acta*, **669**, 21 (1981).
6. Hitomi, Y. and Fujii, S. *Int. Archs Aller. Appl. Immun.*, **69**, 262 (1982).
7. Hong, K., Kinoshita, T., Miyazaki, W., Izawa, T., and Inoue, K. *J. Immunol.*, **122**, 2418 (1979).
8. Ikari, N., Sakai, Y., Hitomi, Y., and Fujii, S. *Biochim. Biophys. Acta*, **742**, 318 (1983).
9. Kaneko, I., Fearon, D. T., and Austen, K. F. *J. Immunol.*, **124**, 1194 (1980).
10. Lesavre, P., Gaillard, M., and Halbwachs-Mecarelli, L. *Eur. J. Immunol.*, **12**, 252 (1982).
11. Lukas, T. J., Munoz, H., and Erickson, B. W. *J. Immunol.*, **127**, 2555 (1981).

Proteinase Inhibitors: Medical and Biological Aspects, eds. N. Katunuma et al., pp. 45–53 (1983).
Japan Sci. Soc. Press, Tokyo/Springer-Verlag, Berlin

Legume Protease Inhibitors: Inhibition Mechanism of Peanut Protease Inhibitors

Tokuji IKENAKA and Shigemi NORIOKA

*Department of Chemistry, Osaka University College of Science**

Legume seeds contain various protein protease inhibitors. Some of them have been isolated in pure form and characterized in terms of their chemical, physicochemical, and inhibitory properties. Based on these properties, legume protease inhibitors were classified into two families; one is the Kunitz soybean trypsin inhibitor family and the other the Bowman-Birk protease inhibitor family.

An example of the first family is the soybean trypsin inhibitor (Kunitz), so-called STI, which has been crystallized and characterized by Kunitz (*10*, *11*). This inhibitor consists of 181 amino acids including two disulfide bridges, and the complete amino acid sequence of STI was determined by Koide and Ikenaka (*9*). STI was used by Laskowski *et al.* for the investigation of the inhibition mechanism of serine protease inhibitors, especially for the development of the concept of the reactive sites of serine protease inhibitors (*4*, *18*). The three-dimensional structure of the porcine trypsin-STI complex was determined by Blow's group (*20*). Homologous inhibitors have also been isolated from several legume seeds, barley and rice. However, the only complete amino acid sequence reported to date is that for STI.

* Machikaneyama 1-1, Toyonaka, Osaka 560, Japan.

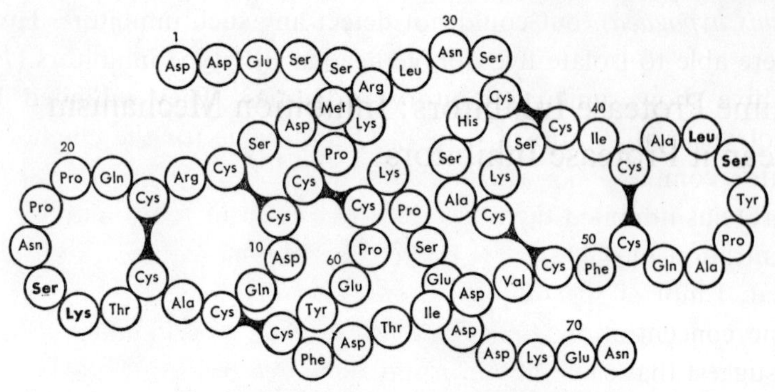

Fig. 1. Covalent structure of BBI (*14*).

An example of the second family is soybean Bowman-Birk protease inhibitor, so-called BBI, which was isolated by Bowman in 1946 (*3*) and characterized by Birk (*2*). The covalent structure of this inhibitor elucidated by Odani and Ikenaka (*14*) is shown in Fig. 1. This inhibitor possesses 71 amino acid residues and is characterized by a high cystine content, namely, seven disulfide bridges. Preliminary X-ray crystallographic analysis of a Bowman-Birk soybean trypsin inhibitor was reported by Wei *et al.* (*21*), and recently Gaier's group succeeded in crystallizing the ternary complex consisting of α-chymotrypsin, β-trypsin, and BBI, and reported preliminary crystallographic data of the complex (*5*).

BBI inhibits two proteases simultaneously and independently, trypsin being bound to one reactive site, Lys_{16}-Ser_{17}, and chymotrypsin to the other reactive site, Leu_{43}-Ser_{44}. BBI is, therefore, called double-headed inhibitor. Homologous inhibitors were found in seeds of almost all leguminous plants. Many of them were isolated in pure form and their amino acid sequences were determined (*1*, *6–8*, *13*, *15*, *16*, *19*, *22–24*). BBI consists of two homologous regions. Each containing three loops, and linked together by two polypeptide chains. The outer loop consists of nine amino acid residues and possesses a reactive site. These data may suggest that these double-headed inhibitors are evolved from a single-headed inhibitor by gene duplication.

PEANUT TRYPSIN-CHYMOTRYPSIN INHIBITORS

Recently we attempted to isolate single-headed inhibitors from peanuts

(*Arachis hypogaea*), but could not detect any such inhibitors. However, we were able to isolate five double-headed BBI type inhibitors (*12*) and these five show similar inhibitory properties. They inhibited bovine chymotrypsin at a ratio of 1 to 1. The K_i value for the chymotrypsin-inhibitor complex was calculated to be 1.2×10^{-8} M. Furthermore, all five proteins inhibited bovine trypsin at a ratio of 1 to 1.4 at an enzyme concentration of 4.4×10^{-7} M. When the enzyme concentration was increased, 1 mol of the inhibitor bound 2 mol of the enzyme. At a low enzyme concentration, only 1 mol of enzyme was inhibited. These results suggest that each inhibitor may have two trypsin reactive sites with different K_i values.

In contrast to BBI, the peanut inhibitor-trypsin complex showed very weak anti-chymotryptic activity. On the other hand, it is interesting to note that when the inhibitor-chymotrypsin complex was mixed with an equimolar amount of trypsin, trypsin was inhibited rapidly with a relatively slower release of a final value of 70% of the chymotrypsin after 10 min. This suggests that the binding of trypsin to a reactive site

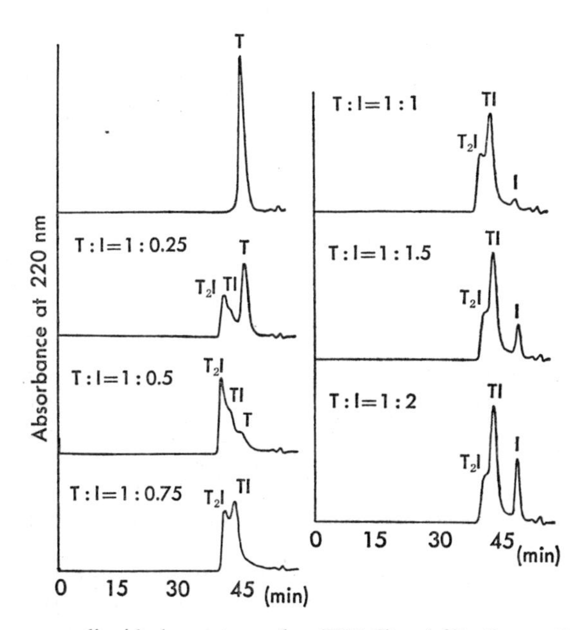

Fig. 2. High pressure liquid chromatography (HPLC) gel-filtration pattern of the formation of complexes (T_2I and TI) by mixing trypsin (T) with different molar ratios of inhibitor B-III (I).

is preferential to that of chymotrypsin. In order to release all of the chymotrypsin from the chymotrypsin-inhibitor complex, 1.3 molar equivalent of trypsin was required.

The nature of the inhibitor-trypsin complex was investigated by gel-filtration at different molar ratios of trypsin and inhibitor. As shown in Fig. 2, the formation of two types of complexes, namely T_2I and TI, was observed. For example, the TI complex was predominant when the trypsin to inhibitor was used at a ratio of 1 : 1. In contrast to these findings, chymotrypsin was found to form only one complex with the

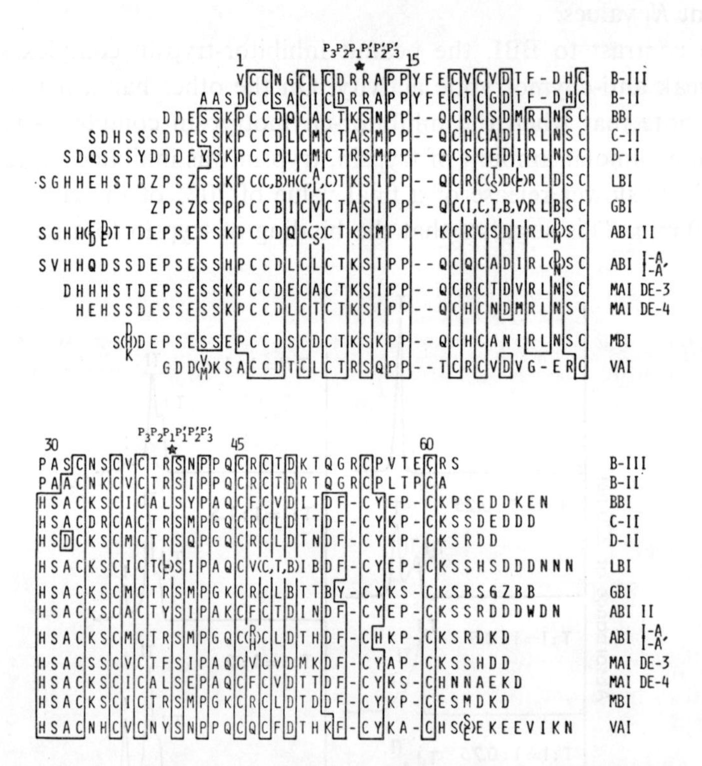

Fig. 3. Comparison of amino acid sequences of various Bowman-Birk family protease inhibitors.

B-III, peanut inhibitor B-III: B-II, peanut inhibitor B-II; BBI (*13*); C-II, soybean inhibitor C-II (*15*); D-II, soybean inhibitor D-II (*16*); LBI, lima bean inhibitor IV (*19*); GBI, garden bean inhibitor II′ (*22*); ABI II, azuki bean inhibitor II (*6, 23*); ABI I-A and I-A′, azuki bean inhibitor I-A and I-A′ (*8*); MAI DE-3 and DE-4, *Macrotyloma axillare* inhibitor DE-3 and DE-4 (*7*); MBI, mung bean inhibitor (*24*); VAI, *Vicia angustifolia* inhibitor (*1*).

inhibitor (CI type). These results agree with the data of inhibition study, namely, that the peanut inhibitors possess two reactive sites for trypsin with different K_i values and one reactive site for chymotrypsin.

Subsequently, the complete amino acid sequences of the five inhibitors were determined. Figure 3 shows the amino acid sequences of peanut inhibitors B-II and B-III with those of other BBI-family inhibitors. Peanut inhibitors A-I, A-II, and B-I have the same sequence as that of B-III except for different N-terminals. Therefore, we speculated that A-I (with N-terminal of Ser-Ser-Ser-Asp-Asp-Asn-Val-), B-I (Asp-Asn-Val-), and B-III (Val-) are derived from A-II (with N-terminal of Glu-Ala-Ser-Ser-Ser-Ser-Asp-Asp-Asn-Val-) or from an inhibitor with a longer N-terminal amino acid sequence. However, the sequence of B-II differs at various positions from that of inhibitor B-III.

The two asterisks point to the positions of the reactive site peptide bonds of inhibitors. These data suggest that the reactive sites of peanut inhibitors are Arg-Arg and Arg-Ser. There are several significant differences between peanut inhibitors and the other BBI family inhibitors. There are four amino acid insertions (Tyr_{15}, Phe_{16}, Arg_{54}, and Glu or Pro_{59}) and one deletion (No. 25). The P_1' position of the first reactive sites of the peanut inhibitors is replaced by arginine residue instead of serine residue. In all BBI-family double-headed inhibitors hitherto known the P_1' position is occupied by serine. Earlier we assumed that the P_1' serine residues of the inhibitors were invariant, because the replacement of this serine residue of BBI by other amino acids would significantly decrease the inhibitory activity over that of the original inhibitor (*17*). The P_2 position of the first reactive sites of the peanut inhibitors is replaced by aspartic acid instead of threonine residue. Both the P_2 and P_1' positions of the first reactive sites of other BBI-family inhibitors were occupied by neutral amino acids. But all the peanut inhibitors have negatively charged aspartic acid and positively charged arginine in the P_2 and P_1' positions of their first reactive sites, respectively, and these charges might neutralize each other. This neutralization may be very important for the inhibition of proteases. B-III, the most abundant inhibitor in peanuts, has only one amino acid residue, valine, N-terminal to the first half cystine residue. In addition, there are only two amino acids, arginine and serine, C-terminal to the last half cystine residue. The size of the N- and C-terminal portions of the polypeptide

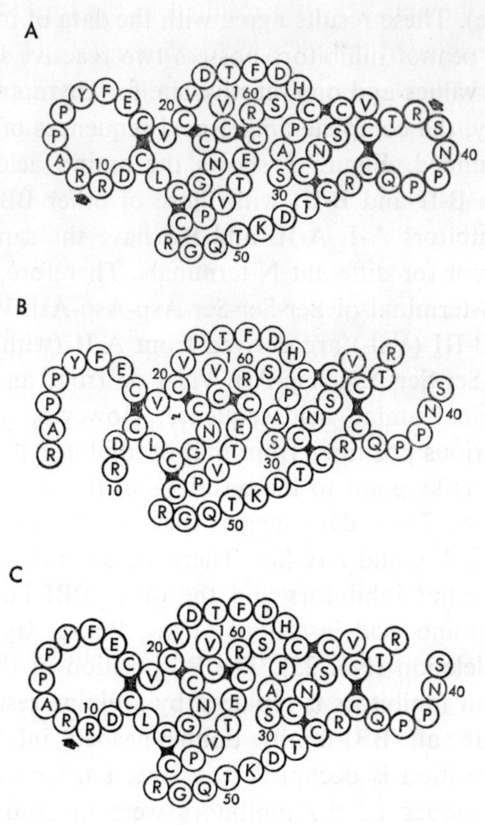

Fig. 4. Covalent structure of native peanut inhibitor B-III and its enzymatically modified inhibitors.
A: native B-III (I). B: trypsin modified B-III (I^{**}_{RS}). C: trypsin regenerated B-III (I^{*}_{S}).

chain seems to be important to explain the difference of K_i values of the first and second reactive sites.

Figure 4A shows the covalent structure of peanut inhibitor B-III including seven disulfide bridges, the locations of which were deduced from the sequence of BBI (Fig. 1). The most significant difference between B-III and the other BBI family inhibitors is the insertion of two amino acids, probably Tyr and Phe, in the first reactive site loop, resulting in an eleven-membered ring rather than the usual nine-membered ring.

The reactive sites of B-III were determined by the methods of Finkenstadt and Laskowski (*4*). Incubation of the inhibitor with catalytic amounts of bovine β-trypsin resulted in the formation of a modified inhibitor, designated I_{RS}^{**}, of which two peptide bonds, Arg_{10}-Arg_{11} and Arg_{38}-Ser_{39}, were cleaved as shown in Fig. 4B. This finding suggests that these two bonds are the reactive sites for trypsin. The trypsin modified B-III still inhibits 2 mol of trypsin, but its chymotrypsin inhibitory activity is extremely low.

The resynthesis of the cleaved reactive site peptide bonds was performed by the Laskowski technique (*18*). The trypsin modified inhibitor, I_{RS}^{**}, was adsorbed on a trypsin-Sepharose column and then eluted with 4 M urea, pH 2.3. Determination of the N-terminal amino acids of the inhibitor thus obtained demonstrated that 60% of the Arg-Arg bond was resynthesized, but not the Arg-Ser bond. The structure of trypsin regenerated B-III, I_{S}^{*}, is presented in Fig. 4C. This means that the two reactive sites react differently with trypsin.

To study the chymotrypsin reactive site, peanut inhibitor B-III was incubated with chymotrypsin at pH 3.0 for 3 weeks, however, no peptide bond was cleaved. Therefore, the reactive site for chymotrypsin could not be defined. To obtain further information, the resynthesis of the cleaved reactive site peptide bonds of I_{RS}^{**} (Fig. 4B) by incubation with chymotrypsin was carried out. After 60 hr of incubation, about 20% of I_{RS}^{**} had changed to the CI-complex. The chymotrypsin regenerated inhibitor was separated from the CI complex. Its determined structure was the same as that of the trypsin regenerated B-III, in which the reactive site Arg-Arg was synthesized. This chymotrypsin regenerated B-III can inhibit chymotrypsin similarly to the native inhibitor. From these results, the Arg_{10}-Arg_{11} bond was suggested to be the reactive site for chymotrypsin.

INHIBITION MECHANISM OF PEANUT PROTEASE INHIBITORS

The inhibition by B-III of trypsin and chymotrypsin may be explained as shown in Fig. 5. B-III has two reactive sites, A and B. Site A is Arg-Arg and site B is Arg-Ser. The K_i value of the complex with trypsin at site A is much smaller than that of site B. The addition of an equimolar amount of trypsin to B-III leads to the formation of mainly TI complex, in which most of the trypsin is bound to site A. If more

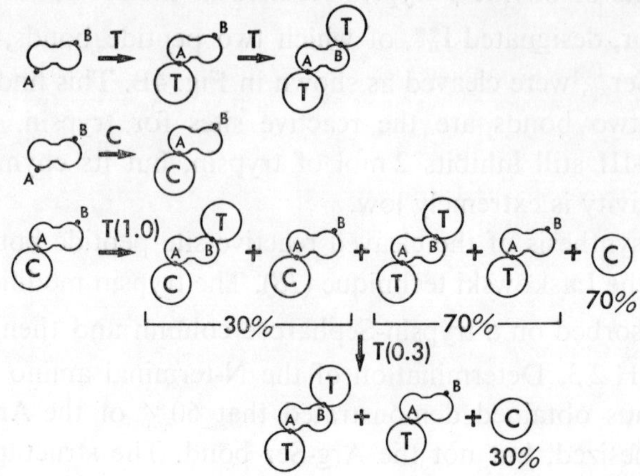

Fig. 5. Schematic explanation of the inhibition mechanism of peanut inhibitor B-III. T, trypsin; C, chymotrypsin; A, B, reactive sites A and B.

trypsin is added, the T_2I complex is formed. Chymotrypsin also binds at site A, but more weakly than trypsin. Therefore, when an equimolar amount of trypsin is added to the CI complex, trypsin binds first at site B and then replaces chymotrypsin at site A, releasing 70% of chymotrypsin with the formation of the CIT, CI, T_2I, and TI complexes in the ratios indicated in the figure. Further addition of 0.3 molar equivalent of trypsin releases all chymotrypsin resulting in T_2I and TI complexes.

The peanut is thought to be one of the most evolved legumes, and its inhibitors are characterized by many amino acid replacements as described in this paper. In particular, the Ser→Arg and Thr→Asp replacements at the P_1' and P_2 positions of the first reactive site and also the extension of the ring structure to an eleven-membered loop without loss of the inhibitory activity represents an important landmark in the study of the structure and function of protease inhibitors. The role of these inhibitors in legume seeds is still obscure, however, these protease inhibitors may fulfill many important biological functions.

REFERENCES

1. Abe, O., Shimogawa, Y., Kuromizu, K., Ohata, J., and Araki, T. Abstr. of 32th Symposium on Protein Structure, p. 77 (1981) (in Japanese).
2. Birk, Y., Gertler, A., and Khalef, S. *Biochem. J.*, 87, 281 (1963).

3. Bowman, D. E. *Proc. Soc. Exp. Biol. Med.*, **63**, 547 (1946).
4. Finkenstadt, W. R. and Laskowski, M., Jr. *J. Biol. Chem.*, **240**, PC962 (1965).
5. Gaier, J. R., Tulinsky, A., and Liener, I. E. *J. Biol. Chem.*, **256**, 11417 (1981).
6. Ishikawa, C., Nakamura, S., Watanabe, K., and Takahashi, K. *FEBS Lett.*, **99**, 97 (1979).
7. Joubert, F. J., Kruger, H., Townshend, G. S., and Botes, D. P. *Eur. J. Biochem.*, **97**, 85 (1979).
8. Kiyohara, T., Yokota, K., Masaki, Y., Masaki, O., Iwasaki, T., and Yoshikawa, M. *J. Biochem.*, **90**, 721 (1981).
9. Koide, T. and Ikenaka, T. *Eur. J. Biochem.*, **32**, 417 (1973).
10. Kunitz, M. *J. Gen. Physiol.*, **29**, 149 (1946).
11. Kunitz, M. *J. Gen. Physiol.*, **30**, 291 (1947).
12. Norioka, S., Omichi, K., and Ikenaka, T. *J. Biochem.*, **91**, 1427 (1982).
13. Odani, S. and Ikenaka, T. *J. Biochem.*, **71**, 839 (1972).
14. Odani, S. and Ikenaka, T. *J. Biochem.*, **74**, 697 (1973).
15. Odani, S. and Ikenaka, T. *J. Biochem.*, **82**, 1523 (1977).
16. Odani, S. and Ikenaka, T. *J. Biochem.*, **83**, 737 (1978).
17. Odani, S. and Ikenaka, T. *J. Biochem.*, **84**, 1 (1978).
18. Ozawa, K. and Laskowski, M., Jr. *J. Biol. Chem.*, **241**, 3955 (1966).
19. Stevens, F. C., Wuerz, S., and Krahn, J. *In* Bayer Symposium V "Protease Inhibitors," eds. H. Fritz, H. Tschesche, L. J. Greene, and E. Truscheit, p. 344 (1974). Springer-Verlag, Berlin.
20. Sweet, R. M., Wright, H. T., Janin, J., Chothia, C. H., and Blow, D. M. *Biochemistry*, **13**, 4212 (1974).
21. Wei, C. H., Basu, S. P., and Einstein, J. R. *J. Biol. Chem.*, **254**, 4892 (1979).
22. Wilson, K. A. and Laskowski, M., Sr. *J. Biol. Chem.*, **250**, 4261 (1975).
23. Yoshikawa, M., Kiyohara, T., Iwasaki, T., Ishii, Y., and Kimura, N. *Agric. Biol. Chem.*, **43**, 787 (1979).
24. Zhang, Y., Luo, S., Tan, F., Chi, C., Xu, L., and Zhang, A. *Sci. Sinica*, **25**, 268 (1982).

Proteinase Inhibitors: Medical and Biological Aspects, eds. N. Katunuma et al., pp. 55–68 (1983).
Japan Sci. Soc. Press, Tokyo/Springer-Verlag, Berlin

Relationship between the Amino Acid Sequence and Inhibitory Activity of Protein Inhibitors of Proteinases

M. LASKOWSKI, Jr., M. TASHIRO, M. W. EMPIE, S. J. PARK, I. KATO, W. ARDELT, and M. WIECZOREK

*Department of Chemistry, Purdue University**

PROTEIN INHIBITORS OF SERINE PROTEINASES

In contrast to the great excitement surrounding the early exploration of protein inhibitors of cysteine, aspartic acid, and metallo-proteinases, where relatively little was as yet established, the study of protein inhibitors of serine proteinases is a field approaching maturity. A huge number of serine proteinase inhibitors have been described, about 200 inhibitory domains were sequenced, at least 10 relevant 3-dimensional structures were determined, and a substantial outline of a mechanism of action—the standard mechanism has emerged (*8*). The inhibitors have evolved to fit the enzyme surface very well and make many favorable contacts with the enzyme. Few conformational changes occur on interaction—this is an example of an old-fashioned E. Fischer "lock and key" fit. Most inhibitors differ from most substrates in that the reactive site is held within disulfide bridges and thus after the hydrolysis the resultant peptide fragments are still held together. Furthermore, and at first surprisingly, the equilibrium constant for the peptide bond hy-

* W. Lafayette, IN 47907, U.S.A.

drolysis is not very large but in general it is close to unity near neutral pH.

Another surprising fact about inhibitors is that in each heavily studied inhibitor family, the reactive site amino acid P_1 appears to change during evolution. This finding is in sharp contrast with the usual expectation of workers who study protein evolution. In most proteins, the amino acid residues comprising the active site are conserved, but in inhibitors these residues vary. A partial explanation of this reactive site variability has been provided (*8*).

INHIBITORS AS A SUITABLE SUBSTRATE FOR THE SEQUENCE TO REACTIVITY ALGORITHM (*3*, *7*)

The major problem of protein chemistry is that while it has been thoroughly established that the amino acid sequence dictates the three dimensional structure and therefore the reactivity and ultimately the function of a protein molecule, we do not know the predictive algorithms. Most authors now focus on the determination of the sequence to conformation algorithm. We have chosen, instead, to focus upon the sequence to reactivity algorithm by comparing the reactivity (thus far K_{assoc} for interaction with various enzymes) with the sequence of many homologous (therefore we assume possessing the same main chain conformation) proteins (avian ovomucoid third domains).

The choice of protein proteinase inhibitors is justified by their small size (in our case 56 residues or less), relatively small number of residues in enzyme inhibitor contact, great rigidity, very small conformational changes on association and by the extensive knowledge of X-ray structures and of the mechanism of action. Avian ovomucoid third domains were chosen for all of these reasons and for two more—one is relatively easy availability, the second is great variability in contact region. The first reason can best be summarized by pointing out that all avian egg whites contain ovomucoid and that it is present at an approximate concentration of 10 mg/ml (*4*). Ovomucoid is quite easy to isolate. The ovomucoid third domain can be obtained by specific hydrolysis of the connecting peptide between the second and third domains with either staphyloccocal proteinase or thermolysin (*6*). This procedure is not completely general, but it succeeded for us for 60 species and (possibly

TABLE I. The Amino Acid Sequence of Turkey Ovomucoid Third Domain and the Alternatives from 50 Additional Avian Species

Val Ile	6	136	13	
Asp	7	137	12	
Cys	8	138	11	
Ser	9	139	10	
Glu Gly Asp	10	140	9	
Tyr His	11	141	8	
Pro	12	142	7	
***Lys Gln Arg	13	143	6	P_n
***Pro	14	144	5	
***Ala Asp Thr Gly Val	15	145	4	
***Cys	16	146	3	
***Thr Pro Met Arg Leu Ser	17	147	2	
***Leu Ala Thr Met Lys Ser Val Gln	18	148	1	
***Glu Asp	19	149	1	
***Tyr Asp Glu His Arg Gln Phe	20	150	2	
***Arg Leu Met Lys Phe	21	151	3	
Pro	22	152	4	
Leu Val Ile	23	153	5	
Cys	24	154	6	
Gly	25	155	7	
Ser	26	156	8	
Asp Asn	27	157	9	
Asn Ser	28	158	10	
Lys Glu Gln	29	159	11	
Thr Ile	20	160	12	
Tyr	31	161	13	
***Gly Ser Asp Ala Val Asn	32	162	14	
Asn Ser Asp	33	163	15	
Lys	34	164	16	
Cys	35	165	17	
***Asn Asp Ser Ala	36	166	18	
Phe	37	167	19	P_n'
Cys	38	168	20	
Asn	39	169	21	
Ala	40	170	22	
Val Ala	41	171	23	
Val Leu	42	172	24	
Glu Lys Asp	43	173	25	
Ser Lys	44	174	26	
Asn Ser	45	175	27	
Gly Val	46	176	28	
Thr	47	177	29	
Leu	48	178	30	
Thr Asn	49	179	31	
Leu Val	50	180	32	
Ser Asn Arg Gly	51	181	33	
His Arg	52	182	34	
Phe Ile Leu	53	183	35	
Gly	54	184	36	
Lys Gln Glu Thr	55	185	37	
Cys	56	186	38	

Three numbering systems are presented: turkey third domain, entire chicken ovomucoid, and Schechter-Berger. The residues which are in contact with the enzyme in the turkey ovomucoid third domain—*Streptomyces griseus* proteinase B complex (*5*, *10*), are starred (ref. *6* and unpublished results).

temporarily) failed for 3. In most species, about half of the third domain is glycosylated at Asn[45], residue P_{27}' (see Table I), the remainder is not. Most of our work was done on carbohydrate-free material. The yield of this is about 10% of crude ovomucoid or 1 mg/ml of avian egg white. It is most fortunate that the third domain is small enough to determine its entire sequence in a single sequencer run. Table I shows the sequence of turkey ovomucoid third domain and the various alternative amino acids obtained from sequencing ovomucoids from 50 avian species. Of these 50 ovomucoids, 40 turned out to be distinct, the remainder were duplicates of other sequences in the set.

The second reason for choosing ovomucoid third domains is apparent by inspection of Table I. Note that those positions where there are a large number of alternatives are almost without exception positions which are in contact with the enzyme in enzyme-inhibitor complex. This result was anticipated from early work of Feeney and associates (*4*, *11*) who carried out extensive comparative studies of avian ovomucoids prior to the advent of rapid sequencing. Feeney and coworkers found that ovomucoid molecules from various species could inhibit in some cases one (singleheaded), in other cases two (doubleheaded), and in yet others three (tripleheaded) enzyme molecules. This finding is nicely explained by our finding that each avian ovomucoid has three tandem homologous domains. The second finding of the Feeney group is that ovomucoids from even closely related avian species have very distinct specificity. It turns out that a large part of this variability in specificity is localized in the ovomucoid third domains and that it is explained by high variability of the residues in contact with the enzyme.

WHAT ARE THE USES OF THE ALGORITHM?

The determination of a sequence to reactivity algorithm first of all would show us that protein sequences can be "read." In this case, what we have in mind is the ability to predict the equilibrium constants for association with various enzymes from the sequence alone. While the algorithm would be determined empirically, it would still have a very large predictive power. The answer might encourage protein chemists to find similar algorithms for other proteins. The most obvious example that comes to mind are not only the inhibitors but the serine proteinases themselves.

The second purpose of determining the algorithm is to provide to protein chemists very detailed information about energies of interactions. Since $\Delta G^\circ = -RT \ln K_a$ we are measuring changes in free energy of enzyme-inhibitor association resulting from specified amino acid substitutions making contacts that can be examined in detail by X-ray crystallography. Such data will allow us to calibrate the potential functions used in theoretical protein chemistry.

However, there are two purposes directly related to inhibitors. The first is to understand inhibitory specificity. At present, most workers in the field associate the specificity with the P_1 residue, *e.g.*, if P_1 is Lys or Arg then trypsin will be inhibited. It turns out, however, that the other 10 residues of the inhibitor which make contact with the enzyme also have an effect on specificity. This conclusion—differential effects of residues other than P_1—was not completely anticipated. It will allow us, when the algorithm is completely worked out, to design an inhibitor with maximal discrimination for any specified pair of enzymes. Thus, since the use of inhibitors as drugs is contemplated, it would be a method of truly rational drug design.

Another aspect of the algorithm may make an important contribution to the understanding of the mechanism of enzyme-inhibitor interaction. The current puzzle is to explain how inhibitors differ from excellent substrates. Both have high k_{cat}/K_m, but substrates and inhibitors partition this differently. When the algorithm is extended to measurement of the various rate constants in the mechanism—the partitioning of k_{cat}/K_m into its components may become clearer.

SINGLE SUBSTITUTIONS ARE EASIEST TO UNDERSTAND

At the start of building of a sequence to reactivity algorithm, it is very dangerous (and incorrect) to make comparisons between variants, which differ from one another by more than a single amino acid replacement. Therefore, our first problem was to generate a set of variants which are connected to one another by a single amino acid replacement. All such replacements are shown in Fig. 1. There are 20 entries. Seventeen are

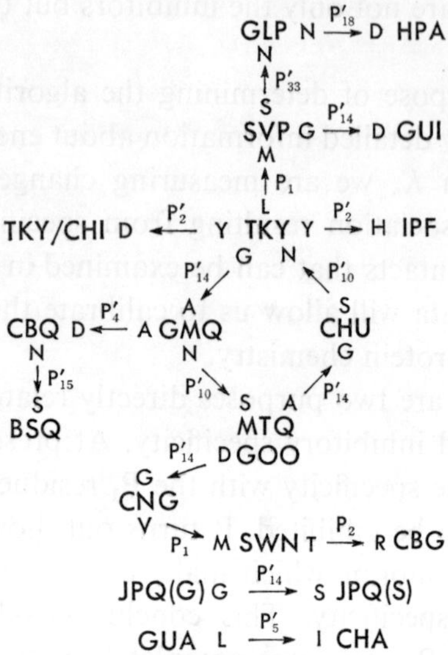

GLP N $\xrightarrow{P'_{18}}$ D HPA
N
$\uparrow P'_{33}$
S
SVP G $\xrightarrow{P'_{14}}$ D GUI
M
$\uparrow P_1$
L
TKY/CHI D $\xleftarrow{P'_2}$ Y TKY Y $\xrightarrow{P'_2}$ H IPF
G N
P'_{14} P'_{10}
A S
CBQ D $\xleftarrow{P'_4}$ A GMQ CHU
N N G
$\downarrow P'_{15}$ P'_{10} P'_{14}
S S A
BSQ MTQ
P'_{14} D GOO
G
CNG
V
P_1 M SWN T $\xrightarrow{P_2}$ R CBG
JPQ(G) G $\xrightarrow{P'_{14}}$ S JPQ(S)
GUA L $\xrightarrow{P'_5}$ I CHA

Fig. 1. A scheme showing the relationship between those avian ovomucoid third domains which can be related to one another by a single amino acid replacement (the position and nature of the replacements indicated by arrows).
GLP, golden pheasant (*Chrysolophus pictus*); HPA, Hungarian partridge (*Perdix perdix*); SVP, silver pheasant (*Lophura nycthemera*); GUI, helmet guineafowl (*Numida meleagris*); CHI, chicken (*Gallus gallus*), TKY, turkey (*Meleagris gallopavo*); IPF, Indian peafowl (*Pavo cristatus*); CBQ, chestnut bellied scaled quail (*Callipepla squamata castanogastris*); GMQ, Gambel's quail (*Lophortyx gambelii*); CHU, Chukar partridge (*Alectoris chukar*); BSQ, blue scaled quail (*Callipepla squamata pallida*); MTQ, mountain quail (*Oreortyx pictus*); GOO, goose (*Anser anser*); CNG, Canada goose (*Branta canadensis*); SWN, black swan (*Cygnus atratus*); CBG, cape-barren goose (*Cereopsis novaehollandiae*); JPQ, Japanese quail (*Coturnix coturnix*); GUA, Spix's guan (*Penelope jacquacu*); CHA, plain chachalaca (*Ortalis vetula*). TKY/CHI is a hybrid of residues 1–18 of turkey, 19–56 of chicken; JPQ(G) and JPO(S) are genetic variants within the Japanese quail species.

main products from different avian species; one comparison is from different products of two subspecies: chestnut bellied scaled quail (*Callipepla squamata castanogastris*) and blue scaled quail (*Callipepla squamata pallida*). Another comparison is a consequence of a polymorphism in the Japanese quail population. Finally, one variant, the TKY/CHI or turkey/chicken hybrid, was obtained by covalent combination of residues 1–18 of turkey with residues 19–56 of chicken (*13*). This hybrid is a forerunner of a new source of variants which can be obtained by various types of protein engineering.

HOW MANY VARIANTS ARE NEEDED?

A third domain is 51 amino acid residues long.*[1] Therefore, there are 20^{51} or 2×10^{66} sequences of this length. It is clear that this astronomical number of variants must be reduced prior to the start of extensive experimental work. Two*[2] methods of reduction we have come up with are (a) to assume that only the positions in contact with the enzyme affect K_{assoc} and (b) to assume that the effects of residue changes upon the free energy of interaction are strictly additive. The first simplification reduces the number of positions to be concerned with from 51 to 11 (see Table I) and therefore the number of needed variants to $20^{11}=2\times10^{14}$, a very much smaller number than before but still formidable.*[3] The number can only be significantly reduced if the combinatorial aspects of the problem are eliminated, *i.e.*, when we assume additivity. This is equivalent to saying that having determined the effect of the P_4 change from Ala to Asp by comparing Gambel's quail with chestnut bellied scaled quail (see Table I and Fig. 1), we can confidently use this K_{assoc} factor (or difference in ΔG°_{assoc}) for all other inhibitor pairs where this difference occurs, no matter what the remainder of the sequence is. Such a simplification reduces us to requiring 11 (variable contact positions)$\times$(20$-$1) (alternative amino acids)$+$1$=$210 variants. Of course, additional variants will be needed to prove (1) that the residues do not in contact matter and (2) additivity assumption is correct. It is already clear that both of these assumptions are not strictly correct and therefore a somewhat larger number of variants is needed. However, if the number required remains in the hundreds, it will be exceedingly difficult to attain but manageable. For the sake of compari-

*[1] The third domains we use are in fact longer. However, their length differs depending upon how much of the connecting peptide between second and third domains (five residues in turkey third domain cut with staphyloccocal proteinase) is included in the domain. We have fairly good evidence that these residues do not affect K_{assoc}. These residues are not shown in Fig. 1.

*[2] Another reduction in numbers is also possible. We wish to compare only those domains which have the same main chain conformation. Thus, the *structural* residues should not be allowed to vary. Precise calculation of this restriction is complex.

*[3] Of the 11 contact residues, two, Pro at P_5 and Cys at P_3, do not have alternatives and probably should be considered structural. Thus, the number of positions to be studied can probably be reduced to 9.

son, we currently have 40 variants, 20 of which allow single substitution comparison (Fig. 1). Of the 210 required terms in the simplest possible algorithm we can now evaluate 13, if all the simplifying assumptions are allowed.

In the paper by Empie and Laskowski (*3*), we report comparison of K_{assoc} with chymotrypsin, elastase, and subtilisin. Here, most of the data listed involve *Streptomyces griseus* proteinases A and B, (from here on we call them SGPA and SGPB, respectively) because the 3-dimensional structure of the SGPB complex with turkey ovomucoid third domain was recently determined. It should be emphasized here that an algorithm cannot possibly be solved without the aid of X-ray crystallography. The determinations of the structure of free Japanese quail ovomucoid third domain (*9*, *12*) the SGPB-turkey third domain complex (*5*, *10*) as well as that of pancreatic secretory trypsin inhibitor-trypsinogen complex (*2*) are incorporated implicitly or explicitly into all that is said here.

RESIDUES NOT IN CONTACT DO NOT MATTER

We have already shown (*3*) that attachment of carbohydrate to Asn^{45} or residue P_{27}' has no effect upon K_{assoc}. Furthermore, substitutions at positions P_5', P_{10}', P_{33}' have been examined and in all cases found to be precisely without effect. On the other hand, there are exceptions. Position P_{15}' of turkey ovomucoid makes no contact with SGPB in complex. However, the side chain of P_{15}' which is Asn, not only in most ovomucoid third domains but also in most Kazal family inhibitors donates hydrogen bonds to the main chain oxygens of P_2 and P_1' both in free inhibitor and in complex (*5*, *9*, *10*, *12*). Replacement of P_{15}' by Ser (comparison of chestnut bellied with blue scaled quail) weakens K_{assoc} with SGPB by a factor of 5. Serine has only one hydrogen to donate and therefore this replacement is likely to eliminate one of the two hydrogen bonds to the reactive site and thus change its geometry. Thus, we need to consider not only residues in contact with the enzyme in complex but also those whose *side chains* can interact with contact residues. Examination of side chain hydrogen bonds in free Japanese quail inhibitor, turkey ovomucoid and in silver pheasant third domain (free) (O. Epp, W. Bode, and R. Huber, personal communication) suggests that there are only three such residues Asn[33] (P_{15}'), Asn[39] (P_{21}') and possibly Lys[55] (P_{37}'). Of these, Asn[33] was just discussed, Asn[39] is unvaried in all sequences we have (see Table I). The penultimate position, P_{37}', varies a good deal, but we do not have a good single substitution valid comparison. We infer that the change of this residue from Lys[55] to Glu[55] increases the K_{assoc} with several enzymes by a factor of 2–4.

In conclusion, we have shown that changing amino acids or introducing large substituents (carbohydrate) in most positions not in contact has no effect at all (within our measuring error of about 10%) upon K_{assoc}. A few positions do affect K_{assoc}. These effects are relatively small (less than a factor of 10) and can be rationalized on the basis of known 3 dimensional structures. However, at the present stage it is not totally safe to ignore all positions not in contact without a test.

REACTIVE SITE RESIDUE P_1 IS MOST IMPORTANT BUT OTHER CONTACTS MATTER A GREAT DEAL

The development of the reactive site model described earlier focused a great deal of attention on the reactive site residue, P_1. This was especially so because most of the early work was with trypsin inhibitors where the P_1 interactions are especially dominant. The reactive site is still extremely important as can be seen from the comparison of black swan and Canada goose ovomucoid third domains. In this and subsequent comparisons, there is only one difference between the compared sequences.

The three order of magnitude difference for chymotrypsin is quite striking. It becomes even more striking when we learn (data not shown) that P_1 Met and Leu are closely similar, although for chymotrypsin, SGPA and SGPB Leu is preferred (for subtilisin Met is preferred over

		P_4	P_3	P_2	P_1	P_1'	P_2'	P_3'	
Black swan	···	Ala	Cys	Thr	Met	Glu	Tyr	Met	···
Canada goose	···	Ala	Cys	Thr	Val	Glu	Tyr	Met	···

	(K_{assoc} (M^{-1}))			
	Chymotrypsin	SGPA	SGPB	Subtilisin
Black swan	2.4×10^9	2.5×10^{11}	2.7×10^{10}	5.1×10^{10}
Canada goose	2.7×10^6	2.8×10^9	3.6×10^8	6.1×10^8

Leu). Thus, in changing from Leu to Val—a subtraction of a single $-CH_2-$ group K_{assoc} declines by a factor of more than 1,000, much more than might be expected from simple hydrophobic transfer consideration.

However, large changes resulting from a single residue replacement are not limited to the P_1 position. The most dramatic example we have obtained recently is

	P_4	P_3	P_2	P_1	P_1'	P_2'	P_3'
Turkey	··· Ala	Cys	Thr	Leu	Glu	Tyr	Arg ···
Turkey/chicken hybrid	··· Ala	Cys	Thr	Leu	Glu	Asp	Arg ···

(K_{assoc} (M^{-1}))	Chymotrypsin	SGPA	SGPB
Turkey	1.8×10^{11}	2.3×10^{11}	5.6×10^{10}
Turkey/chicken hybrid	7.6×10^{6}	7.7×10^{8}	4.8×10^{8}

where the second entry was obtained by covalent hybrid formation of residues 1–18 of turkey and 19–56 of chicken. The factor of 2×10^4 in K_{assoc} for chymotrypsin is the largest we have obtained thus far for any single residue replacement. Replacements at many other contact positions, *e.g.*, P_4, P_2, P_2' (shown above), P_3' and P_{14}' produce really dramatic effects (changes of about two orders of magnitude for some enzymes). It is thus clear that a good deal of specificity resides in contact residues other than P_1.

RESIDUES OTHER THAN P_1 EXERT LARGE DIFFERENTIAL EFFECTS

It is now clear that residues other than P_1 are important—the question is are they important in the same way to all serine proteinases or do they exert important differential effects? This can best be illustrated by an example. It is well known that several trypsin inhibitors with Arg P_1 inhibit chymotrypsin. If there were no differential effects of residues other than P_1, the ratio of K_{assoc} for trypsin and for chymotrypsin would always be the same. In practice, this would mean that *all* very strong trypsin inhibitors would be efficient inhibitors of chymotrypsin, while weak inhibitors of trypsin would never inhibit chymotrypsin. With limited data, it is easy to believe such a statement, and for some time one of us (ML) did. However, the statement is quite wrong.

Another example of differential effects can be provided for two very similar enzymes, SGPA and SGPB. From the structure of SGPB-turkey ovomucoid complex, one can deduce what residues of SGPA are in contact with the inhibitor. It turns out that they are essentially the same in both enzymes. One might thus expect that either (a) (the binding constants of various third domains will be essentially the same for SGPA and SGPB) or (b) (while the constants for SGPA and SGPB are not the same, the ratio of these constants might be the same independent of the ovomucoid) are true. After we screened about 15 variants, all of which had Thr at P_2 position, we decided that statement (b) is approximately true and that SGPA is approximately 10 times more strongly inhibited than SGPB by any variant. We were wrong. After variants at the P_2 position were considered, it became clear that the inhibition order can be reversed even for this closely similar pair. A particularly dramatic example is given by

		P_4	P_3	P_2	P_1	P_1'	P_2'	P_3'	
Black swan	···	Ala	Cys	Thr	Met	Glu	Tyr	Met	···
Cape barren goose	···	Ala	Cys	Arg	Met	Glu	Tyr	Met	···

	(K_{assoc} (M^{-1}))		
	Chymotrypsin	SGPA	SGPB
Black swan	2.4×10^{9}	2.5×10^{11}	2.7×10^{10}
Cape barren goose	2.4×10^{6}	1.8×10^{8}	9.7×10^{8}

It thus becomes clear that even for very similar enzymes the secondary specificity differs. Furthermore, it is clear that inhibitors can be designed such that they inhibit only one of several serine proteinases present in some biologically or medically important system. Of course, prior to such a design all serine proteinases that are present must be well characterized.

ADDITIVITY OF RESIDUE CONTRIBUTIONS

While we have now acquired a relatively large body of data on the relationship of K_{assoc} to sequence, and while we have already learned a great deal from it, this research will be an "endless game" unless we can show that the effect of a specific substitution at one position is independent of the nature of amino acid residues present in other positions. This is best illustrated by an example shown in Fig. 2.

If the effect upon K_{assoc} of the P_{14}' Gly→Asp change were strictly

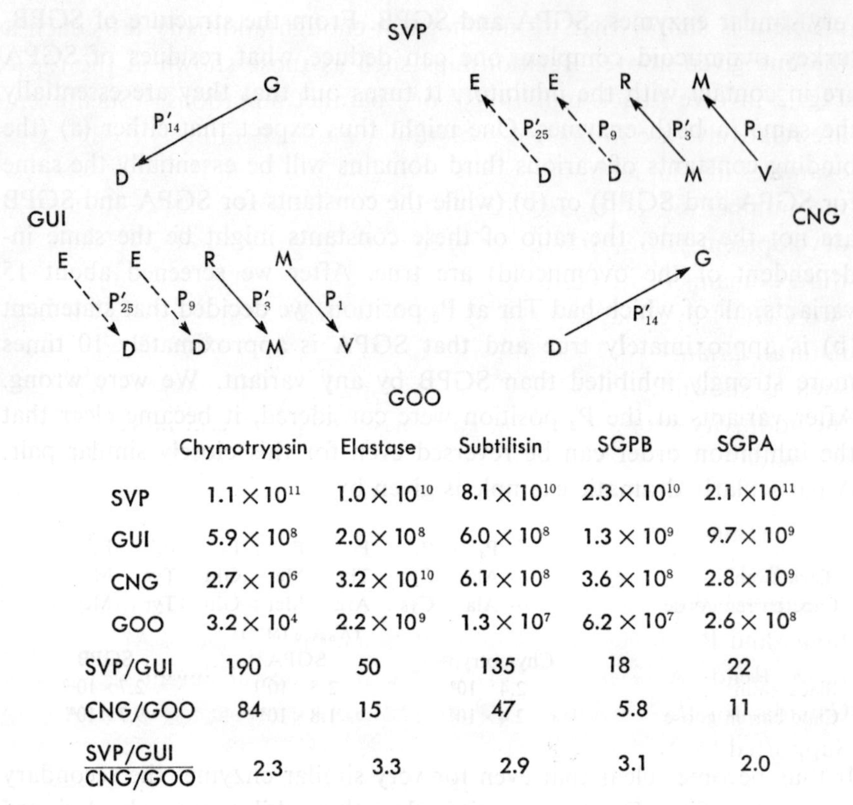

	Chymotrypsin	Elastase	Subtilisin	SGPB	SGPA
SVP	1.1×10^{11}	1.0×10^{10}	8.1×10^{10}	2.3×10^{10}	2.1×10^{11}
GUI	5.9×10^{8}	2.0×10^{8}	6.0×10^{8}	1.3×10^{9}	9.7×10^{9}
CNG	2.7×10^{6}	3.2×10^{10}	6.1×10^{8}	3.6×10^{8}	2.8×10^{9}
GOO	3.2×10^{4}	2.2×10^{9}	1.3×10^{7}	6.2×10^{7}	2.6×10^{8}
SVP/GUI	190	50	135	18	22
CNG/GOO	84	15	47	5.8	11
$\frac{\text{SVP/GUI}}{\text{CNG/GOO}}$	2.3	3.3	2.9	3.1	2.0

Fig. 2. Scheme for testing additivity of residue contributions.
The symbols SVP, GUI, CNG, and GOO are defined in Fig. 1. The arrows indicate changes in sequences. It is essential to see that the same change occurs between SVP and GUI and between CNG and GOO and also that the same changes occur between SVP and CNG and between GUI and GOO. Solid arrows indicate changes in contact positions, while broken arrows indicate changes not in contact.

the same independently of the rest of the sequence, then the "defect factor" would be exactly one and the equilibrium constants for one of the four variants would be *exactly* predictable from the other three. The conclusion of our single try at the additivity question is that the "missing" equilibrium constant is approximately but not exactly predictable. The defect factor is about 3, and essentially the same in each case. It is a tribute to the current accuracy of K_{assoc} determination that we can claim that the "defect factor" differs from one. Since it is constant for all enzymes, it appears that it can be explained. Our current problem is simply that we did not run enough additivity tests such as the one given above. Relationships such as those shown at the top of Fig. 2 involve "back mutations" or "parallel mutations." Such events are relatively rare in protein evolution. Thus, it is unlikely to find good examples in a limited sample of variants. As the sample grows, however, the number of potential tests increases. Probably the most hopeful development is the formation of covalent hybrids between ovomucoid third domains of two different species. If the two NH_2 terminal and the two COOH terminal peptides differ in any pair of species, then the two natural variants as well as the two possible hybrids always form a cycle such as shown in Fig. 2 and allow for an additivity test. We hope that the additivity problem can be understood when several more tests of it are made.

Acknowledgment
This work could not be seriously attempted without 3-dimensional structures. We are deeply grateful to E. Weber, E. Papamokos, W. Bode, and R. Huber, Martinsried, West Germany, and to M. Fujinaga, R. J. Read, A. Sielecki, and M. N. G. James, Edmonton, Alberta, Canada, for their willingness to undertake the crystallographic studies. Supported by National Institutes of Health Grant GM 10831.

REFERENCES

1. Ardelt, W. and Laskowski, M., Jr. *Acta Biochim. Pol.*, **30** (1983), in press.
2. Bolognesi, M., Gatti, G., Menegatti, E., Guarneri, M., Marquart, N., Papamokos, E., and Huber, R. *J. Mol. Biol.*, **162**, 839 (1982).
3. Empie, M. W. and Laskowski, M., Jr. *Biochemistry*, **21**, 2274 (1982).
4. Feeney, R. E. and Allison, R. G. *Evolutionary Biochemistry of Proteins.* John Wiley & Sons, New York (1969).
5. Fujinaga, M., Read, R. J., Sielecki, A., Ardelt, W., Laskowski, M., Jr., and James, M. N. G. *Proc. Natl. Acad. Sci. U.S.*, **79**, 4868 (1982).
6. Kato, I., Kohr, W. J., and Laskowski, M., Jr. *Proc. FEBS Meet.*, **47**, 197 (1978).
7. Laskowski, M., Jr., Empie, M. W., Kato, I., Kohr, W. J., Ardelt, W., Bogard, W. C., Jr., Weber, E., Papamokos, E., Bode, W., and Huber, R. *32nd Mosbach Colloquium*, eds. H. Eggerer and R. Huber, p. 136 (1981). Springer-Verlag, Berlin.
8. Laskowski, M., Jr. and Kato, I. *Annu. Rev. Biochem.*, **49**, 593 (1980).
9. Papamokos, E., Weber, E., Bode, W., Huber, R., Empie, M. W., Kato, I., and Laskowski, M., Jr. *J. Mol. Biol.*, **158**, 515 (1982).
10. Read, R. J., Fujinaga, M., Sielecki, A. R., and James, M. N. G. *J. Mol. Biol.* (1983), submitted for publication.

11. Rhodes, M. B., Bennet, N., and Feeney, R. E. *J. Biol. Chem.*, **235**, 1686 (1960).
12. Weber, E., Papamokos, E., Bode, W., Huber, R., Kato, I., and Laskowski, M., Jr. *J. Mol. Biol.*, **149**, 109 (1981).
13. Wieczorek, M. and Laskowski, M., Jr. *Biochemistry*, **22** (1983), in press.

Proteinase Inhibitors: Medical and Biological Aspects, eds. N. Katunuma et al., pp. 69–75 (1983).
Japan Sci. Soc. Press, Tokyo/Springer-Verlag, Berlin

Ca-activated Neutral Protease (CANP) and Its Inhibitors in Pathological States

Hideo SUGITA, Shoichi ISHIURA, and Ikuya NONAKA

*Division of Neuromuscular Research, National Center for Nervous, Mental and Muscular Disorders**

The muscular dystrophies in man and animal are characterized by progressive muscle atrophy and weakness, presumably with the continuous loss of both soluble and myofibrillar proteins from muscle cells. The decrease in muscle proteins in muscular dystrophies is largely ascribable to an increased protein degradation, decreased protein synthesis or both.

There are still many controversies concerning the enzymatic mechanisms involved in these processes, but recent studies have suggested that in Duchenne muscular dystrophy (DMD), an increased intracellular Ca ion due to an excess influx of extracellular Ca ion through a disrupted plasma membrane may activate Ca-activated neutral protease (CANP) to initiate the degradation of the structural proteins (*2*).

LOCALIZATION OF CANP IN MYOFIBRIL

The most susceptible protein to CANP is an unknown Z-band protein which rendered α-actinin free from Z-band. In view of this fact, the localization of CANP in glycerinated chicken myofibril was determined

* Ogawa Higashi 2620, Kodaira, Tokyo 187, Japan.

by an immunofluorescent method. Z-band under a phase microscope was definitely stained with a fluorescent labeled antibody against CANP (*4*). The amount of CANP bound to myofibrils (bound CANP) was approximately 4% of that contained in whole muscle homogenate. Ouchterlony's immunodiffusion procedure proved that the bound CANP showed the same precipitation line as the soluble one. The preliminary result suggested that the bound CANP is more sensitive to Ca ion than the soluble one, but still requires an unphysiologically high Ca-ion concentration for its activation.

CALCIUM INDUCED DEGENERATION OF MUSCLE PROTEIN IN EXCISED MUSCLE AND EFFECT OF E-64-C

To support the Ca hypothesis of the degradation of the structural proteins in DMD, we experimentally increased the intracellular Ca-ion concentration of excised intact rat muscle using Ca ionophore, A23187 *in vitro* and observed the degeneration of muscle fibers morphologically and biochemically (*6*, *12*). The effect of protease inhibitor, E-64-C was also investigated.

Intact rat soleus and extensor digitorum longus (EDL) were removed with tendons and incubated in Krebs-Ringer bicarbonate solution containing 25 μg/ml of A23187, 5 mM of glucose and 0.5 mM cycloheximide under 95% O_2-5% CO_2 for 3 hr at 37°C in the presence or absence of 30 μM of E-64-C. The control buffer consisted of the same solution except than 1 mM EGTA was substituted for Ca ion. After 3 hr, the muscles were removed and used for electron microscopy and SDS-gel electrophoresis. The released proteins were also analyzed by SDS-gel electrophoresis.

The protein released from the muscle into the medium was more abundant when incubated with Ca-buffer than in control, 23% more in EDL and 63% in soleus. As the released soluble protein measured by creatine kinase (CK) was almost equal, the relative increase in liberated protein in the presence of Ca ion was supposed to originate mainly from the solubilization of the structural proteins from the muscle into the medium. On electron microscopy, the muscles incubated with Ca ion clearly lost their Z-band structure in strong contrast

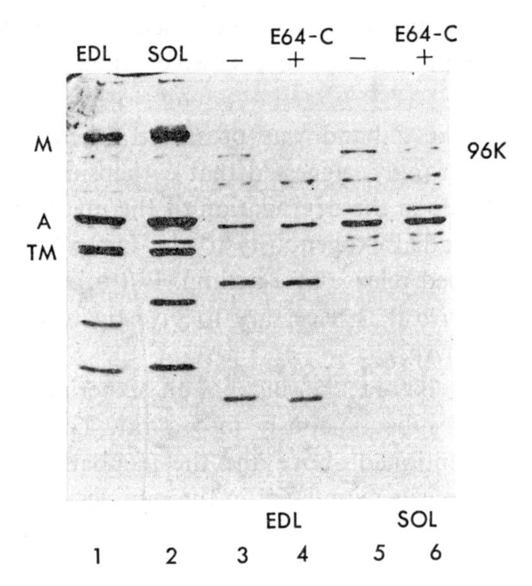

Fig. 1. Effect of protease inhibitor on Ca^{2+}-induced release of α-actinin from muscle. SOL, soleus; M, heavy chain of myosin; 96K, α-actinin; A, actin; TM, tropomyosin. Lanes 1 and 2 are gel electrophoretic patterns of structural proteins incubated in Ca^{2+}-buffer. Lanes 3, 4 (EDL) and 5, 6 (SOL) are solubilized proteins in the medium in the absence (−) and presence (+) of E-64-C.

to their preservation in control, as already mentioned by many investigators (*1*, *13*).

Concomitant with the disappearance of Z-band, there was a marked decrease of α-actinin in Ca-treated muscle, more prominent in EDL than in soleus.

Now the question may arise whether Z-band loss is caused by Ca-induced proteolysis. If this is the case, the release of α-actinin should be inhibited by specific inhibitors of Ca-dependent proteases. We utilized E-64-C for this purpose.

As shown in Fig. 1, the addition of 30 μM of E-64-C into the medium decreased or inhibited the release of α-actinin from the muscle cell, more clearly in EDL than in soleus. Decrease in tyrosine release has been reported by Kameyama and Etlinger using leupeptin (*7*).

As the loss of α-actinin from the Z-band was suppressed by an inhibitor of CANP, E-64-C, the degradation of Z-band is supposed to be caused not by the direct effect of increased Ca, but by a proteolytic action of CANP, activated by an excessive influx of extracellular Ca ion.

The fact that the Z-band was preserved when incubated in the absence of Ca ion strongly suggested that cathepsins in the muscle did not play an initial role in the degradation of the myofibril.

E-64-C, when added exogenously to the incubation medium, suppressed the Ca-induced release of α-actinin. But to expect the beneficial effect of E-64-C *in vivo* it is necessary to investigate whether this drug enters the muscle *in vivo*.

Twenty-four hr before the incubation experiment, 10 mg/kg of E-64-C was injected subcutaneously into a rat. The soleus and EDL were removed as mentioned above and the incubation experiment was repeated. If E-64-C was accumulated in the muscles, release of α-actinin should be inhibited. However, the release of α-actinin was not influenced by the earlier injection of E-64-C (*5*).

The distribution of E-64-C in various organs of the rat was studied using radioactive [^{14}C]-E-64-C by a subcutaneous route. Twenty-four hr after a single shot of 10 mg/kg of [^{14}C]-E-64-C, the activity was highest in kidney (7.46 μg/g) and liver (1.43 μg/g) and the muscle was relatively inaccessible to penetration (0.088 μg/g). But successive injections for 10 days increased this amount to 7 times that contained in a single shot, suggesting that this drug has a tendency to accumulate in the muscle (*9*).

In any event, to expect a beneficial effect by a subcutaneous route it is necessary to increase the doses more than 10-fold, as CANP is approximately 8 times more resistant to E-64-C than cathepsin B (*5*).

EFFECT OF E-64-C ON HEREDITARY DYSTROPHIC CHICKEN *IN VIVO* (*11*)

Taking the above findings into consideration, studies of the *in vivo* effect of E-64-C on hereditary dystrophic chicken line 413 have been made to clarify whether this protease inhibitor has any demonstrable influence on functional disability, serum enzymes, muscle protease, and muscle pathology. Application of protease inhibitor to dystrophic chicken and other domestic animals has been reported by many inves-

TABLE I. Outline of *In Vivo* Administration of E-64-C to Dystrophic Chicken

				♂	♀
1) Period:	10–89 days *ex ovo*				
2) Amount:	150 mg or 300 mg/kg/day, divided twice a day				
3) Route:	subcutaneously, back of the chicken				
4) Numbers:	line 413	untreated	22	12	10
		150 mg	21	11	10
		300 mg	21	11	10
	line 412	normal	11	4	7
5) Parameters:					
Clinical:	growth curve, flip number				
Biochemical:	serum CK, PK, LDH, muscle proteases				
Morphological:	number of necrotic fibers, histogram of fiber diameters				

PK, pyruvate kinase; LDH, lactate dehydrogenase.

TABLE II. CANP Activities of the Superficial Pectoralis Muscle of the Dystrophic Chicken Treated with E-64-C

	CANP activity (cpm+S.E.)			
	n	Male	*n*	Female
Control	12	6,686± 538	10	4,743±240
E-64-C 150 mg/kg	11	5,222± 663	10	3,451±303**
300 mg/kg	11	5,389± 508	10	3,751±269*
Normal	4	4,167±1,390	7	2,593±443

CANP activity was expressed as cpm of radioactivity per 0.1 ml of supernatant. Asterisks shows the decreasd CANP activity with statistical significance (** $p<0.01$; * $p<0.05$).

tigators. However, the results have so far been controversial. The outline of the experiment is shown in Table I.

1. *Clinical Study*

Males were heavier than females and a normal animal was heavier than a dystrophic. The drug treatment did not influence the development of body weight. Males lost their righting ability earlier and at a faster rate than females. In both sexes, the drug treatment had no influence on the flip number at all.

2. *Biochemical Study*

In general, serum enzymes were more markedly increased in dystrophic chicken than in normal throughout the experiment. Drug treatment had no effect on the progressively increasing level of serum CK in either sex, contrary to the findings of Stracher's group (*10*) and Hudecki's group (*3*). Table II shows the dose response of CANP activity after administration of E-64-C. The dystrophic chicken showed higher activity than normal. Females treated with E-64-C showed a 30% reduction in CANP activity compared to control, but still higher than normal. However, the male dystrophic chicken did not show a significant reduction in CANP activities. No difference was observed between the 150 mg and 300 mg groups.

The cathepsins B, H, and D in this series of experiments has been reported by Noda *et al.* (*8*). According to them, both cathepsins B and H were reduced to the level of a normal chicken but cathepsin D, which is insensitive to E-64-C, was unchanged. The discrepancy of the effect of E-64-C on cathepsins and CANP may be explained by the difference in sensitivity of the two proteases to E-64-C.

3. *Morphological Study*

The number of necrotic fibers and histograms of the fiber diameters were not influenced at all by the drug treatment.

In spite of the high treatment doses, we could obtain no beneficial effect, clinically, biochemically or morphologically. One of the reasons for this negative result might be the drug's low permeation into the muscle cell. Another possibility is that the protein degradation in dystrophic chicken was not triggered by a surface membrane contrary to its proposal in the case of DMD. Further fundamental studies are necessary in terms of the route of drug administration, modification of the structure of the drug or exploration of a better solvent or liposome to obtain more effective permeability of the drug into the muscle cells.

REFERENCES

1. Busch, W. A., Stromer, M. H., Goll, D. E., and Suzuki, A. *J. Cell Biol.*, **52**, 367 (1972).
2. Ebashi, S. and Sugita, H. *Curr. Top. Nerve Muscle Res.*, 73 (1979).
3. Hudecki, M. S., Pollina, C. M., and Heffner, R. R. *J. Clin. Invest.*, **67**, 969 (1981).
4. Ishiura, S., Sugita, H., Nonaka, I., and Imahori, K. *J. Biochem.*, **87**, 343 (1980).
5. Ishiura, S., Hanada, K., Tamai, M., Kashiwagi, K., and Sugita, H. *J. Biochem.*, **90**, 1557 (1981).
6. Ishiura, S., Nonaka, I., and Sugita, H. *J. Biochem.*, **90**, 283 (1981).
7. Kameyama, T. and Etlinger, J. D. *Nature*, **279**, 344 (1979).
8. Noda, T., Isogai, K., Katunuma, N., Tanimoto, Y., and Ohzeki, M. *J. Biochem.*, **90**, 893 (1981).

9. Ohzeki, M., Nozu, T., Fukushima, K., Kono, Y., Urano, H., and Yoshida, H. *In* "Annual Report of New Drug Development (E-64)," ed. K. Imahori, p. 163 (1981).
10. Stracher, A., McGowan, E. B., and Shafiq, S. A. *Science*, **200**, 50 (1978).
11. Sugita, H., Kimura, M., Tarumoto, Y., Tamai, M., Hanada, K., Ishiura, S., Nonaka, I., Ohzeki, M., and Imahori, K. *Muscle and Nerve*, **5**, 738 (1982).
12. Sugita, H., Ishiura, S., and Kohama, K. *In* "Proceedings of the Fifth International Congress of Neuromuscular Diseases," ed. G. Serratrice, D. Cross, and C. Desnuelle. Raven Press, New York, in press.
13. Uchino, M. and Chou, S. M. *Proc. Japan Acad.*, **56**, 480 (1980).

9. Ohzeki, M., Noru, T., Fukushima, K., Kono, Y., Urano, H., and Yoshida, H. *In* "Annual Report of New Drug Development (E-64)," ed. K. Imahori, p. 163 (1981).
10. Stracher, A., McGowan, E. B., and Shafiq, S. A. *Science*, 200, 50 (1978).
11. Sugita, H., Kimura, M., Tarumoto, Y., Tamai, M., Hanada, K., Ishiura, S., Nonaka, I., Ohzeki, M., and Imahori, K. *Muscle and Nerve*, 5, 738 (1982).
12. Sugita, H., Ishiura, S., and Kohama, K. *In* "Proceedings of the Fifth International Congress of Neuromuscular Diseases," ed. G. Serratrice, D. Cross, and C. Desnuelle, Raven Press, New York, in press.
13. Uchino, M. and Chou, S. M. *Proc. Japan Acad.*, 56, 480 (1980).

Proteinase Inhibitors: Medical and Biological Aspects, eds. N. Katunuma et al., pp. 77–83 (1983).
Japan Sci. Soc. Press, Tokyo/Springer-Verlag, Berlin

Inhibition of Sister Chromatid Exchange and Mitogenesis by Microbial Proteinase Inhibitors

Kazuo UMEZAWA

*Department of Molecular Oncology, Institute of Medical Science, University of Tokyo**

Proteinase inhibitors have been shown to inhibit skin tumorigenesis in experimental animals (*3*). They have also inhibited blood-borne lung metastases in rats injected intravenously with Yoshida ascites hepatoma cells (*8*). Tumorigenic cells often produce a higher amount of proteinases than normal cells. 12-*O*-Tetradecanoylphorbol-13-acetate, a tumor promoter in mouse skin, induces plasminogen activator in cell cultures (*15*). Involvement of proteinases in the mechanism of carcinogenesis has been suggested by these observations.

Several low molecular weight peptides with specific inhibitory effects on various proteinases were isolated from culture filtrates of *Streptomyces* (*10*). Leupeptin is an inhibitor of trypsin, papain, plasmin, and cathepsin B; antipain, of trypsin, papain, and cathepsin B; chymostatin, of chymotrypsin; elastatinal, of elastase; pepstatin, of carboxylproteinases such as pepsin, cathepsin D, and renin; phosphoramidon, of metalloproteinases such as thermolysin and collagenase; bestatin, of aminopeptidase B and leucine aminopeptidase; and amastatin, of aminopeptidase A and leucine aminopeptidase.

In the course of our carcinogenesis research we studied the type

* 4-6-1 Shirokanedai, Minato-ku, Tokyo 108, Japan.

of proteinases involved in carcinogen-induced sister chromatid exchange (SCE) and mitogenesis of tumor cells using microbial proteinase inhibitors.

INHIBITION OF SISTER CHROMATID EXCHANGE BY PROTEINASE INHIBITORS

Chemical mutagens and carcinogens were shown to induce SCE in various cell lines including cultured human lymphocytes (*4*). Measurement of SCE has been developed as a cytological method for detecting chemical mutagens and carcinogens. SCE is a phenomenon which may be related to the repair of damaged DNA, although its etiology is uncertain.

Elastatinal was found to inhibit chemically induced mutagenesis in *Salmonella typhimurium* (*11*) and also chemically induced SCE in human lymphocytes transformed by Epstein Barr virus (*12*). From scoring 40 human peripheral lymphocytes, an average of 11.0 spontaneous SCEs/cell were found. *N*-Methyl-*N'*-nitro-*N*-nitrosoguanidine (MNNG) at concentrations of 10^{-7} M and 10^{-6} M induced 5.7 and 12.1 SCEs/cell, respectively, while 10^{-5} M MNNG inhibited mitosis. The addition of 10 μg/ml of elastatinal did not affect spontaneous SCEs, but reduced the SCEs induced by MNNG by about 50%, as shown in Fig. 1D. The differences at both concentrations of MNNG were significant by the Student's *t*-test ($p<0.001$). Similarly, *N*-ethyl-*N'*-nitro-*N*-nitrosoguanidine (ENNG) at concentrations of 10^{-7} M and 10^{-6} M induced 6.9 and 13.0 SCEs/cell, respectively, while 10^{-5} M ENNG inhibited mitosis. Elastatinal at 10 μg/ml also reduced SCEs induced by ENNG by about 65%. The differences were also significant at both concentrations of ENNG. Elastatinal inhibited induction of SCE by 4-nitroquinoline *N*-oxide (4NQO), but did not inhibit it by vinblastin, vincristin or nicotinamide, an inhibitor of poly(ADP-ribose)polymerase. Elastatinal has been used to classify the mechanism of SCE induction (*14*).

SCE is not specifically inhibited by elastatinal. Leupeptin, antipain, chymostatin, bestatin, and amastatin at a concentration of 10 μg/ml also inhibited SCEs induced by MNNG as shown in Fig. 1 (*13*). At both 10^{-7} M and 10^{-6} M MNNG, the proteinase inhibitor-induced de-

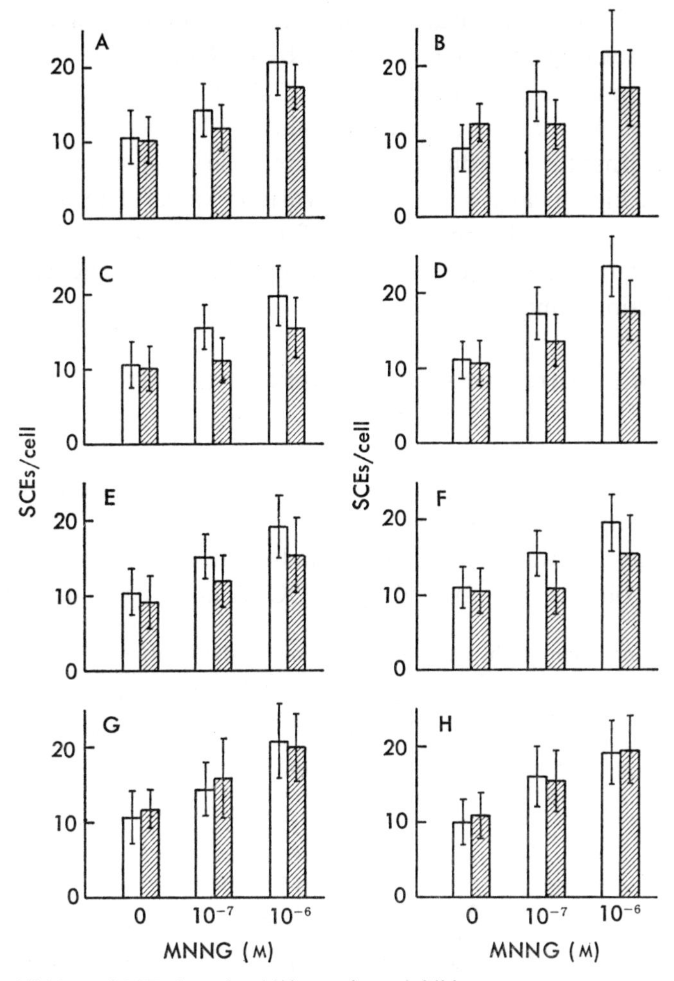

Fig. 1. Inhibition of SCEs by microbial proteinase inhibitors.
A: leupeptin. B: antipain. C: chymostatin. D: elastatinal. E: bestatin. F: amastatin. G: phosphoramidon. H: elasnin. ▭ SCEs/cell without proteinase inhibitors; ▨ SCEs/cell with 10 μg/ml of ptoteinase inhibitors. Vertical lines indicate standard deviations.

creases in SCE induction were all statistically significant ($p<0.001$). Except for antipain, the proteinase inhibitors did not significantly influence the spontaneous rate of SCE. Addition of 10 μg/ml of antipain induced SCEs without MNNG ($p<0.001$), but inhibited SCE induction by MNNG ($p<0.001$). Elasnin is an inhibitor of granulocyte elastase isolated from *Streptomyces noboritoensis* (*6*), while elastatinal is an inhibitor of pancreatic elastase. Elasnin and phosphoramidon at 10 μg/ml did not influence SCE induction by MNNG.

Incubation of 10^{-6} M MNNG with 10 μg/ml of proteinase inhibitors did not significantly alter the stability of MNNG in Roswell Park Memorial Institute medium. Therefore, these proteinase inhibitors should act biologically on the cells. Various kinds of proteinases seem to be involved in the mechanism of carcinogen-induced SCEs. The cellular location of the proteinases involved in SCE and their role in SCE-induction remain to be clarified.

INHIBITION OF MITOGENESIS BY PROTEINASE INHIBITORS

Proteinases are known to induce mitogenesis of cultured fibroblasts (*9*). Microbial proteinase inhibitors inhibited serum-induced mitogenesis of tumorigenic fibroblasts.

Mitogenesis was assayed by thymidine incorporation with hamster dermal fibroblasts transformed by 4NQO. Among proteinase inhibitors

TABLE I. Inhibition of Thymidine Incorporation by Microbial Proteinase Inhibitors

Chemical	Dose (μg/ml)	TdR incorporation (%) Serum-induced[a]	TdR incorporation (%) Thrombin-induced[b]
Control	—	100	100
Leupeptin	10	65.4	38.2
	50	51.8	8.0
Antipain	10	85.7	41.1
	50	65.8	26.7
Elastatinal	50	93.2	106.6
Phosphoramidon	50	93.4	107.4
Chymostatin	20[c]	92.9	97.8
Pepstatin	20	91.6	110.2
Bestatin	20	93.1	101.7
Amastatin	20	95.4	126.7

[a] 4NQO-transformed hamster dermal fibroblasts were incubated with the indicated chemicals for 18 hr in Dulbecco's modified Eagle's medium containing 1% fetal bovine serum; then ^{3}H-thymidine (1 μCi, 1 μg/ml) was added for 1 hr. The control value was 10,965 cpm/dish.

[b] Secondary cultures of chick embryonic fibroblasts were incubated with the indicated chemicals for 12 hr in medium containing 3 unit/ml thorombin; then ^{3}H-thymidine was added for 1 hr. The control value was 2,838 cpm/dish.

[c] Chymostatin, pepstatin, bestatin, and amastatin were dissolved in 0.2% dimethyl sulfoxide.

of microbial origin, only leupeptin and antipain inhibited thymidine incorporation in the cells as shown in Table I. Leupeptin had a stronger effect than antipain. Mitogenesis of the cells was enhanced by fetal bovine serum; leupeptin was most inhibitory for mitogenesis at about a 1% serum concentration. The inactive structural analogues of leupeptin (CH_3CO-Leu-Leu-Arg-CHO), leupeptinol (CH_3CO-Leu-Leu-Arg-CH_2OH), and leupeptinic acid (CH_3CO-Leu-Leu-Arg-COOH) did not inhibit serum-induced thymidine incorporation in the cells. Therefore, inhibition of mitogenesis by leupeptin should be due to inhibition of proteinases, but not to a cytotoxic effect.

Since the argininal moiety of leupeptin is essential for inhibition of proteinases, leupeptin derivatives in which various chemical groups are introduced into leupeptin in place of the *N*-acetylleucine moiety were prepared. These leupeptin derivatives showed different inhibitory activities on trypsin, papain, kallikrein, and plasmin as shown in Table II. Dansyl-Leu-Arg-CHO inhibited thrombin, while leupeptin and other leupeptin derivatives did not. Inhibition of thymidine incorporation by these leupeptin derivatives was roughly proportional to their trypsin or plasmin inhibitory activities, but not to the papain, kallikrein, or thrombin inhibitory activities. Leupeptin, antipain, and carbobenzyloxy-Phe-Leu-Arg-CHO inhibited multiplication of the cells in the presence of 1% fetal bovine serum as shown in Table III.

Leupeptin and its derivatives also inhibited thymidine incorpora-

TABLE II. Inhibition of Thymidine Incorporation by Leupeptin Derivatives

R-Leu-Arg-CHO	IC_{50} (μg/ml)				IC_{35} (μg/ml)
R	Trypsin	Papain	Kallikrein	Plasmin	TdR incorporation[a]
Pyroglutamyl	0.50	0.75	1.1	0.50	2.7
Z-pyroglutamyl	0.15	0.17	14.0	0.53	2.5
Z-Phe	0.80	0.40	1.0	0.73	2.4
Mandelyl	0.70	0.30	14.0	1.4	3.5
Acetyl-Leu	1.0	0.40	35	4.7	11
m-Chlorobenzoyl	4.5	0.85	1.4	6.0	12
Nicotinyl	5.0	0.14	0.75	9.0	40
Dansyl	0.85	0.14	40	12.3	17
Benzoyl	14	0.05	3.0	18.3	>50
2-Ethyl-*n*-butanoyl	9.0	0.40	3.3	30	>50
Cyclopropane-CO	26	0.17	12	>200	>50

[a] Serum-induced thymidine incorporation of 4NQO-transformed hamster dermal fibroblasts.

TABLE III. Inhibition of Cell Multiplication by Proteinase Inhibitors

Chemical	Dose (μg/ml)	Cell number (cells$\times 10^5$/dish) mean$\pm$S.D.[a]
Control	—	4.56$\pm$0.32
Leupeptin	10	3.36$\pm$0.06
	50	1.93$\pm$0.32
Antipain	10	3.18$\pm$0.44
	50	2.55$\pm$0.25
Z-Phe-Leu-Arg-CHO	10	2.53$\pm$0.19
	50	1.86$\pm$0.12

[a] 4NQO-transformed hamster dermal fibroblasts (6×10^4 cells/dish) were incubated with the indicated chemicals for 3 days in medium containing 1% fetal bovine serum; then the cell number was counted after trypsinization. Means for 4 samples.

tion in a duck embryonic cell line (ATCC CCL 141). However, the inhibitory effect of leupeptin on thymidine incorporation was modest, being only up to 50% even at higher concentrations of leupeptin. Serum-induced thymidine incorporation in chick embryo fibroblasts, C3H10T1/2 cells, and HeLa cells was not inhibited by leupeptin.

Leupeptin strongly inhibited throbmin-induced mitogenesis of chick embryonic fibroblasts. Thrombin induces DNA synthesis in chick embryonic fibroblasts in their quiescent state, and this induction is blocked by thrombin inhibitors (*2*). Leupeptin and antipain inhibited thrombin-induced thymidine incorporation as shown in Table I, although they do not inhibit thrombin. Inhibition by semisynthetic leupeptin derivatives was again proportional to their trypsin and plasmin inhibitory activities. Leupeptin-sensitive proteinases, above all trypsin and plasmin, may be involved in the mechanism of both serum-induced mitogenesis of tumorigenic hamster fibroblasts and thrombin-induced mitogenesis of chick embryonic fibroblasts.

The membrane-enriched fraction was isolated from tumorigenic hamster fibroblasts by the method described by Lesco *et al.* (*5*). This fraction showed serine proteinase activity with ^{3}H-tosylarginine methyl ester as substrate. The proteinase activity was solubilized by 0.1% Triton X-100, and the solubilized enzyme activity was inhibited by leupeptin and antipain. It was reported that the microfilament structure of rat embryonic cells is dissociated by plasmin and trypsin, but not by chymotrypsin, thrombin, or urokinase (*7*). Furthermore, cytoplasmic microfilaments are thought to be connected to cell surface structures

(*1*). Therefore, leupeptin may stabilize microfilament organization by protecting cell surface structures from proteolytic attack, thus inhibiting mitogenesis.

REFERENCES

1. Abercrombie, M. *In* "Cell Behavior," eds. R. Bellairs, A. Curtis, and G. Dunn, p. 19 (1982). Cambridge Univ. Press, Cambridge.
2. Chen, L. B. and Buchanan, J. M. *Proc. Natl. Acad. Sci. U.S.*, **72**, 131 (1975).
3. Hozumi, M., Ogawa, M., Sugimura, T., Takeuchi, T., and Umezawa, H. *Cancer Res.*, **32**, 1725 (1972).
4. Latt, S. A. *Proc. Natl. Acad. Sci. U.S.* **71**, 3162 (1974).
5. Lesco, L., Donlon, M., Marinetti, G. V., and Hare, J. D. *Biochim. Piophys. Acta*, **311**, 173 (1973).
6. Omura, S., Ohno, H., Saheki, T., Yoshida, M., and Nakagawa, A. *Biochem. Biophys. Res. Commun.*, **83**, 704 (1978).
7. Pollack, R. and Rifkin, D. *Cell*, **6**, 495 (1975).
8. Saito, D., Sawamura, M., Umezawa, K., Kanai, Y., Furihata, C., Matsushima, T., and Sugimura, T. *Cancer Res.*, **40**, 2539 (1980).
9. Sefton, B. M. and Rubin, H. *Nature*, **227**, 843 (1970).
10. Umezawa, H. *Methods Enzymol.*, **45**, 678 (1976).
11. Umezawa, K., Matsushima, T., and Sugimura, T. *Proc. Japan Acad.*, **53B**, 30 (1977).
12. Umezawa, K., Sawamura, M., Matsushima, T., and Sugimura, T. *Chem.-Biol. Interact.*, **24**, 107 (1979).
13. Umezawa, K., Sawamura, M., Matsushima, T., and Sugimura, T. *Chem.-Biol. Interact.*, **30**, 247 (1980).
14. Utakoji, T., Hosoda, K., Umezawa, K., Sawamura, M., Matsushima, T., Miwa, M., and Sugimura, T. *Biochem. Biophys. Res. Commun.*, **90**, 1147 (1979).
15. Wigler, M. and Weinstein, I. B. *Nature*, **259**, 232 (1976).

Proteinase Inhibitors: Medical and Biological Aspects, eds. N. Katunuma et al., pp. 85–95 (1983).
Japan Sci. Soc. Press, Tokyo/Springer-Verlag, Berlin

Proteinases and Their Inhibitors in Inflammation: Basic Concepts and Clinical Implication

M. JOCHUM,[*1] K.-H. DUSWALD,[*2] S. NEUMANN,[*3]
J. WITTE,[*2] H. FRITZ,[*1] and U. SEEMÜLLER[*1]

Department of Clinical Chemistry and Biochemistry, University of Munich,[*1] *Surgical Clinic, University of Munich,*[*2] *and Department of Biochemical Research, E. Merck*[*3]

THE ROLE OF LYSOSOMAL ENZYMES IN INFLAMMATION

During the inflammatory response various systemic or local tissue cells are activated thereby releasing internal, mostly lysosomal enzymes. They trigger the activation of the clotting, fibrinolysis and complement cascades, the disruption of cell membranes and tissue structure, and the release of toxic peptides (Fig. 1).

Phagocytes, especially the granulocytes and monocytes or macrophages, but also fibroblasts, endothelial cells, and mast cells are known to be very rich in such internal or lysosomal enzymes. So far, only the properties and pathobiochemical effects of enzymes of the azurophilic and specific lysosomes of polymorphonuclear granulocytes (neutrophils) have been investigated in more detail. Such enzymes, for example the neutrophil elastase and cathepsin G as well as the acidic cathepsins are preformed and stored in the lysosomes in fully active form (*7*, *11*). In this way, they can respond immediately to perform their biological

[*1],[*2] D-8000 München 2, FRG.
[*3] D-6100 Darmstadt 1, FRG.

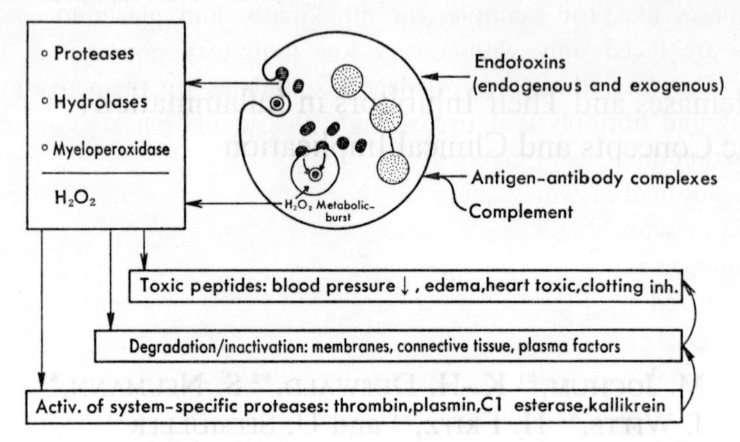

Fig. 1. Liberation and effects of lysosomal factors. For details see text.

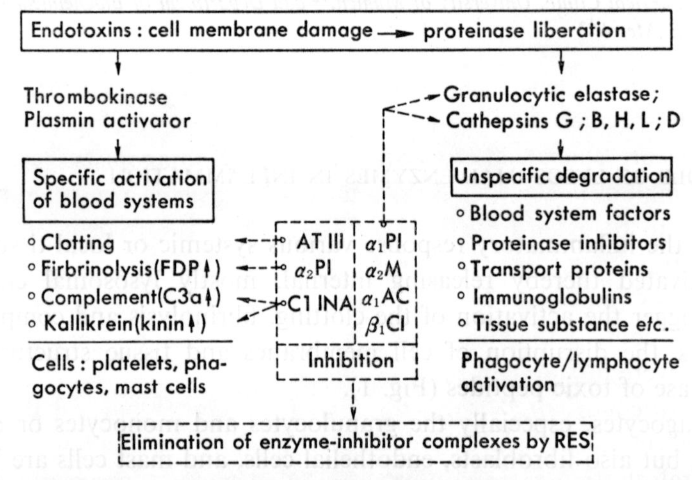

Fig. 2. Consumption of plasma factors during inflammation.
AT III, antithrombin III; α_2PI, α_2-plasmin inhibitor; C1 INA, C1 inactivator; α_1PI, α_1-proteinase inhibitor (formerly α_1-antitrypsin); α_2M, α_2-macroglobulin; α_1AC, α_1-antichymotrypsin; β_1CI, β_1-collagenase inhibitor; FDP, fibrin(-ogen) degradation products; C3a, complement factor. For details see text.

function, namely, the degradation of extracellular and intracellular material after phagocytosis.

Outside the phagocytes, proteolytic action of the lysosomal enzymes is normally prevented or balanced by proteinase inhibitors present in plasma, interstitial fluid and body secretions. Non-lysosomal proteinases like, for example, thrombokinases and plasminogen activators are faced only with a very low inhibitory potential in body fluids. They are, therefore, privileged candidates for the activation of clotting and fibrinolysis if released into the circulation after increased production due to an inflammatory stimulus.

Lysosomal enzymes liberated during severe inflammation like septicemia or septic shock can enhance, together with thrombokinases and plasminogen activators, the inflammatory response *via* two major routes characterized by either substrate-specific or substrate-unspecific proteolysis (Fig. 2).

The system-specific proteinases, thrombokinases and plasminogen activators, trigger the activation of the clotting, fibrinolysis and complement cascades (summarized as 'blood systems') by *substrate-specific* proteolysis of proenzymes and cofactors. The activated enzymes are subsequently inhibited by their natural inhibitors; the enzyme-inhibitor complexes thus formed are rapidly eliminated from the circulation by the reticuloendothelial system (RES). Hence, in a series of steps based on highly specific interactions not only the proteinases respectively their zymogens are consumed but also their natural antagonists, the system-specific inhibitors. Until recently, the given sequence of reactions was assumed to be exclusively responsible for the development of disseminated intravascular coagulation (DIC).

Results obtained very recently indicate that an additional reaction path may contribute considerably to the consumption of plasma factors during severe inflammations. This implies inactivation of plasma factors by *substrate-unspecific* proteolysis due to liberated lysosomal proteinases. Egbring and coworkers observed in animals a significant decrease in several clotting factors after infusion of endotoxin or human neutrophil elastase (*4*). Similar results were obtained by Ohlsson and coworkers (*1*, *2*) in canine endotoxemia. Moreover, in patients suffering from sepsis or septic shock a striking consumption of blood system factors, immunoglobulins and proteinase inhibitors has been found by the teams of Egbring (*5*), Aasen (*3*), Gallimore (*6*), and Witte (*14*).

NEUTROPHIL ELASTASE AND PLASMA FACTORS IN SEPSIS

In the following study neutrophil elastase was chosen as a marker

enzyme in order to demonstrate the release of lysosomal enzymes during the development of septicemia.

1. Assay of Liberated Elastase

Due to the presence of an excess of the endogenous inhibitors α_1-proteinase inhibitor (α_1PI) and α_2-macroglobulin (α_2M), direct measurement of the neutrophil-derived proteinase activities in plasma or other body fluids is not feasible. However, increased levels of the elastase-α_1PI (E-α_1PI) complex would be already a clear indication for elastase liberation. Quantitative estimation of the plasma levels of the E-α_1PI complex was carried out with a highly sensitive enzyme-linked immunoassay (*12*). Briefly, the E-α_1PI complex of the plasma sample was bound to surface-fixed antibodies directed against neutrophil elastase. After washing, a second alkaline phosphatase-labelled antibody directed against α_1PI was fixed to the complex. Under suitable conditions, the activity of fixed alkaline phosphatase towards *p*-nitrophenylphosphate is proportional to the concentration of the E-α_1PI complex in the sample.

In a first approach, we were interested to see whether a relationship exists between the plasma levels of E-α_1PI and the severity of postoperative infections. To achieve this purpose, besides other factors the levels of factor XIII (Faktor XIII-Schnelltest, Behringwerke AG Marburg) and antithrombin III (AT III) (S-2238, Deutsche Kabi Munich) were continuously monitored because both clotting factors are known to be easily degraded by neutrophil elastase *in vitro* (*9*).

2. Patients

In the clinical trial, patients subjected to major abdominal surgery were included if the operation time exceeded 120 min. Diagnosis of septicemia in the postoperative course was confirmed by prospectively established septic criteria:

Defined infection site and pos. bact. culture
Body temperature $>$ 38.5°C
Leukocytosis with $>$ 15,000 cells/mm^3 or
Leukocytopenia with $<$ 5,000 cells/mm^3
Platelets $<$ 100,000/mm^3 or drop $>$ 30%
(Positive blood culture)

In the prospective study more than 120 patients were included. Thirty of them fulfilled the defined septic criteria during the postoperative course. Of these patients, fourteen survived the infection (group B) whereas sixteen died as a direct result of septicemia (group C). Eleven patients being without infection after abdominal surgery served as controls (group A).

3. *E-α_1PI Levels*

With the enzyme-linked immunoassay elastase levels between 60 and 110 ng/ml were found in 153 healthy individuals. In patients without preoperative infection (groups A and B), the operative trauma was followed by an increase of the E-α_1PI level up to 3-fold of the normal value. Patients suffering from preoperative infections (6 out of 16 in group C) showed already clearly elevated preoperative E-α_1PI levels. Immediately after surgery a slight decrease was observed, probably due to elimination of the infection focus. Before onset of sepsis, the E-α_1PI concentrations of group B and C showed a moderate elevation but no significant changes compared to the postoperative levels. However, at the beginning of septicemia a highly significant increase of the E-α_1PI

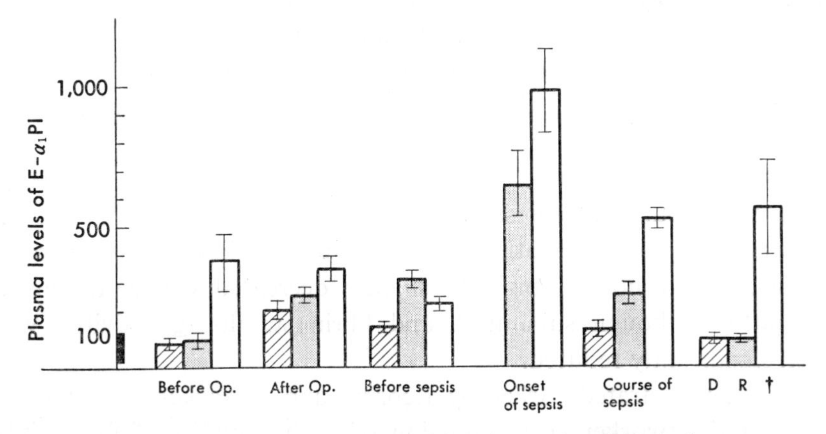

Fig. 3. Plasma levels of E-α_1PI in patients subjected to major abdominal surgery. (A) patients (n=11) without postoperative infection. (B) patients (n=14) surviving postoperative septicemia. (C) patients (n=16) dying as a result of septicemia. The E-α_1PI levels are given as mean values (±S.E.M.) for the day before operation, the day after operation as well as for the postoperative phase before sepsis, at onset of sepsis and during septicemia. Last determinations were done on day of discharge (D) for group A, on day of recovery (R) for group B, and before death (†) for group C.

levels could be detected: up to 6-fold in group B and up to 10-fold in group C. Peak levels were found above 2,500 ng/ml in both groups. The E-α_1PI levels of septic patients who recovered showed a clear tendency towards normal values. In patients with persisting septicemia, high levels of E-α_1PI were measured until death (Fig. 3).

4. AT III Activity

In non-infected patients the activity of AT III, the most important inhibitor of the clotting system, was in the normal range during the whole observation period. In infected patients, however, the AT III activity was found already below the clinically critical concentration of about 75% of the standard mean value before onset of septicemia. This low value normalized in all patients overcoming the infection, whereas a further significant decrease (up to 45% of the norm) was found in group C patients with lethal outcome. Probably, the extremely low AT III activity in the latter patients, having permanently elevated E-α_1PI levels, may be due to a significant degree to degradation by lysosomal enzymes and especially by elastase.

5. Factor XIII Activity and A and S Subunit Levels

Similar results were obtained for factor XIII, the fibrin stabilizing coagulation factor. In plasma of patients who did not survive septicemia, the factor XIII activity decreased up to 28% of the standard mean value. As measured by immunoelectrophoresis, these patients also had very low concentrations of both subunit A, comprising the active enzyme, and subunit S, representing the carrier protein (data not shown). In contrast, group A patients with an uncomplicated postoperative course showed normal or only slightly decreased concentrations of subunit S, although subunit A and fibrin stabilizing activities were often significantly reduced.

As demonstrated earlier by Egbring and coworkers (*5*) and Ikematsu and coworkers (*8*), reduction of both subunits of factor XIII cannot be due to activation of the clotting cascade alone. During clotting, that means by the action of thrombin only subunit A is consumed simultaneously with the factor XIII activity but not subunit S. Elastase, however, is able to degrade both subunits to a similar degree. These data and the results presented in our clinical trial suggest that in the

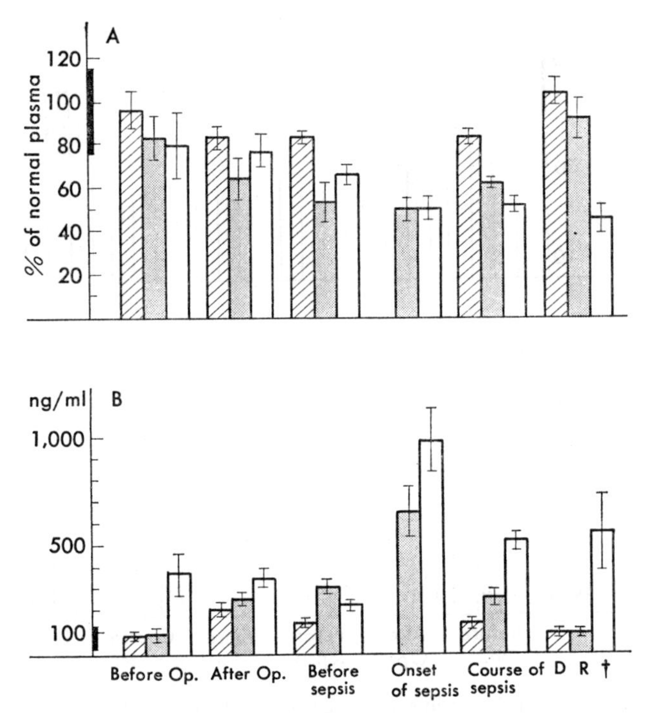

Fig. 4. Plasma levels of the inhibitory activity of AT III (A) compared to the amount of the E-α_1PI (B) in patients subjected to major abdominal surgery. For details, see legend to Fig. 3.

patients suffering from septicemia, unspecific proteolytic degradation by granulocytic elastase and/or other lysosomal proteinases is involved to a significant degree in the depletion of factor XIII.

6. *Conclusion from the Clinical Studies*

The results of the clinical studies show that in inflammatory diseases a correlation exists between the release of a lysosomal enzyme marker, the neutrophil elastase, and the clinical situation of the patient, respectively, the consumption of selected plasma factors. We take this as a clear indication that liberated lysosomal factors and especially neutrophil proteinases contribute significantly to the inflammatory response of the organism by substrate-unspecific degradation of plasma and other factors. Early application of suitable and potent exogenous proteinase

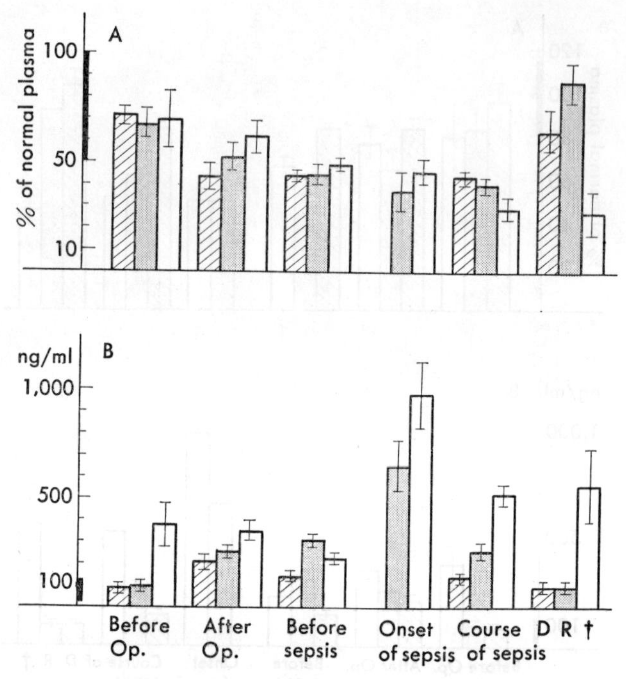

Fig. 5. Plasma levels of the fibrin stabilizing activity of factor XIII (F XIII) (A) compared to the amount of the E-α_1PI (B) in patients subjected to major abdominal surgery. For details, see legend to Fig. 3.

inhibitors should prevent or at least diminish, therefore, such destructive proteolytic processes.

INHIBITOR THERAPY IN EXPERIMENTAL ENDOTOXEMIA

To confirm this assumption, we established an endotoxemia model in dogs by intravenous infusion of *Escherichia coli* endotoxin for 2 hr. Thereby, a significant decrease was observed in the plasma levels of the clotting factors AT III, prothrombin, and factor XIII, of the fibrinolysis factors plasminogen and α_2-antiplasmin, and of the complement factor C3. The levels were followed up over an experimental period of 14 hr and their alterations checked for statistical significance (*10*) (Fig. 6).

Simultaneous intravenous administration of a relatively specific inhibitor of neutrophil elastase and cathepsin G, the Bowman-Birk

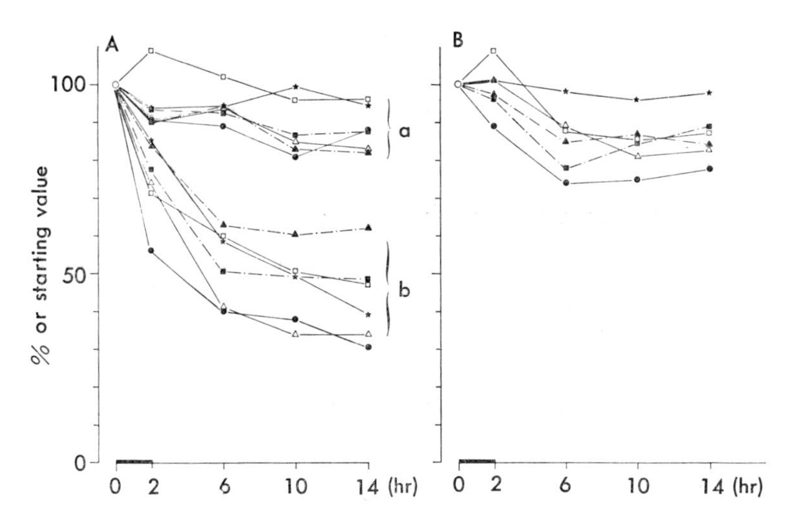

Fig. 6. Changes in the plasma levels of various plasma factors during the acute phase of experimental endotoxemia in dogs (n=6 for each group) without or with inhibitor treatment.
Applied inhibitor dosages: 10–25 trypsin inhibiting units (Bowman-Birk inhibitor) per kg body weight over the observation period of 14 hr. Data are given as mean values in percentage of the individual starting values. Thick line at the abscissa=endotoxin infusion period. For further details see text and ref. *10*. A: a, controls, b, endotoxin-treated. B: inhibitor and endotoxin-treated. ● complement C3; ★ factor XIII; △ prothrombin; ▲ AT III; □ plasminogen; ■ antiplasmin.

inhibitor from soybeans, clearly reduced the endotoxin-induced decline of the tested plasma factors. The inhibitor (*Mr* 7,000) was effective in dosages ranging from 3–8 mg (*i.e.*, 10–25 trypsin inhibiting units) per kg body weight. A reasonable assumption would be, therefore, that the exogenous inhibitor was able to prevent or reduce the neutrophil proteinase-induced consumption reactions very effectively.

EGLIN, A POTENT INHIBITOR OF NEUTROPHIL PROTEINASES

Another promising candidate for an inhibitor therapy in inflammatory processes might be eglin, an inhibitor present in the leech *Hirudo medicinalis*. It is a mini inhibitor protein with a molecular weight of 8,100 consisting of a single peptide chain with 70 amino acid residues (Fig. 7) (*13*). Eglin inhibits very strongly the neutral granulocytic proteinases elastase and cathepsin G, as well as chymotrypsin and sub-

1 10 20
NH_2-Thr-Glu-Phe-Gly-Ser-Glu-Leu-Lys-Ser-Phe-Pro-Glu-Val-Val-Gly-Lys-Thr-Val-Asn-Gln-

30 40
Ala-Arg-Glu-Tyr-Phe-Thr-Leu-His-Tyr-Pro-Gln-Tyr-Asp-Val-Tyr-Phe-Leu-Pro-Glu-Gly-

50 60
Ser-Pro-Val-Thr-Leu-Asn-Leu-Arg-Tyr-Asn-Arg-Val-Arg-Val-Phe-Tyr-Asn-Pro-Gly-Thr-

70
Asn-Val-Val-Asn-His-Val-Pro-His-Val-Gly-COOH

Fig. 7. Primary structure of eglin, an inhibitor of neutrophil elastase and cathepsin G from the leech *H. medicinalis.*

tilisin. The dissociation constants are in the range of 10^{-10} M. Eglin might be a useful drug in inflammatory processes which are at least partially caused by neutral granulocytic proteinases.

GENERAL CONCLUSIONS

In severe inflammatory processes, multiple trauma or shock, various cells such as neutrophils, macrophages, endothelial cells, and mast cells are stimulated or disintegrated. In this way a high potential of lysosomal enzymes is released of which the proteinases are of special pathogenetic effectiveness. Recent studies in our laboratory and by others indicate strongly that *substrate-unspecific proteolysis* by lysosomal proteinases and especially by the neutrophil elastase contributes to a significant degree to the consumption and/or degradation of extracellular substances in such diseases. On the other hand, early administration of convenient exogenous inhibitors directed against the lysosomal enzymes should have a positive therapeutic effect also in humans.

Acknowledgments
This work was supported by the Deutsche Forschungsgemeinschaft, Sonderforschungsbereich 51. The excellent technical assistance of Mrs. U. Hof and Mrs. C. Seidl is greatly appreciated.

REFERENCES

1. Aasen, A. O., Ohlsson, K., Larsbraaten, M., and Amundsen, E. *Eur. J. Surg. Res.*, **10**, 63 (1978).
2. Aasen, A. O. and Ohlsson, K. *Hoppe-Seyler's Z. Physiol. Chem.*, **359**, 683 (1978).
3. Aasen, A. O., Smith-Erichsen, N., Gallimore, M. J., and Amundsen, E. *Adv. Shock Res.*, **4**, 1 (1980).

4. Egbring, R., Gramse, M., Heimburger, N., and Havemann, K. *Diath. Haemorrh.*, **38**, 222 (1977).
5. Egbring, R., Schmidt, W., Fuchs, G., and Havemann, K. *Blood*, **49**, 219 (1977).
6. Gallimore, M. J., Aasen, A. O., Lyngaas, K., Larsbraaten, M., Smith-Erichsen, N., and Amundsen, E. *Microvasc. Res.*, **18**, 292 (1979).
7. Havemann, K. and Janoff, A. *In* "Neutral Proteases of Human Polymorphonuclear Leukocytes," eds. K. Havemann and A. Janoff (1978). Urban und Schwarzenberg Verlag, Baltimore-Munich.
8. Ikematsu, S., McDonagh, R. P., Reisner, H. M., Skrzynia, C., and McDonagh, J. *J. Lab. Clin. Med.*, **97**, 662 (1981).
9. Jochum, M., Lander, S., Heimburger, N., and Fritz, H. *Hoppe-Seyler's Z. Physiol. Chem.*, **362**, 103 (1981).
10. Jochum, M., Witte, J., Schiessler, H., Selbmann, H. K., Ruckdeschl, G., and Fritz, H. *Eur. Surg. Res.*, **13**, 152 (1981).
11. Klebanoff, S. J. and Clark, R. A. *In* "The Neutrophil Function and Clinical Disorders," eds. Klebanoff and Clark (1978). Elsevier/North-Holland Biomedical Press, Amsterdam.
12. Neumann, S., Hennrich, N., Gunzer, G., and Lang, H. *In* "Progress in Clinical Enzymology-II," eds. D. M. Goldberg and M. Werner (1983). Masson Publishing USA, Inc., in press.
13. Seemüller, U., Fritz, H., and Eulitz, M. *Methods Enzymol.*, **80**, 804 (1981).
14. Witte, J., Jochum, M., Scherer, R., Schramm, W., Hochstrasser, K., and Fritz, H. *Intensive Care Med.*, **8**, 215 (1982).

Proteinase Inhibitors: Medical and Biological Aspects, eds. N. Katunuma et al., pp. 97–109 (1983).
Japan Sci. Soc. Press, Tokyo/Springer-Verlag, Berlin

An Investigation of Intracellular Proteinases during Differentiation of Cultured Muscle Cells

F. J. ROISEN,[*1] H. KIRSCHKE,[*2] R. COLELLA,[*2] L. WOOD,[*2] A. C. St. JOHN,[*2] E. FEKETE,[*2] Q-S. LI,[*2] G. YORKE,[*1] and J. W. C. BIRD[*2]

Department of Anatomy, University of Medicine and Dentistry of New Jersey-Rutgers Medical School[*1] *and Bureau of Biological Research, Rutgers University*[*2]

Evidence supporting a role for the lysosome in muscle development and in terminal myofibrillary degradation in pathological states has steadily increased in recent years (*4*). However, uncertainties concerning the precise lysosomal intracellular location and enzymatic activity exist because of the heterogeneous cellular nature of muscle. Recent studies in our laboratory have utilized established rat myogenic lines and primary chick and rat embryonic muscle cultures to examine the role of the lysosomal apparatus in myogenesis. Relatively homogeneous myoblast populations were grown under defined culture conditions to provide two morphologically distinct developmental stages: pre-fusion and post-fusion (Fig. 1). The presence of a prominent lysosomal apparatus in each of these stages has been demonstrated by: acridine orange staining; cytochemical localization of acid phosphatase; immunocytochemical localization of the cathepsins B and H; determination of the activity levels of cathepsins B, D, H, and L; an examination of protein degradation in the presence of several proteinase inhibitors; and an evaluation of the effects of these inhibitors on myogenesis.

[*1,*2] Piscataway, N. J. 08854, U.S.A.

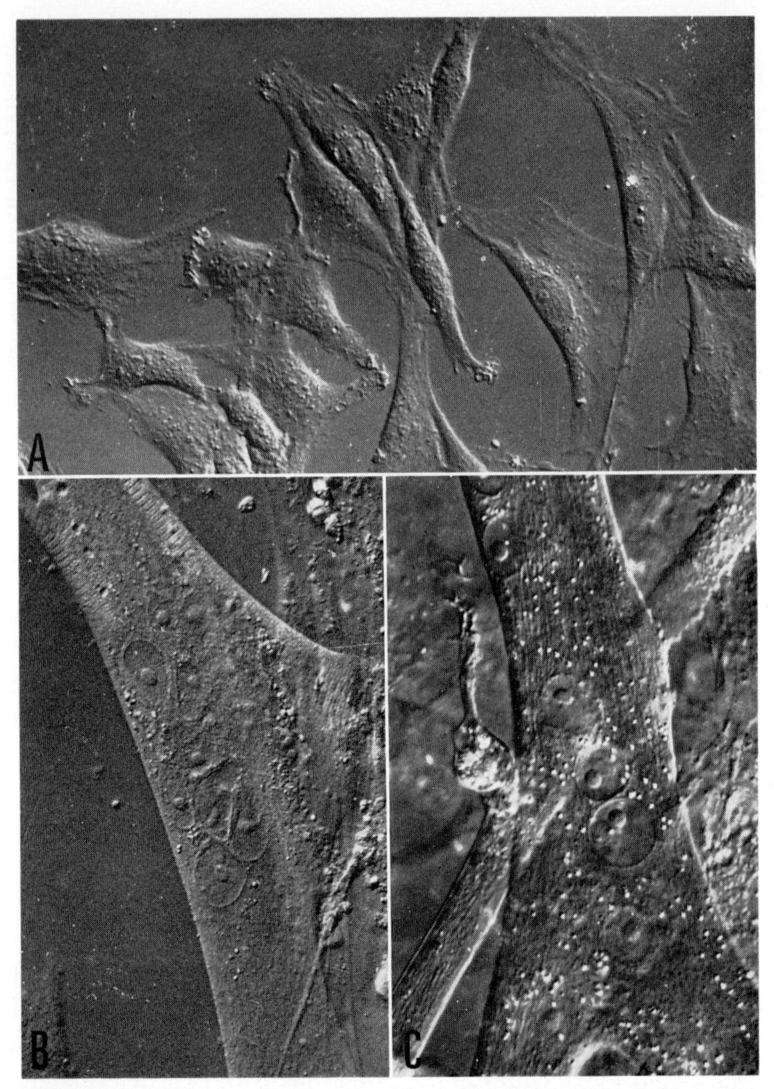

Fig. 1. Rat myoblasts were cultured to provide two developmentally distinct cell populations; Nomarski optics.
A: pre-fusion. L_6 myoblasts were plated at a low density to minimize cell-cell contact and prevent fusion. Four days *in vitro* (DIV). ×450. B: post-fusion. L_6 myoblasts plated at high density to insure fusion. Nuclear chains and prominent striations characterize this contractile myotube. 17 DIV. ×600. C: post-fusion. Primary muscle culture with nuclear chains and well developed myofibrillary apparatus. 6 DIV. ×600.

These studies suggested that the lysosomal proteinases play a prominent role in muscle differentiation.

MORPHOLOGICAL AND CYTOCHEMICAL STUDIES

Vital staining with acridine orange, a metachromatic dye used as a fluorescent marker for presumptive lysosomal material, revealed many prominent orange staining vesicles in pre-fusion and post-fusion cultures (*3*). Similarly we observed positive reaction-product to the known lysosomal marker enzyme acid phosphatase in representative cultures of both developmental stages. Generally less reaction-product was found in the post-fusion myotubes than in pre-fusion myoblasts (Fig. 2). The enzyme was also found to undergo a redistribution from the Golgi apparatus and smooth endoplasmic reticulum prior to fusion, to lysosomes and autophagic vesicles during fusion which eventually localized in the smooth endoplasmic reticulum (SER) and occasional lysosomes in highly differentiated post-fusion myotubes. This specific intracellular redistribution during the process of differentiation might facilitate a conservative lysosomal-mediated re-utilization of cytoplasmic and organellar components during muscle development.

The characterization and availability of monospecific rabbit antisera to rat liver cathepsins B, H, and L (*8*) has provided a probe for localizing these lysosomal enzymes in primary and established muscle culture. Indirect immunofluorescent studies with an fluorescein isothiocyanate (FITC) labeled second antibody of semi-synchronous pre-fusion populations of the rat L_6 myogenic line demonstrated that cathepsins B and H had a similar if not identical perinuclear distribution (Fig. 3A–B). Positive fluorescence was observed as discrete punctate areas asymmetrically distributed in the immediate perinuclear sarcoplasm which frequently appeared as two capped regions on opposite sides of the nucleus (Fig. 3A inset). This pattern of fluorescence was obtained consistently under a variety of culture conditions and was never observed in pre-immune, nonimmune, or second antibody controls. The pattern of perinuclear fluorescence found for the cathepsins B and H always occurred within the region of acridine orange-positive material. The antiserum to cathepsin B also uniformly decorated an area immediately

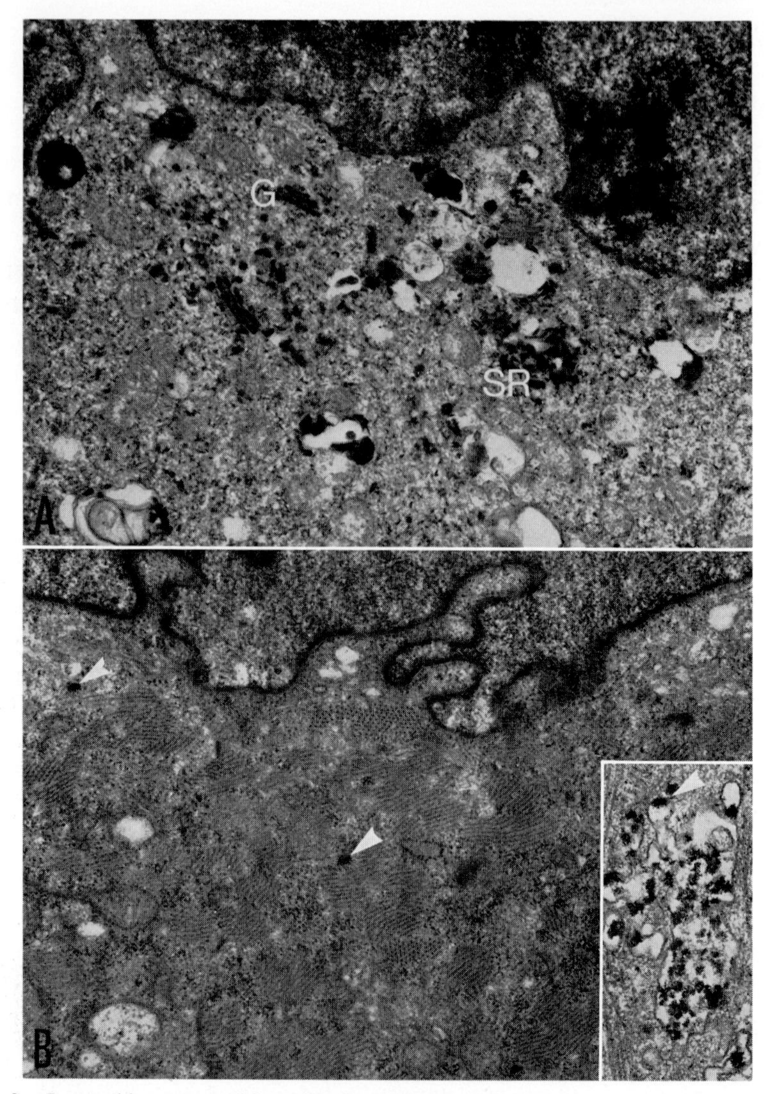

Fig. 2. L_6 myoblasts assayed for acid phosphatase.
A: pre-fusion. Reaction product is localized in a perinuclear region within the Golgi (G) and elements of the sarcoplasmic reticulum (SR). ×22,600. B: post-fusion. Actively contracting myotubes exhibit a high level of myofibrillary organization. Reaction product (arrowheads) was observed in the sarcoplasmic reticulum and occasional lysosomes. ×20,400. Inset. ×100,000.

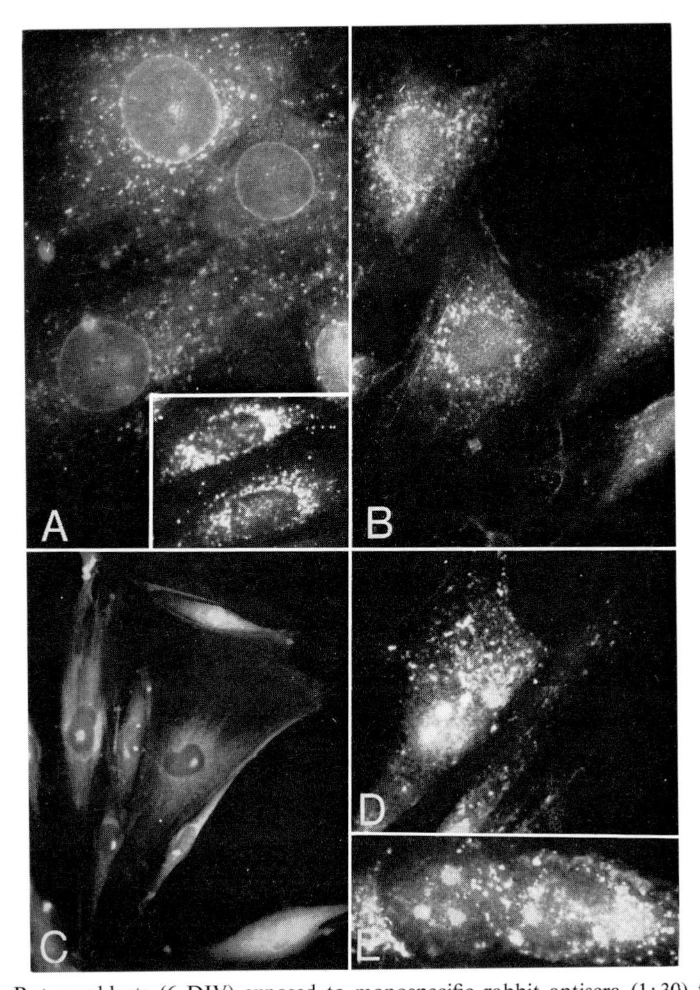

Fig. 3. Rat myoblasts (6 DIV) exposed to monospecific rabbit antisera (1:30) for 18 hr at 4°C followed by 1 hr exposure to FITC labeled goat anti-rabbit IgG.
Epifluorescence optics. A–C: pre-fusion L_6 cells. D–E: primary rat muscle. A: cathepsin B-positive perinuclear immunofluorescence highlighting the nuclear envelope. ×1,600. Inset illustrates asymmetric nuclear capping. ×500. B: cathepsin H-positive immunofluorescence with a similar pattern as that for cathepsin B. ×1,400. C–E: the cathepsin L-positive fluorescence appears more diffuse than the distribution seen for cathepsins B and H in pre-fusion. ×450, ×500, ×1,400, respectively.

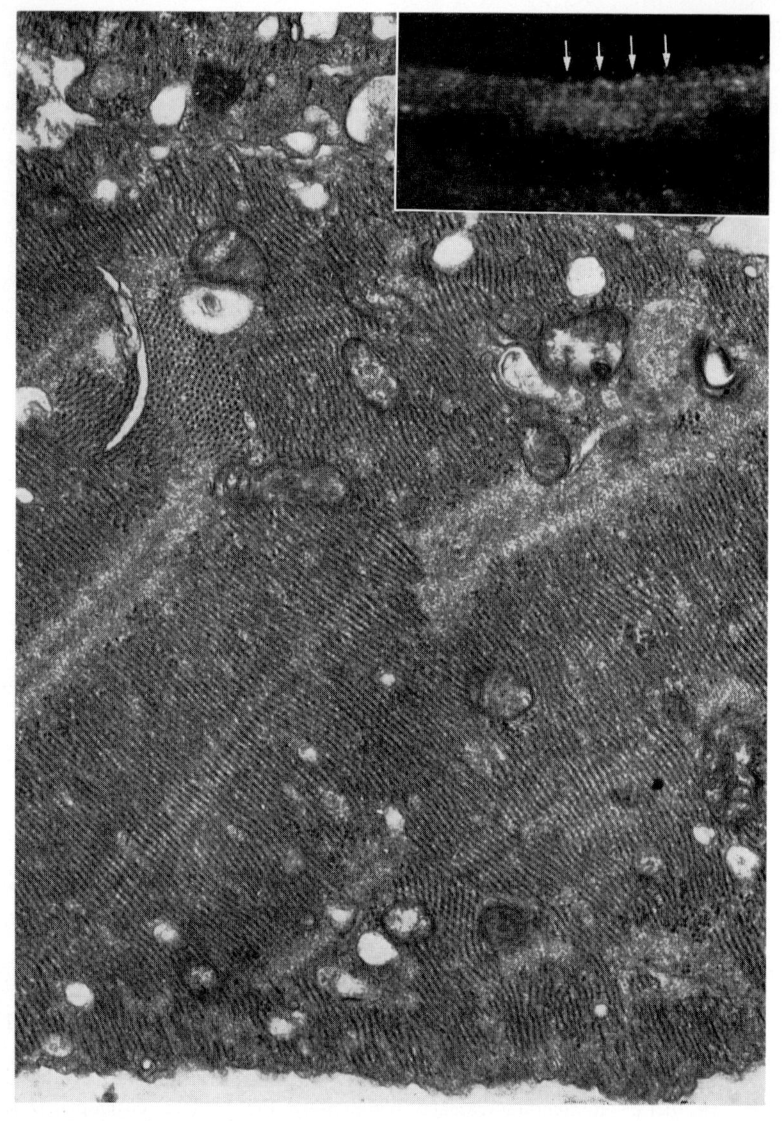

Fig. 4. Rat post-fusion myotube illustrating a high level of sarcomeric development and tubular system.
×38,000. Inset: cathepsin B-positive fluorescence. The periodic (arrowheads) fluorescence observed with antibody to rat liver cathepsin B reflects the sarcomeric arrangement of enzyme-containing structures. Epifluorescence optics, ×900.

adjacent to the nuclear membrane as evidenced by the highlighted nuclear envelopes (Fig. 3A).

In contrast to the distribution found for cathepsins B and H, the antiserum against cathepsin L produced a variable, less abundant, randomly dispersed pattern of fluorescence (Fig. 3C–E). These differences may reflect the relatively low levels of cathepsin L found in pre-fusion myoblasts (*11*).

Indirect immunofluorescent studies on highly differentiated post-fusion myotubes formed in cultures of either primary rat thigh muscle or the established rat L_6 myogenic line frequently exhibited a localized periodic fluorescence with antisera to cathepsins B and H that often appeared cross-streated (Fig. 4). This striated appearance probably reflected a specific alignment of the enzyme within a sarcomerically arranged component of the myofibrillary apparatus. Our studies on the cytochemical localization of acid phosphatase suggest further that the sarcoplasmic reticulum is the likely site of this fluorescence (*14*).

The pre-fusion mononuclear primary presumptive myoblasts never exhibited a uniform periodic fluorescence but did display a highly localized perinuclear fluorescence morphologically similar to that found in the pre-fusion rat L_6 myoblasts. Antisera to the rat liver cathepsins B, H, and L were all found to be species specific. They did not cross-react with any of the cells in our primary chick embryonic muscle cultures.

Our continuing immunofluorescent studies have demonstrated a specific intracellular distribution for cathepsins B and H in developing myoblasts. Future refinements in these techniques should identify the precise location of the lysosomal enzymes during myogenesis.

ENZYME ACTIVITIES IN PRE- AND POST-FUSION POPULATIONS

Recently we completed detailed studies on the concentrations of cathepsins B, H, and L in cultured myoblasts (*7*). A combination of very sensitive fluorescent substrates and relatively specific enzyme inhibitors were used to determine which substrates were most specific for each of the cysteine proteinases in our crude cell extracts. A summary of the reactivity of the specific substrates is found in Table I. Z-Arg-Arg-2-naphthylamide(NNap) and Z-Arg-Arg-4-methyl-7-coumaryl-amide

TABLE I. Cysteine Proteinase Substrates

Assay substrates	Lysosomal cysteine proteinases		
	Cathepsin B	Cathepsin H	Cathepsin L
Bz-Arg-NNap	+	+	−
Z-Arg-Arg-NNap	+	−	−
Arg-NNap	−	+	−
Azocasein	+	+	+
Z-Arg-Arg-NMec	+	−	−
Arg-NMec	−	+	−
Z-Phe-Arg-NMec	+	−	+
Z-Phe-Arg-NMec +Z-Phe-Phe-CHN_2[a]	+	−	−

-NNap, -2-naphthylamide; -NMec, -4-methyl-7-coumarylamide. + indicates reaction of enzyme with substrate.

[a] Only valid for the enzymes from rat. Inhibitor must be used in a specific concentration.

(NMec) (*1*) were highly specific substrates for cathepsin B. The-NMec substrates have also been shown to be much more sensitive than the corresponding naphthylamides, as illustrated for cathepsin B by a greater K_{cat}/K_m for Z-Arg-Arg-NMec ($3{,}419 \cdot 10^3 \cdot \text{M}^{-1} \cdot \text{sec}^{-1}$) than for Z-Arg-Arg-NNap ($887 \cdot 10^3 \cdot \text{M}^{-1} \cdot \text{sec}^{-1}$) (*1*).

No specific substrate has yet been found for cathepsin L. We used Z-Phe-Arg-NMec as a substrate for cathepsin L, and believe it to be superior when compared to azocasein or radio-labeled cytosol proteins, because of its high sensitivity (about 10 ng enzyme/ml can be determined). Z-Phe-Arg-NMec is rapidly hydrolyzed by cathepsins B and L (*1, 8*). The activities of cathepsins B and L against Z-Phe-Arg-NMec may be distinguished by judicious use of the inhibitor Z-Phe-Phe-CHN_2. This inhibitor reacts irreversibly and rapidly with cathepsin L, and reversibly with human and rat liver cathepsin B though irreversibly with rat cathepsin B from bovine spleen (*8*). Therefore, the reaction of Z-Phe-Phe-CHN_2 as a specific inhibitor of rat cathepsin L is very much dependent on the conditions of the assay. We found that purified rat liver cathepsin L was inhibited 100% after a preincubation of 5 min with 5×10^{-7} M Z-Phe-Phe-CHN_2. Z-Phe-Phe-CHN_2 produced a negligible (<2%) inhibition of cathepsin B purified from either rat liver or muscle.

Arg-NMec and Arg-NNap are suitable substrates for cathepsin H and arginyl aminopeptidase (*10*), but caution should be taken since they can be hydrolyzed by soluble alanyl aminopeptidase and leucyl aminopeptidase (*10*). The latter enzymes can be inhibited by puromycin and bestatin, respectively (*6*). Under our assay conditions, rat liver cathepsin H was completely inhibited by 10^{-4} M E-64; at 2×10^{-6} M E-64 85–90% inhibition was achieved. The lower concentration also completely inhibited the enzyme if the preincubation time was longer than 5 min.

The -NMec substrates allow the determination of cathepsins L, B, and H in solutions containing the enzymes at concentrations of 5–20 ng/ml, and therefore make it possible to measure these enzymes in an extract of cultured cells from a single 60 mm diameter petri dish.

We found that the fluorescent coumarylamide substrates were approximately 20 times more sensitive than the 2-naphthylamide substrates for the cysteine proteinases in muscle. The extracts of cultured myoblasts contained concentrations of the cysteine proteinases that were occasionally two orders of magnitude greater than the *in vivo* extracts (*7*). The reason for the difference in changes of the acid hydrolases is not clear, for example, cathepsin L was 500×, cathepsin B was 100×, and cathepsin H was 13× higher in *in vitro* myoblasts than in *in vivo* muscle extracts. The reason may have been that the different cell types in muscle contained different amounts of cysteine proteinases in their lysosomes; they also may have had different concentrations of endogenous inhibitors which might exhibit different actions on the various cysteine proteinases. The presence of potent endogenous inhibitor(s) to these enzymes in muscle tissue as well as in cultured myoblasts has been demonstrated (see below).

During differentiation, the mononucleated spindle-shaped, embryonic myoblasts proliferate for 3–4 days before they fuse and differentiate into multinucleated myotubes. One to two days later these fibers begin to contract spontaneously in the culture dishes. We demonstrated morphologically the presence of lysosomes in rat myoblasts, myotubes, and in a fusion-deficient mutant clone (*7*). Our data demonstrated an increase in lysosomal enzyme activities during differentiation. Cathepsin D had no change in activity, but the cysteine proteinase specific activities were increased 2–7 times in the myotubes as compared to the myoblasts. In view of the increase in protein concentration after fusion, the increase in the cysteine prteinases was especially high.

ENDOGENOUS INHIBITORS

We reported previously an endogenous inhibitor to cathepsin B in the cytoplasmic extract of muscle (*12*) which has been found in skeletal and heart muscles of the rat, cow, rabbit, pig, and human (*2*). This inhibitor has recently been purified from muscle (*5*, *9*) and found to be active against cathepsins B and H. We have examined the level of this inhibitory activity in extracts of pre- and post-fusion muscle cultures. Different degrees of inhibition were found for each of the three cysteine proteinases. Under our experimental conditions the maximal inhibition for these enzymes was 50%. Our studies provide important evidence that the myoblast (as opposed to other cell types in muscle) contains the endogenous inhibitors. The ability to measure the intracellular levels of endogenous inhibitors would allow calculation of the true intracellular levels of enzyme activities even in the presence of the inhibitors. Unfortunately calculations of this nature await further studies of the inhibitor kinetics.

PROTEIN TURNOVER

Our studies have shown that dramatic morphological changes associated with differentiation of pre-fusion, actively dividing myoblasts into multinucleated myotubes are accompanied by changes in the specific activities of lysosomal enzymes. The extensive development of the lysosomal apparatus in these cultures suggests that active protein turnover occurs during this development. To study the characteristics of this process, the rate of intracellular protein degradation was examined during development of L_6 myoblast cultures (*13*).

Pre-fusion L_6 myoblasts demonstrated a rapid rate of protein degradation; after 24 hr 30–60% of the labeled proteins were degraded. The degradation of labeled proteins in post-fusion cultures ranged from 35–45% (*13*). Therefore the average proteins in developing L_6 myoblasts had a half-life of 1–2 days.

We investigated the effects of inhibitors of lysosomal proteinases on intracellular protein degradation in these cells. Ammonium chloride is a general inhibitor of lysosomal activities since it is a weak base which accumulates within lysosomes and increases the intralysosomal pH thereby disrupting the functioning of lysosomal enzymes *in vivo*. At a concentration of 5–10 mM, ammonium chloride led to a consistent decrease in protein catabolism in pre-fusion, post-fusion and in a fusion-deficient subclone of the L_6 rat myoblasts. Approximately 20% of the proteins degraded over a 5 hr period were hydrolyzed by ammonium-sensitive enzymes.

To identify the lysosomal enzymes that were involved in protein catabolism, we examined the effects of well-characterized proteinase inhibitors on protein turnover. Leupeptin, pepstatin, and phenylmethyl sulfonyl fluoride (PMSF) were added to cultured L_6 myoblasts for a 24-hr incubation period. Leupeptin (50 μg/ml) inhibited proteolysis a small (11–15%) but significant extent, suggesting that a leupeptin-sensitive component was involved in the degradation rates (*11*). Vandenburgh and Kaufman (*15*) reported that leupeptin caused a similar inhibition of protein degradation in primary chick myotubes. Our studies with primary chick cultures have further confirmed this report. Since leupeptin is a reversible inhibitor of cathepsins B, H, and L, we were unable to assay cell extracts to identify which enzyme activities decreased and caused the lowered protein degradation.

The addition of PMSF was toxic to the cells, especially after fusion, so no assessment of the role of PMSF-sensitive proteases in intracellular protein catabolism could be made. The addition of pepstatin (50 μg/ml), a cathepsin D inhibitor, had no significant effect on the measured rates of intracellular protein catabolism. Caution must be taken in the interpretation of this data since this inhibitor does not readily cross the cell membrane.

Our recent studies have examined the effects of the irreversible cysteine proteinase inhibitor, E-64, on protein degradation in L_6 myoblasts and in primary cultures of chick embryonic muscle cells. This inhibitor was not toxic to the myoblasts at concentrations up to 250 μM. When L_6 myoblasts were treated for 24 hr with 125 μM E-64, protein degradation was reduced by approximately 20% in both pre-fusion and post-fusion populations. The same concentration of the inhibitor produced a 10–19% decrease in protein degradation in primary chick muscle cultures.

Since E-64 reacts stoichiometrically with cysteine proteinases, it is possible to determine the extent of inhibition of lysosomal proteinases after the 24 hr treatment period. The level of E-64 used in these studies was sufficient to inhibit 95–100% of the activities of cathepsins B, H, and L compared to untreated cultures. Thus we were able to correlate directly the changes in these lysosomal proteinase activities with the changes in intracellular protein degradation.

EFFECTS OF PROTEINASE INHIBITORS ON MYOBLAST FUSION

Primary chick myoblasts grown in medium containing one of several proteinase inhibitors had a slower rate of fusion when compared to control cultures. By 72 hr *in vitro*, cells grown in medium containing the lysosomotrophic amines NH_4Cl (2 mM) and chloroquine (3 μM) had 27% and 23% less nuclei found in myotubes when compared to control cultures, respectively. The cysteine proteinase inhibitors, leupeptin (20 μM), E-64 (25 μM), and antipain (20 μM) also inhibited fusion at 72 hr by 32%, 15%, and 64%, respectively. The aspartyl proteinase inhibitor, pepstatin (20 μM) and the aminopeptidase inhibitor bestatin (20 μM), had no apparent effect on fusion.

These results suggested that the cysteine proteinases were actively involved in fusion. However, at 25 μM, E-64 inhibited fusion only 15% whereas the cysteine proteinases, cathepsins B, H, and L were completely inhibited in cultures grown in medium containing E-64. The effects of E-64 on fusion were transient; by 6 days *in vitro* cultures grown in the presence of E-64 regained the same level of fusion as control cultures of the same age. An examination of cathepsins B, H, and L levels at days 3–6 revealed a 95–100% inhibition of these enzymes when E-64 was present throughout the culture period. Therefore, a primary role for the lysosomal cysteine proteinases in the fusion process seems unlikely since total inhibition of the cysteine proteinase activity by E-64 was not accompanied by a comparable inhibition of fusion.

Acknowledgments
This research was supported by a grant from the Muscular Dystrophy Association, Charles and Johanna Busch Endowment, and NIH NS

11299. The authors thank Janet Baxter for her expert technical assistance.

REFERENCES

1. Barrett, A. J. and Kirschke, H. *Methods Enzymol.*, **80**, 535 (1981).
2. Bird, J. W. C., Spanier, A. M., and Schwartz, W. N. *In* "Protein Turnover and Lysosome Function," eds. H. Segal and D. Doyle, p. 589 (1978). Academic Press, New York
3. Bird, J. W. C., Roisen, F. J., Yorke, G., Lee, J. A., McElligott, M. A., Triemer, D. F., and St. John, A. *J. Histochem. Cytochem.*, **29**, 431 (1981).
4. Bird, J. W. C. and Roisen, F. J. *In* "Myology," eds. A. Engel and D. Banks. McGraw Hill, New York, in press.
5. Brooks, R. M. and Bird, J. W. C. *Fed. Proc.*, **40**, 912 (1981).
6. Kawata, S., Takayama, S., Ninomiya, K., and Makisumi, S. *J. Biochem.*, **88**, 1601 (1980).
7. Kirschke, H., Wood, L., Roisen, F. J., and Bird, J. W. C. *Biochem. J.*, in press.
8. Kirschke, H. and Shaw, E. *Biochem. Biophys. Res. Commun.*, **101**, 454 (1981).
9. Lenney, J. F., Tolan, J., Sugai, W., and Lee, A. *Eur. J. Biochem.*, **101**, 153 (1979).
10. McDonald, J. K. and Barrett, A. J. *In* "Mammalian Proteases," Vol. 2, ed. A. J. Barrett (1977). North-Holland Publ. Co., Amsterdam.
11. McElligott, M. A., Roisen, F. J., Keaton, K. S., Triemer, D. F., Li, Q-S., St. John, A. C., Yorke, G., and Bird, J. W. C. *Acta Biol. Med. Germ.*, **40**, 1333 (1981).
12. Schwartz, W. N. and Bird, J. W. C. *Biochem. J.*, **167**, 811 (1977).
13. St. John, A. C., McElligott, M. A., Lee, J. A., Keaton, K. S., Yorke, G., Roisen, F. J., and Bird, J. W. C. *In* "Proteinases and Their Inhibitors: Structure, Function and Applied Aspects," eds. V. Turk and L. Vitale, p. 13 (1981). Pergamon Press, New York.
14. Triemer, D. F., Bird, J. W. C., and Roisen, F. J. Submitted.
15. Vandenburgh, H. and Kaufman, S. *J. Biol. Chem.*, **255**, 5826 (1980).

ENDOGENOUS PROTEINASE INHIBITORS

Proteinase Inhibitors: Medical and Biological Aspects, eds. N. Katunuma et al., pp. 113–123 (1983).
Japan Sci. Soc. Press, Tokyo/Springer-Verlag, Berlin

Cysteine Proteinase Inhibitors in Mammalian Plasma

James F. LENNEY

*Department of Pharmacology, School of Medicine, University of Hawaii**

Until recently, cathepsin B was the only known mammalian cysteine proteinase (endopeptidase). In the last decade about ten additional cysteine proteinases have been described. Most of these enzymes are lysosomal; they constitute a very important part of the machinery for intracellular proteolysis. Under certain conditions these proteinases are secreted into the extracellular space where they are presumably active. Because of the potency of the cysteine proteinases, it is important to determine how these enzymes are controlled or regulated.

Within cells, in every tissue that has been analyzed, there is a low molecular weight inhibitor of cysteine proteinases (*11*). This inhibitor is in the cytosol and would complex cysteine proteinases should they enter this compartment, for example, by leakage from lysosomes.

Cysteine proteinases outside of cells would encounter extracellular inhibitors such as those present in plasma. In the last 15 years, six plasma cysteine proteinase inhibitors have been described. These are not nearly as well known as the extensively studied plasma serine proteinase inhibitors. However, with the increasing recognition of the importance of cysteine proteinases, it is possible that this family of inhibitors may

* 1960 East West Road, Honolulu, Hawaii 96822, U.S.A.

turn out to be of comparable significance. The total number of proteinase inhibitors that have been reported in plasma is now approaching twenty. Mammals have built up a strong line of defense against any active proteinase which may find its way into plasma!

A CHRONOLOGY OF STUDIES ON PLASMA CYSTEINE PROTEINASE INHIBITORS

The pioneering literature on cysteine proteinase inhibitors in the blood stream of mammals is listed chronologically in Table I. In 1967, Snellman and Sylven (*26*) reported that purified human haptoglobin inhibited cathepsin B. (In the light of recent knowledge, the enzyme preparation employed was a mixture containing many liver lysosomal cysteine proteinases.) The inhibitory activity of haptoglobin is interesting because this molecule is an acute phase protein whose function is usually considered to be the complexing of free hemoglobin.

In 1971, Tokaji (*29*) found that rabbit, guinea pig or cow serum contained a thermostable factor which inhibited papain as well as a skin cysteine proteinase. The molecular weight of the inhibitor was not

TABLE I. Chronology of Salient Studies on Cysteine Proteinase Inhibitors in Mammalian Plasma or Serum

Source of plasma or serum	Inhibitor	Molecular weight of inhibitor	Proteinase employed	Year	Reference
Human	Haptoglobin	100,000	Calf liver cathepsin B	1967	*26*
Rabbit, guinea pig, cow	—	—	Skin cysteine proteinase and papain	1971	*29*
Human	α_2-Macroglobulin	725,000	Human liver cathepsin B	1973	*28*
	IgG	150,000			
Mouse	Haptoglobin	900	Cathepsin B	1974	*27*
Human	α_2TPI	90,000	Papain, ficin	1974	*22, 23*
Rat	I_1	74,000	Papain	1976	*6*
Human	α_1TPI	170,000	Papain	1979	*21*
	α_2TPI	90,000			
Human	Antithrombin III	62,000	Papain	1980	*30*
Rat	Haptoglobin	60,000	Rat liver cathepsin B	1980	*17*
Human	Haptoglobin	100,000	Human liver cathepsin B	1981	*9*
Rat	TPI-L	16,000	Cathepsin H and papain	1982	*5*
Horse, cow	TPI-L	12,000	Thiol cathepsins	1982	*2*
Human	TPI-L	13,000	Cathepsins B, H, and T	1982	*12*

determined. Starkey and Barrett (*28*) conducted a careful study of human serum, searching for inhibitors of human liver cathepsin B. They concluded that α_2-macroglobulin (also an acute phase protein) is the major inhibitor, while immunoglobulin G (IgG) has low potency, and haptoglobin has no activity against cathepsin B. They stated that serum proteins smaller than *Mr* 150,000 did not inhibit or bind cathepsin B. Subsequently, Snellman and Sylven (*27*) reported that an oligosaccharide (*Mr* 900) derived from mouse ascites fluid haptoglobin inhibited pure cathepsin B, papain, and trypsin.

In 1974 and 1977, Sasaki *et al.* (*22, 23*) described an inhibitor (*Mr* 90,000) of papain, ficin, and bromelain in human serum. Because of its electrophoretic mobility, this protein was designated α_2-thiol proteinase inhibitor (α_2-TPI). By gel filtration on Sephadex G-200, it was separated from the aggregated form of haptoglobin (*22*). In 1976, Jarvinen (*6*) showed that rat serum contained a papain inhibitor with a molecular weight of 74,000. In 1979 Ryley (*21*) isolated and purified two papain inhibitors from human plasma. These proteins had molecular weights of 90,000 and 170,000 and appeared to be immunologically identical (presumably monomer and dimer). This inhibitor did not cross-react with human haptoglobin, α_2-macroglobulin, or the well-known serine proteinase inhibitors of human plasma. Human serum contained an average of 43 mg/100 ml of these proteins. The pure inhibitor contained 10% carbohydrate and had a pI of 4.8, with a high acidic amino acid content. Jarvinen (*7*) also purified these inhibitors from human serum and found that α_1TPI had a pI of about 3.4 and α_2TPI had a pI of about 4.7. Both inhibitors were very effective against papain and ficin, less effective against bromelain and human cathepsin B, and were inactive against trypsin, chymotrypsin, and elastase. These inhibitors were shown to be immunologically distinct from 15 human serum proteins, including haptoglobin and antithrombin III.

In 1980, Valeri *et al.* (*30*) reported that purified human antithrombin III inhibited papain. Pagano and coworkers (*17*) studied the inhibition of rat liver cathepsin B by purified rat serum haptoglobin. Inhibition was more pronounced when proteins were used as cathepsin B substrates than when the B chain of insulin or benzoyl arginine-2-naphthylamide were used. Kalsheker *et al.* (*9*) showed that purified human plasma haptoglobin inhibited purified human liver cathepsin B, using benzoyl-Arg-Arg-2-naphthylamide as substrate. This inhibition was reversed by the addition of monospecific antiserum to haptoglobin.

Further studies of human plasma α_1TPI and α_2TPI were conducted by Sasaki and coworkers (*24*). Two forms of α_2TPI were detected, and the three proteins were antigenically indistinguishable. The higher molecular weight form was more stable at 80°C than were the two low molecular weight forms. Pagano and Engler (*18*) showed that αTPI from ascites or pleural fluid inhibited human liver cathepsin L, while Gounaris and Barrett (*3*) found that human plasma αTPI was more effective against cathepsin H than against cathepsin B. Sasaki *et al.* (*25*) found that human α_1TPI and α_2TPI were strong inhibitors of porcine muscle calcium-activated neutral proteinase, a non-lysosomal cysteine proteinase. In whole plasma, αTPI had a greater affinity for this enzyme than did α_2-macroglobulin. Pagano *et al.* (*19*) showed that rat haptoglobin inhibited rat liver cathepsin L as well as cathepsin B. They also demonstrated that asialohaptoglobin and the haptoglobin-hemoglobin complex were effective against cathepsins B and L. (The five references cited in this paragraph are omitted from Table I.)

It is interesting to note that the heavy chains of haptoglobin have a high degree of sequence homology to the family of mammalian serine proteinases (*1*). This region of the haptoglobin molecule is not involved in the binding of hemoglobin (*1*). Since the hemoglobin-binding site is not concerned with the binding of cysteine proteinases (*19*) it will be instructive to determine whether the "serine proteinase region" is.

In 1982, three laboratories independently discovered a low molecular weight cysteine proteinase inhibitor in serum. Iwata and coworkers (*5*) found that rat serum contained a protein (*Mr* 16,000) which was an effective inhibitor of papain and rat liver cathepsin H. These authors stated that this inhibitor (TPI-L) resembled the low molecular weight inhibitor which they had previously isolated from rat liver in its inhibition spectrum, molecular weight, and its pI of 9.2. Christopher *et al.* (*2*) reported that bovine and equine serum contained a *Mr* 12,000 inhibitor of lysosomal cysteine proteinases. This protein was stable at pH 4 when heated at 100°C, but was inactivated under these conditions in the presence of sulfhydryl compounds. Lenney *et al.* (*12*) showed that human serum and plasma contained multiple forms of a low molecular weight protein which inhibited several lysosomal cysteine cathcpsins. Very recently Hirado and coworkers (*4*) purified TPI-L from bovine plasma. This molecule had a molecular weight of 15,500 and was immunologically distinct from a *Mr* 73,000 papain inhibitor which was also isolated from bovine plasma.

There are a number of discrepancies between the findings of the various laboratories that have studied the cysteine proteinase inhibitors in plasma. On several occasions, only one or two of the plasma inhibitors were detected and the others were overlooked. Results obtained in our laboratory indicate that the number of inhibitors detected in human plasma depends upon many factors such as plasma pretreatment procedures, the reference proteinase, the substrate, and the assay procedure. For example, antithrombin III inhibits papain but not cathepsin B, while human TPI-L inhibits cathepsin B but not papain.

To summarize the literature cited, plasma contains six different cysteine proteinase inhibitors: haptoglobin, α_2-macroglobulin, IgG, αTPI, antithrombin III, and TPI-L. These proteins are readily distinguished from one another on the basis of criteria such as molecular weight, chromatographic separation, or immunological identity.

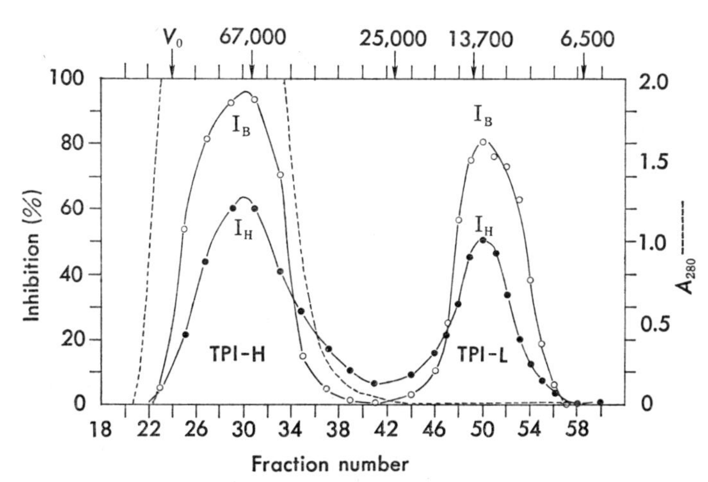

Fig. 1. Sephadex G-75 elution profile of human serum inhibitors of cathepsins B (○) and H (●).

Two ml of serum were placed on the column bed (1.5×114 cm) and 3.0 ml fractions were collected, eluting with 10 mM phosphate buffer pH 7.4 containing 0.9% NaCl and 0.02% NaN_3. The arrows at the top indicate the void volume and the elution volumes of the calibrating proteins.

A CLOSER LOOK AT THE LOW MOLECULAR WEIGHT INHIBITOR IN HUMAN PLASMA

In our laboratory, when human plasma or serum was chromatographed on a Sephadex G-75 column, cysteine cathepsin inhibitors with molecular weights of approximately 80,000 and 13,000 were eluted (*12*). A typical elution pattern is shown in Fig. 1. Cathepsin B and H were assayed using Cbz-Phe-Arg-aminomethylcoumarin and Leu-naphthylamide respectively as substrates, measuring the hydrolysis products fluorometrically. The low molecular weight inhibitor (TPI-L) was purified about 350-fold by this gel filtration step. When TPI-L was subjected to cationic polyacrylamide gel electrophoresis, eight protein bands were detected by Coomassie Blue staining. TPI-L was shown to be a protein; its activity was destroyed when it was treated with crystalline pepsin, trypsin, or subtilisin.

Human tissues contain a cathepsin B inhibitor having approximately the same molecular size as plasma TPI-L (*11*). Therefore it was imperative that the tissue and plasma inhibitors be compared. Plasma TPI-L and a partially purified human lung inhibitor were tested against eleven different proteinases. Both inhibited liver cathepsins B and H and rat kidney cathepsin T. The lung inhibitor was very effective against papain, ficin, and bromelain, whereas plasma TPI-L had little or no effect on these plant cysteine proteinases. Neither preparation inhibited trypsin or chymotrypsin. Both inhibitors were stable when heated at 90°C for 10 min. However, heating in the presence of dithiothreitol destroyed plasma TPI-L only, whereas heating in the presence of oxidized glutathione destroyed the lung inhibitor selectively. Finally, the lung inhibitor was a slightly smaller molecule than plasma TPI-L, having an approximate molecular weight of 11,000. It was concluded that the plasma and lung inhibitors probably represent two different gene products.

Plasma TPI-L is present in multiple molecular forms having isoelectric points ranging from approximately 4 to 12. TPI-L was prepared from a sample containing serum pooled from 40 people. When this TPI-L preparation was chromatographed on a hydroxylapatite column, nine peaks were resolved; each inhibited cathepsins B and H. When plasma from one person was subjected to these same chromatographic procedures, only two forms of the inhibitor were obtained. A plasma sample from another person was analyzed by chromatography on Sephadex G-75 and CM-cellulose, with the resolution of four inhibitor peaks. These preliminary results indicate that there are differences between individuals in regard to the number of molecular forms of the inhibitor. It is also possible that some of the microheterogeneity is attributable to proteolysis by the proteinases activated during clotting.

The plasma TPI-L-cathepsin B complex is a fully effective inhibitor of cathepsin H and the TPI-L-cathepsin H complex is fully effective against cathepsin B (*13*). Therefore it appears that TPI-L is a double headed inhibitor and the two active centers do not overlap.

Human rheumatoid synovial fluid was analyzed by chromatography on a Sephadex G-75 column. The elution pattern resembled that obtained for plasma (Fig. 1), containing the high and low molecular weight peaks. Human cerebrospinal fluid contained only the low molecular weight inhibitor, and this peak had the same inhibition spectrum as serum TPI-L. Therefore the cerebrospinal fluid inhibitor is probably of serum rather than tissue origin.

When plasma is chromatographed on a Sephadex G-75 column, α_2-macroglobulin emerges at the void volume. However, this inhibitor was not detected in Fig. 1 because proteinases "trapped" by α_2-macroglobulin retain their activity against low molecular weight substrates. Haptoglobin and αTPI should both be present in the TPI-H peak. We analyzed highly purified samples of human haptoglobin and α_2TPI and found that they had similar potencies as inhibitors of liver cathepsins B and H. A sample of human antithrombin III was inactive when tested against these two proteinases. We found that the TPI-H peak was very active in inhibiting papain and ficin, whereas TPI-L had little or no activity against these enzymes. The earlier workers who studied the cysteine proteinase inhibitors of serum used papain or ficin as reference proteinases, and therefore did not detect TPI-L when they analyzed gel filtration column eluates (*6*, *23*). Judging from the size of the peaks in Fig. 1, the total inhibitory capacity of TPI-L in plasma probably equals or exceeds that of haptoglobin or αTPI.

Recently Mort and coworkers (*14*) reported that human ascites fluid contained a latent form (*Mr* 41,000) of a cathepsin B-like proteinase which was readily activated by treatment with pepsin. Since ascites fluid is derived from plasma and TPI-L is destroyed by pepsin, it seems likely that the latent proteinase is an enzyme-plasma TPI-L complex.

PHYSIOLOGICAL ROLE OF THE CYSTEINE PROTEINASE INHIBITORS IN PLASMA

If a cysteine proteinase should enter the blood stream, it would be complexed by the inhibitors having the greatest affinity for that enzyme, probably αTPI, TPI-L, haptoglobin, or α_2-macroglobulin. In the case of the serine proteinase inhibitors of plasma, there is evidence that complexed proteinases are transferred to α_2-macroglobulin and this complex is promptly removed from the blood stream by the reticulo-endothelial system. This sequence of events probably occurs to cysteine proteinases as well. Reaction with a proteinase alters the conformation of α_2-macroglobulin and increases its electrophoretic mobility; it is this altered form which is recognized by the reticuloendothelial cells.

Intracellular enzymes are often detectable in extracellular spaces. One source of these enzymes is the process of phagocytosis, which releases lysosomal hydrolases to the outside of the cell. Proteinases are detectable in extracellular spaces as a result of cell death, inflammation, trauma or pathological conditions such as malignancy. Here the cysteine proteinases would encounter plasma proteinase inhibitors. For example, we found plasma TPI-H and TPI-L in rheumatoid synovial fluid and TPI-L in cerebrospinal fluid (*12*). Inhibitor concentration in these fluids was similar to that found in plasma itself. In rheumatoid arthritis, proteinases released by leucocytes are believed to cause damage to the articulating surfaces. In view of the large number of lysosomal cysteine proteinases (at least six are known, with additional ones being reported every year) it is likely that the plasma-derived inhibitors of these powerful enzymes would normally play a protective role. Figure 1 illustrates that serum TPI-H and TPI-L are individually capable of effectively inhibiting cathepsin B, despite a 30-fold dilution of the serum sample during gel filtration and assay.

In recent years there have been many reports that a cathepsin B-like proteinase is selectively secreted by malignant cells, but not by their normal counterparts. This proteinase could play a role in the invasiveness of cancer cells; the plasma inhibitors of cysteine proteinases probably exert a protective effect against this enzyme. As mentioned above, a cathepsin B-like proteinase is present in human ascites fluid in a latent form (*14*) which seems to be an enzyme-plasma TPI-L complex. There have been reports (*e.g.*, *20*) of a cathepsin B-like proteinase measurable in the sera of cancer patients. However, because of the inhibitors in serum and the removal of proteinases by α_2-macroglobulin, cysteine proteinases entering the blood stream should not be detectable. Human serum is very active against Cbz-Phe-Arg-aminomethylcoumarin whereas plasma is not (*13*). This activity is attributable to serum kallikrein, which is activated during clotting. This substrate is also hydrolyzed by cathepsin B, which has a substrate specificity similar to that of kallikrein.

In muscular dystrophy, elevated concentrations of several proteinases have been reported. For example, Kar and Pearson (*10*) found that patients with muscular dystrophy and related diseases had muscle cathepsin B or H levels two to six times higher than normal. It has also been reported that muscle calcium-activated neutral proteinase is elevated in muscular dystrophy (*15*). All three of these cysteine proteinases are inhibited by the plasma inhibitors, should the enzymes find their way into extracellular spaces. Stracher and coworkers (this volume) have demonstrated that leupeptin, a powerful inhibitor of cysteine proteinases, markedly retards the muscle tissue breakdown that occurs in genetically dystrophic animals. It would be interesting to ascertain whether there are any genetic deficiencies of plasma or tissue proteinase inhibitors in muscular dystrophy patients.

It is well known that plasma α_1-antitrypsin provides protection against the development of emphysema in normal individuals. Lysosomal proteinases from alveolar macrophages are believed to be important causative agents, elastase being a prime suspect. In experimental animals, instillation of pure papain into the lungs produces a condition closely resembling human chronic obstructive lung disease. There are several powerful macrophage lysosomal cysteine proteinases and these enzymes resemble papain in substrate specificity. Since these hydrolases are released during phagocytosis, it is logical to assume that they contribute to the lung tissue damage that occurs in emphysema. In fact, a free cathepsin B-like proteinase has been found in lavage fluid aspirated from human lungs (*16*). Furthermore, Johnson and Travis (*8*) have shown that cathepsin B can destroy α_1-antitrypsin. Just as α_1-antitrypsin protects against elastase, it seems likely that the plasma cysteine proteinase inhibitors provide protection against cathepsin B and the other lysosomal cysteine proteinases. Here also, it will be important to determine whether genetic deficiencies of these inhibitors exist, and whether they predispose the individual to emphysema.

REFERENCES

1. Arcoleo, J. P. and Greer, J. *J. Biol. Chem.*, **257**, 10063 (1982).
2. Christopher, C. W., Cullinane, P. M., Stebbins, D. R., and Kelley, C. A. *Fed. Proc.*, **41**, 1017 (1982).
3. Gounaris, A. D. and Barrett, A. J. *Fed. Proc.*, **41**, 1017 (1982).
4. Hirado, M., Niinobe, M., and Fujii, S. *Biochim. Biophys. Acta* (1983), in press.
5. Iwata, D., Hirado, M., Niinobe, M., and Fujii, S. *Biochem. Biophys. Res. Commun.*, **104**, 1525 (1982).
6. Jarvinen, M. *Acta Chem. Scand.*, **B30**, 933 (1976).
7. Jarvinen, M. *FEBS Lett.*, **108**, 461 (1979).
8. Johnson, D. and Travis, J. *Biochem. J.*, **163**, 639 (1977).
9. Kalsheker, N. A., Bradwell, A. R., and Burnett, D. *Experientia*, **37**, 447 (1981).
10. Kar, N. C. and Pearson, C. M. *Biochem. Med.*, **18**, 126 (1977).
11. Lenney, J. F., Tolan, J. R., Sugai, W. J., and Lee, A. G. *Eur. J. Biochem.*, **101**, 153 (1979).
12. Lenney, J. F., Liao, J. R., Sugg, S. L., Gopalakrishnan, V., Wong, H. C. H., Ouye, K. H., and Chan, P. W. H. *Biochem. Biophys. Res. Commun.*, **108**, 1581 (1982).
13. Lenney, J. F. unpublished data.
14. Mort, J. S., Leduc, M., and Recklies, A. D. *Biochim. Biophys. Acta*, **662**, 173 (1981).
15. Neerunjun, J. S. and Dubowitz, V. *J. Neurol. Sci.*, **40**, 105 (1979).
16. Orlowski, M., Orlowski, J., Lesser, M., and Kilburn, K. H. *J. Lab. Clin. Med.*, **97**, 467 (1981).
17. Pagano, M., Engler, R., Gelin, M., and Jayle, M. F. *Can. J. Biochem.*, **58**, 410 (1980).
18. Pagano, M. and Engler, R. *FEBS Lett.*, **138**, 307 (1982).
19. Pagano, M., Nicola, M. A., and Engler, R. *Can. J. Biochem.*, **60**, 631 (1982).
20. Pietras, R. J., Szego, C. M., Mangan, C. E., Seeler, B. J., and Burtnett, M. M. *Gynecol. Oncol.*, **7**, 1 (1979).
21. Ryley, H. C. *Biochem. Biophys. Res. Commun.*, **89**, 871 (1979).
22. Sasaki, M., Yamamoto, H., Yamamoto, H., and Iida, S. *J. Biochem.*, **75**, 171 (1974).
23. Sasaki, M., Minakata, K., Yamamoto, H., Niwa, M., Kato, T., and Ito, N. *Biochem. Biophys. Res. Commun.*, **76**, 917 (1977).
24. Sasaki, M., Taniguchi, K., and Minakata, K. *J. Biochem.*, **89**, 169 (1981).
25. Sasaki, M., Taniguchi, K., Suzuki, K., and Imahori, K. *Biochem. Biophys. Res. Commun.*, **110**, 256 (1983).

26. Snellman, O. and Sylven, B. *Nature*, **216**, 1033 (1967).
27. Snellman, O. and Sylven, B. *Experientia*, **30**, 1114 (1974).
28. Starkey, P. M. and Barrett, A. J. *Biochem. J.*, **131**, 823 (1973).
29. Tokaji, G. *Kumamoto Med. J.*, **24**, 68 (1971).
30. Valeri, A. M., Wilson, S. M., and Feinman, R. D. *Biochim. Biophys. Acta*, **614**, 526 (1980).

Proteinase Inhibitors: Medical and Biological Aspects, eds. N. Katunuma et al., pp. 125–134 (1983).
Japan Sci. Soc. Press, Tokyo/Springer-Verlag, Berlin

Lysosomal Cysteine Proteinases and Their Protein Inhibitors—Structural Studies

V. TURK,[*1] J. BRZIN,[*1] M. KOPITAR,[*1] I. KREGAR,[*1] P. LOČNIKAR,[*1] M. LONGER,[*1] T. POPOVIĆ,[*1] A. RITONJA,[*1] Lj. VITALE,[*2] W. MACHLEIDT,[*3] T. GIRALDI,[*4] and G. SAVA[*4]

Department of Biochemistry, J. Stefan Institute,[*1] *Department of Organic Chemistry and Biochemistry, Rudjer Bošković Institute,*[*2] *Institute of Physiological Chemistry, Physical Biochemistry and Cell Biology, University of Munich,*[*3] *and Institute of Pharmacology, University of Triest*[*4]

It is well known that tissue proteinases play an important role in the degradation of cell proteins under both normal and a variety of pathological conditions. In particular, lysosomal cysteine (thiol) proteinases are considered to be implicated in hormone metabolism, processing of neuropeptides from their precursors, as well as in pathological processes such as pulmonary emphysema, muscle dystrophy and malignancy. Apart from cathepsin B, so far the most investigated cysteine proteinase, more enzymes of this class were recently discovered, such as cathepsins H, L, N, S, T, and similar enzymes present in various tissues and species.

The recent discovery of the presence of endogenous protein inhibitors of cysteine proteinases added new impetus to this field of investigation. These inhibitors were isolated in different degrees of purity from the human epidermis (*6*), human psoriatic scales (*14*), human spleen (*7*), different rat and human tissues (*11*), and rat liver (*5, 9*).

A previous report from our laboratory described the simultaneous isolation of bovine spleen cysteine proteinases cathepsins B, H, and S and their characterization (*12*). From the same organ and from pig

[*1] 61000 Ljubljana, Yugoslavia. [*2] 41000 Zagreb, Yugoslavia. [*3] 8000 Munich, FRG. [*4] 34100 Triest, Italy.

leucocytes (*3*, *10*) we isolated from cytosol several endogenous inhibitors of cysteine proteinases which differed in their ability to inhibit cathepsins B, H, S, and papain, indicating some differences in their properties.

In this paper we present some properties of cysteine proteinases, and the partial amino acid sequence of cathepsins B and L from bovine spleen. Furthermore, the purification and characterization of cysteine proteinase inhibitors from egg white and from human leucocytes is presented. Amino acid sequences and circular dichroism studies show distinct structural differences among them. The possible antimetastatic effect of cysteine proteinases is examined.

PROPERTIES OF BOVINE SPLEEN CATHEPSINS B, H, AND L AND THE PARTIAL PRIMARY STRUCTURE OF CATHEPSINS B AND L

The simultaneous purification procedure of cathepsins B, H, and L is based on slightly modified methods for the isolation of bovine lymph (*17*) and spleen (*12*) cathepsins. All three enzymes were purified by tissue acid extraction, ammonium sulphate precipitation, Sephadex G-50 column chromatography, covalent chromatography on thiol Sepharose 4B, followed by final separation on CM cellulose. Cathepsin B was

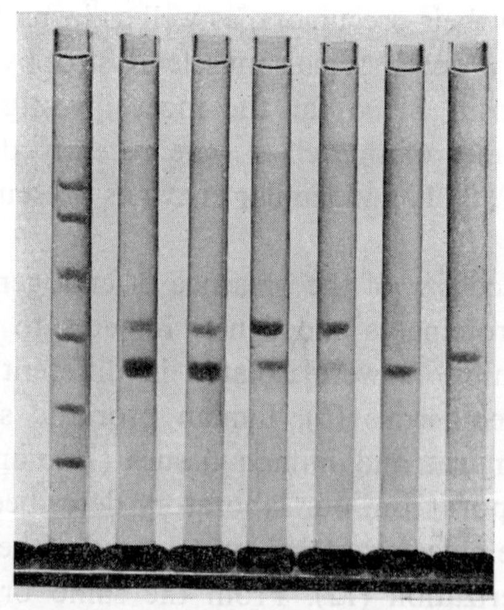

Fig. 1. SDS-PAGE of bovine spleen cysteine proteinases.
From left to right: 1, protein standards; 2–5, fractions of cathepsin B; 6, cathepsin L; 7, cathepsin H.

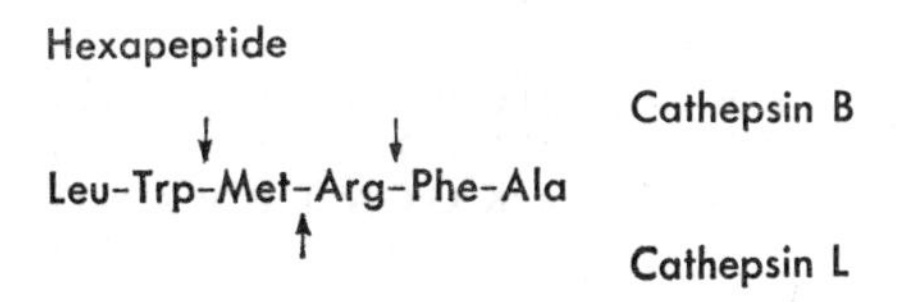

Fig. 2. Substrate specificity of cathepsins B and L toward hexapeptide.

eluted in four active protein fractions which differ in their content of two major polypeptide chains of molecular weight of about 30,000–31,000 and 25,000 (heavy chain), and minor (light chain) of molecular weight of about 5,000 (weakly visible) as detected by SDS-polyacrylamidegel electrophoresis (SDS-PAGE) (Fig. 1). It was suggested that limited proteolytic cleavage might occur to form the heavy and light chain of cathepsin B (*8*). Cathepsins H and L were apparently homogenous.

Some of the properties of cathepsins B, H, and L have already been reported (*12*) and are generally in good agreement with published data (*2*, *8*). Although it is well known that all three enzymes act as endopeptidases, a most interesting observation is the specificity of cathepsin H, which also acts as an aminopeptidase and is therefore described as an aminoendopeptidase (*2*).

Our specificity studies using synthetic hexapeptide (Fig. 2) show that bovine cathepsin H acts on this substrate as an aminopeptidase. Cathepsin B cleaves two peptide bonds, namely Phe-Arg and Met-Trp, whereas cathepsin L acts only on the Met-Arg bond. The same result was obtained by Katunuma and coworkers (*8*), with the exception that we found additional cleavage of the Met-Trp bond by cathepsin B. It has to be added at this point that previously discovered cathepsin S, first isolated in our laboratory and characterized (*12*, *16*), corresponds to cathepsin L on the basis of the substrate specificity already described and similar properties. Small differences are ascribed to species and tissue variations.

Elucidation of the mechanism responsible for specific catalysis of cathepsins B, H, and L requires a knowledge of their structure. Our purification procedure enables us to prepare not only cathepsin B but also cathepsin L in higher amounts enabling the structural characterization of these enzymes. In Fig. 3 the amino terminal sequences of cathepsins B and L are presented and compared with those of papain

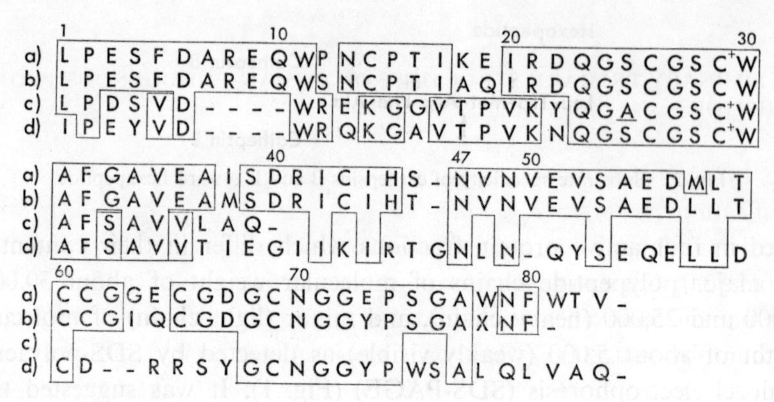

Fig. 3. Comparison of N-terminal sequences of cysteine proteinases.
a) bovine spleen cathepsin B; b) rat liver cathepsin B (*8*); c) bovine spleen cathepsin L; d) papain (*13*). Active site cysteine 29 is marked with a +. Asparagine 47 is C-terminal residue of the light chain of cathepsin B.

(*13*) and rat liver cathepsin B (*8*). A high degree of homology between bovine spleen and rat liver cathepsin B is evident. Among the first 80 amino acid residues there are only 9 amino acids which are not identical. From the data presented, it is also evident that the amino terminal part of cathepsin L is much more homologous to papain. A short loop (residues 7 to 10) is missing in papain as well as in cathepsin L. An additional disulfide bond between Cys^{14} and Cys^{43} was found in cathepsin B which does not exist in papain and cathepsin L. Cys^{29} was found to be in the active site of papain (*13*), cathepsins B and L using active site labelling. The complete homology of residues 26 to 32 in the four enzymes indicates that the structure of the cysteine part of the active site might be the same in all four. A further highly homoregion in cathepsin B and papain was found around Cys^{69} from residues 68 to residue 73, apparently as a consequence of the common disulfide logous bridge Cys^{26}-Cys^{69}.

Recently the light chain of the cathepsin B molecule was isolated from bovine spleen (*15*) and that part of the primary structure (residues 1–47) determined is in complete agreement with our results. The complete primary structure of this important group of proteolytic enzymes will shed additional light on their phylogenetic kinship.

PRELIMINARY PRIMARY STRUCTURE OF HUMAN ENDOGENOUS PROTEIN INHIBITOR

The cytosol of human leucocytes was used as a starting source. For purification, we used a modified version of a method published previously (*3*). In the first purification step we omitted alkaline treatment at pH 12.0. Cytosol was directly applied to a Sephadex G-50 column and papain inhibitory protein was eluted as a single peak. After chromatography on DEAE Sephacel column, the main inhibitory peak comprising about 80% of the total inhibitory activity was applied on papain-Sepharose 4B in 0.01 M Tris buffer, pH 7.8. The inhibitor was eluted with 0.01 M NaOH.

SDS-PAGE showed a single protein band with a molecular weight slightly smaller than cytochrome *c* (Mr=12,400). The pI value of the inhibitor is between 4.7 and 5.0. The inhibitor inhibits papain, human cathepsins B and H, and bovine spleen cathepsin L at equimolar ratios. K_i for papain and cathepsins L and H is about 10^{-9} M, whereas for cathepsin B it is higher.

The preliminary amino acid sequence of this inhibitor was determined by automated solid-phase sequencing and is presented in Fig. 4. Most surprising is the absence of cysteine. There is so far no known primary structure for any of the endogenous protein inhibitors of cysteine proteinases. Therefore we will discuss and compare its properties and structure together with egg white cystatin, the properties and structure of which are described in the next section.

STRUCTURAL PROPERTIES OF PROTEIN EGG WHITE INHIBITOR CYSTATIN

Recently a new isolation procedure was described for cystatin, a protein

```
1         10         20         30
M I P G G L S E A K P A T P E I Q E I V D K V K P Q L E E K T N

          40         50         60
E T Y G K L E A V Q Y K T Q V V A G T N Y Y I K V R A G D N K Y

        70         80         90        98
M H L K V F K S L P G Q N E D L V L T G Y Q V D K N K D D E L T G F
```

Fig. 4. Preliminary complete amino acid sequence of human leucocytes protein endogenous inhibitor.

inhibitor of cysteine proteinases, and partially characterized (*1*). In order to compare this inhibitor with endogenous protein inhibitor, we isolated it using a modified procedure.

The whites of 100 eggs were diluted with distilled water. The homogenate was adjusted to pH 12.0 for 30 min, then acidified to pH 5.5 and the resulting precipitate was removed by centrifugation. The total inhibitory activity preserved was in this step, whereas the specific inhibitory activity increased several fold. The supernatant was immediately adjusted to pH 7.8, concentrated by ultrafiltration and applied to a Sephadex G-50 column. Fractions containing inhibitory activity were pooled and further chromatographed on DEAE Sephacel column in 0.01 M Tris buffer pH 8.0. Three protein peaks (designated as forms I, II, and III) containing inhibitory activity were eluted with a NaCl gradient (0.0–0.1 M). The major protein peak (form II) was rechromatographed on a Sephadex G-50 column in order to remove remaining impurities. SDS-PAGE showed that all three forms are apparently homogenous, having molecular weights corresponding to the protein

TABLE I. Amino Acid Composition of Inhibitors from Egg White Cystatins and Human Leucocyte Endogenous Inhibitor

Amino acid	Egg white form III pI 5.6	Inhibitor form II pI 6.5	Leucocyte inhibitor
Asp	10	9	11
Thr	5	5	7
Ser	12	10	2
Glu	14	13	15
Pro	4	4	5
Gly	5	5	8
Ala	7	7	5
Cys	4	4	0
Val	9	9	9
Met	2	2	2
Ile	6	6	4
Leu	10	8	8
Tyr	5	5	6
Phe	3	3	2
His	1	1	1
Lys	7	7	12
Trp	1	1	0
Arg	8	6	1
Total	113	105	98

standard cytochrome *c*. The pI's of forms I and II are about 6.5, whereas from III has a pI of about 5.6. The amino acid composition of forms II and III shows slight differences (Table I), which can be explained by the differences in the amino-terminal part of the two forms (Fig. 5). It was reported (*1*) that forms with pI 5.6 and 6.5 have the same amino-terminal serine. At this stage it is difficult to explain these differences. Also, it is possible that cleavage of the larger form during the purification procedure might occur. Our preliminary data show that both forms inhibit papain and cathepsins B, H, and L at an equimolar ratio.

Comparing the amino acid composition and amino-terminal sequences of egg white cystatin and human endogenous protein inhibitor,

```
    1               10                  20                  30
a)  S E D R S R L L G A P V P V D E N D E G L Q R A L Q F A M A - -
b)  ————————————————G A P V P V D E N D E G L Q R A L Q F A M A - -
```

Fig. 5. Comparison of N-terminal sequences of two forms of egg white cystatin. a) form III, pI 5.6; b) form II, pI 6.5.

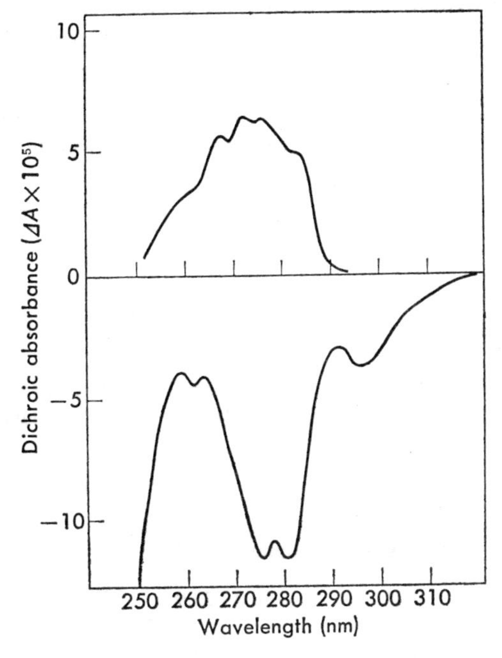

Fig. 6. Near UV circular dichroism spectra of human leucocytes protein endogenous inhibitor (top) and egg white cystatin form II (bottom).

we can conclude that these inhibitors are different. The most striking difference so far obtained is absence of cysteine in human endogenous inhibitor, whereas cystatin presumably contains two S-S bridges. There are also differences in the molecular weight. Molecular weights of egg white cystatin, forms II and III, are about 11,800 and 12,800, respectively, whereas molecular weight of human leucocyte protein endogenous inhibitor is 10,860. Additionally, circular dichroism spectra in the near UV region of egg white inhibitor (form II) and human protein inhibitor are different (Fig. 6). The absence of Trp in human endogenous inhibitor is evident. On the basis of differences between egg white cystatin and endogenous protein inhibitor from human leucocytes we can conclude that they are probably not members of the homologous family of cysteine proteinase inhibitors. More studies to elucidate their structure and mechanism are needed.

PHYSIOLOGICAL ROLE

There is enough evidence that cathepsin B and very likely other cysteine proteinases play an important role in the degradation of proteins. Their activity has been implicated also in various aspects of cancer, including the invasion of host tissue and the formation of metastases. Discovery of their endogenous inhibitors indicated their possible regulatory function, also under pathological conditions. We reported previously that neutral proteinase inhibitors have been found capable of inhibiting metastases formation in mice bearing solid malignant tumors (*4*). Also, the presently tested endogenous protein inhibitor of bovine spleen and egg white cystatin proved active in causing a selective inhibition of

TABLE II. Effects of Protein Inhibitors on Primary Tumor Growth and on the Formation of Spontaneous Pulmonary Metastases in Mice Bearing Subcutaneous Lewis Lung Carcinoma

Inhibitor	Dose (μmol/kg/day)	Percent inhibition of		
		s.c. primary tumor growth	Metastasis	
			Number	Weight
Bovine spleen inhibitor	0.08	−16.0±22.9	31.1±10[a]	71.9±9.1[a]
Egg white	0.5	9.2±7.8	41.3±13.8[a]	34.3±24.5

[a] Means significantly different, Student-Neumann-Keuls test, $p=0.05$.

lung metastases formation in mice bearing subcutaneously (s.c.) Lewis lung carcinoma, since they significantly reduced the number of metastases (also their weight, in the case of protein endogenous inhibitor), whereas no effect whatsoever was caused on the grown of the s.c. primary tumor (Table II).

These results appear to encourage further studies on the use of proteinase inhibitors in experimental systems which may eventually lead to their future use as drugs.

Acknowledgments

We thank Prof. Dr. H. Fritz and Dr. U. Seemüller for their help and stimulating discussions. The excellent technical assistance of Mrs. M. Božič, Mrs. A. Burkeljc, Mrs. S. Košir, and Mrs. U. Borchart is gratefully acknowledged. This work was supported by the Research Council of Slovenia and in part by the Deutsche Forschungsgemeinschaft—Sonder—forschungsbereich 027, München, FRG, and by the NSF, U.S.A.

REFERENCES

1. Barrett, A. J. *Methods Enzymol.*, **80**, 771 (1981).
2. Barrett, A. J. and Kirschke, H. *Methods Enzymol.*, **80**, 535 (1981).
3. Brzin, J., Kopitar, M., Ločnikar, P., and Turk, V. *FEBS Lett.*, **138**, 193 (1982).
4. Giraldi, T., Sava, G., Kopitar, M., Brzin, J., Suhar, A., and Turk, V. *Eur. J. Cancer Clin. Oncol.*, **17**, 1301 (1981).
5. Hirado, M., Iwata, D., Niinobe, M., and Fujii, S. *Biochim. Biophys. Acta*, **669**, 21 (1981).
6. Järvinen, M. *J. Invest. Dermatol.*, **71**, 114 (1978).
7. Järvinen, M. and Rinne, A. *Biochim. Biophys. Acta*, **708**, 210 (1982).
8. Katunuma, N., Towatari, T., Kominami, E., Hashida, S., Takio, K., and Titani, K. *Acta Biol. Med. Germ.*, **40**, 1419 (1981).
9. Kominami, E., Wakamatsu, N., and Katunuma, N. *Biochem. Biophys. Res. Commun.*, **99**, 568 (1981).
10. Kopitar, M., Brzin, J., Zvonar, T., Ločnikar, P., Kregar, I., and Turk, V. *FEBS Lett.*, **91**, 355 (1978).
11. Lenney, J. F., Tolan, J. R., Sugai, W. S., and Lee, A.D. *Eur. J. Biochem.*, **101**, 153 (1979).
12. Ločnikar, P., Popović, T., Lah, T., Kregar, I., Babnik, J., Kopitar, M., and Turk, V. *In* "Proteinases and their Inhibitors. Structure, Function and Applied Aspects," ed. V. Turk and Lj. Vitale, p. 109 (1981). Mladinska Knjiga-Pergamon Press, Ljubljana-Oxford.

13. Mitchel, R. E. J., Chaiken, I. M., and Smith, E. L. *J. Biol. Chem.*, **245**, 3485 (1970).
14. Okitani, O., Fukuyama, K., and Epstein, W. I. *J. Invest. Dermatol.*, **82**, 280 (1982).
15. Pohl, J., Baudyš, M., Tomašek, V., and Kostka, V. *FEBS Lett.*, **142**, 23 (1982).
16. Turnšek, T., Kregar, I., and Lebez, D. *Biochim. Biophys. Acta* **403**, 514 (1975).
17. Zvonar, T., Kregar, I., and Turk, V. *Croat. Chem. Acta*, **52**, 411 (1979).

Proteinase Inhibitors: Medical and Biological Aspects, eds. N. Katunuma et al., pp. 135–145 (1983).
Japan Sci. Soc. Press, Tokyo/Springer-Verlag, Berlin

Structure, Function, and Regulation of Endogenous Thiol Proteinase Inhibitor

Nobuhiko KATUNUMA,[*1] Nobuaki WAKAMATSU,[*1] Koji TAKIO,[*2] Koichi TITANI,[*2] and Eiki KOMINAMI[*1]

Department of Enzyme Chemistry, Institute for Enzyme Research, School of Medicine, The University of Tokushima[*1] *and The Haward Huges Medical Institute Laboratory and Department of Biochemistory, S-70, University of Washington*[*2]

Cathepsins B, H, and L, well characterized lysosomal thiol proteinases, have been purified from various tissues of many animals (*10*, *14*). Recently, cathepsins B and H from rat liver have been sequenced completely (*23*), and the importance of these enzymes in intracellular protein degradation has been demonstrated (*4*, *21*).

There are many reports on the presence of an endogenous inhibitor(s) of thiol proteinases in mammalian tissues: Järvinen (*9*) found that rat skin contains two inhibitors of a rat skin thiol proteinase that hydrolyzes Bz-Arg-2-naphthylamide. Lenney *et al.* (*18*) found thermostable inhibitors of cathepsins B and H in all rat and human tissues examined. Recently, a thiol proteinase inhibitor was isolated in pure form from rat liver in our laboratory (*15*, *16*) and by Hirado *et al.* (*7*).

STRUCTURE

A thiol proteinase inhibitoı was purified from rat liveı by affinity chromatography on inactivated papain-Sepharose 4B (*16*). Essentially, the same procedure can be used for purification of the inhibitor from other

[*1] Kuramoto-cho 3-18-15, Tokushima 770, Japan.
[*2] Seattle, Washington 98195, U.S.A.

organs of rats. Lenney *et al.* (*18*) reported the presence of two inhibitors of cathepsins B and H and used heat treatment for their purification. These inhibitors from rat lung have molecular weights of 14,000. We found that multiple forms of the thiol proteinase inhibitor were observed on purification by a procedure involving heat treatment, but that a single protein was separated by SDS-polyacrylamide gel electrophoresis without heat treatment. The inhibitor purified has two different pI values, as described later, but these inhibitors show the same inhibition spectrum against thiol proteinases and differ from inhibitors I^B and I^H described by Lenney *et al.* (*18*).

The purified inhibitor is a monomeric protein with a molecular weight of 12,500, and contains 9.3% aromatic, 50.5% polar amino acid residues and 2 mol cysteine (Cys-3 and Cys-64) but no tryptophan

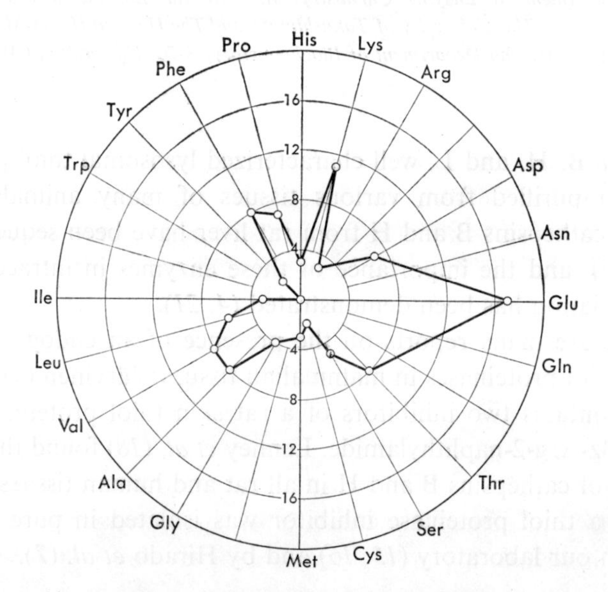

Fig. 1. Amino acid composition of thiol proteinase inhibitor from rat liver.

```
     5         10        15        20        25        30        35        40
X-MM[C]GAPSATMPATTETQEIADKVKSQLEEKANQKFDVFKA
      45        50        55        60    65        70        75        80
 ISFRRQVVAGTNFFIKVDVGEEK[C]VHLRVFEPLPHENKPL
      85        90        95  98
 TLSSYQTDKEKHDELTYF
```

Fig. 2. Total amino acid sequence of native thiol proteinase inhibitor (TPI-B) from rat liver.

(Fig. 1). Its minimum molecular weight, calculated assuming that it had 98 amino acid residues/molecule, is 12,400. This value is in agreement with the molecular weight determined by gel filtration and by SDS-polyacrylamide gel electrophoresis. Its carbohydrate content is less than 0.5 %, indicating that it is not a glycoprotein.

We reported previously on the aminoterminal sequence (42 residues) of the thiol proteinase inhibitor (*23*) and we present here the complete amino acid sequence. For this, it was subjected to sequence analysis after reduction and carboxymethylation of its disulfide bonds and then further purification by reverse-phase high performance liquid chromatography. The complete amino acid sequence of preparation purified by affinity chromatography on inactivated papain-Sepharose with sodium tetracyanate is shown in Fig. 2. The N-terminal amino acid of the preparation purified on papain-Sepharose (TPI-A) was cysteine, but this terminal (methionine) was blocked by acetylation in the inhibitor purified on inactivated papain-Sepharose (TPI-B). These results suggest that TPI-B is the native inhibitor ($Mr \fallingdotseq 12{,}500$) and that TPI-A ($Mr \fallingdotseq 11{,}000$) is formed by limited proteolysis of the N-terminal region of TPI-B. TPI-A might lose Ac-Met-Met residues.

INHIBITION SPECTRUM

Tissue thiol proteinase inhibitor inhibits most thiol proteinases (cathepsins B, H, L, and C papain, and ficin) but not serine proteinases (trypsin, chymotrypsin, mast cell proteases I and II, and cathepsin A), or a carboxyl proteinase, cathepsin D. Thus its inhibitory action is restricted to thiol proteinases. Although calcium-dependent neutral protease is a type of thiol proteinase, it was not inhibited by addition of five times as much inhibitor as required for complete inhibition of papain. The association of most thiol proteinases (papain, ficin, cathepsins H, C, and L) with the inhibitor is stoichiometric and equimolar, but a linear relationship was not obtained for inhibition of cathepsin B by the inhibitor, and its inhibitory effect was much less on cathepsin B than on other thiol proteinases. Moreover, the optimum pH for inhibition of cathepsin B is different from the optimum pH for cathepsin B itself, whereas those for other inhibitor-sensitive thiol proteinases coincide with the pH optima of the proteinases. Recently, we (*23*, *25*) demonstrated the complete amino acid sequences of rat liver cathepsins B and H. Comparison of the sequences of these cathepsins with each other and with those of papain and actinidin showed a striking homology among their primary structure. However, the results also indicated that cathepsin H is more closely related to the plant enzymes, papain and actinidin, than to cathepsin B of the same origin. This may explain the marked difference in inhibition of cathepsin H and cathepsin B by the thiol proteinase inhibitor, although information on the 3-D structure of thiol cathepsins containing a binding pocket for the substrate and on the molecular mechanism of interaction of thiol proteinases and the inhibitor is required for a definite conclusion.

MECHANISM OF INHIBITION

The inhibition of cathepsin H by the inhibitor is rapid, reaching a maximum within 15 sec at 0°C. Kinetical studies indicated that the inhibitor inhibits papain non-competitively and pseudo-irreversibly. When cathepsin H was preincubated with the inhibitor before electrophoresis, a band of a complex of cathepsin H and the inhibitor was observed (Fig. 1). Neither cathepsin H nor inhibitor activity was de-

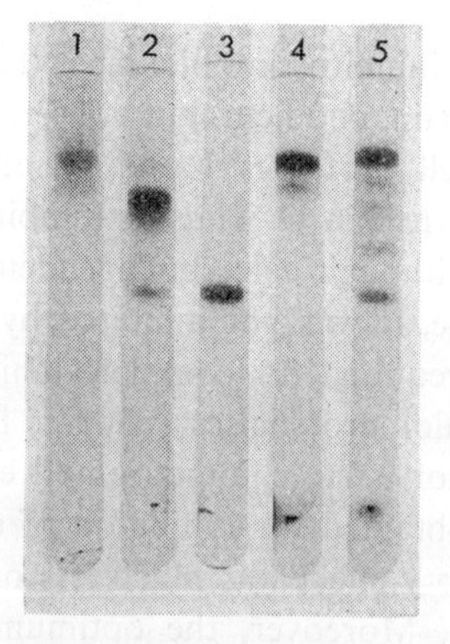

Fig. 3. Formation of complex between cathepsin H and thiol proteinase inhibitor and its retardation by preincubation of cathepsin H with E-64.
1, cathepsin H; 2, cathepsin H plus inhibitor; 3, inhibitor; 4, cathepsin H plus E-64; 5, cathepsin H plus E-64, then plus inhibitor.

tected in the complex extracted from slices of the gel. Inhibitor binds to cathepsin H (one mole to one mole ratio).

Studies by gel filtration on a Sephadex G-25 column showed that when cathepsin H was preincubated with E-64, a strong irreversible inhibitor of thiol proteinases, formation of the enzyme inhibitor complex was inhibited (Fig. 3). On the contrary, when ^{3}H-labelled E-64 was added to the reaction mixture after formation of the enzyme-inhibitor complex, incorporation of ^{3}H-E-64 into cathepsin H was inhibited. However, the inhibitor could still bind to papain-Sepharose 4B inactivated with tetrathionate, as described before, and polyacrylamide gel electrophoresis in the absence of SDS also showed that inactive cathepsin H could still form a complex with the inhibitor. Since cathepsin H is inactive in the absence of a thiol reagent, these results suggest that the thiol proteinase inhibitor does not bind to the active cysteine residue of thiol proteinases. E-64 may not enter the active site of the enzyme in the complex.

On SDS-polyacrylamide gel electrophoresis, the complex between cathepsin H and the inhibitor gave two bands corresponding to cathepsin H and the inhibitor, suggesting that the interaction of thiol proteinase inhibitor and proteinases is non-covalent. The molecular weight of the inhibitor dissociated from the complex in the presence of SDS was about 1,100 daltons less than that of the native inhibitor. This observation seems consistent with the facts that the inhibitor eluted from a column of papain-Sepharose had undergone limited proteolysis, but that eluted from a papain-Sepharose column treated with tetracyanate still had an N-terminal extension. Since the nicked inhibitor purified by affinity chromatography on papain-Sepharose still had inhibitory activity, the N-terminal fragment may not be necessary for such activity.

MODULATION OF THIOL PROTEINASE INHIBITOR ACTIVITY

The intracellular level of thiol proteinase inhibitor is regulated by the rates of its synthesis and degradation. In addition, we demonstrated another mechanism of regulation of the thiol proteinase inhibitor. This inhibitor was purified to the step of Sephadex G-50 chromatography in one day and subjected to isoelectrofocussing. Extraction of the inhibitor from slices of the gel showed that it gave a single peak with a pI of 5.2.

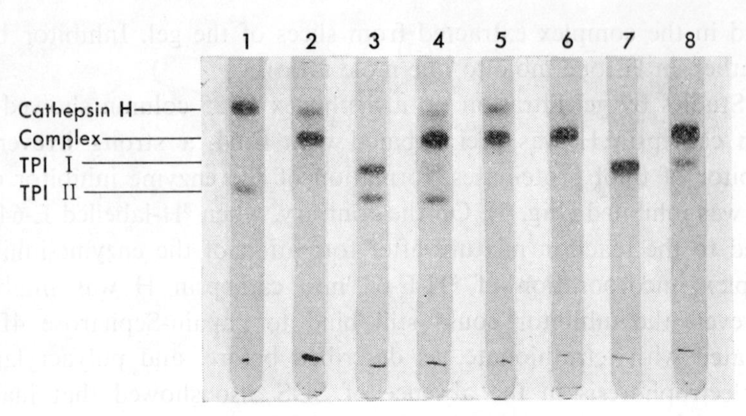

Fig. 4. Effect of cysteine on conversion of TPI-I to TPI-II and on formation of complex with cathepsin H and two types of inhibitors.
1, cathepsin H+TPI-II; 2, cathepsin H+TPI-I; 3, TPI-I+TPI-II; 4, cathepsin H+TPI-I+TPI-II; 5, cathepsin H+inhibitor+cysteine 4 mM; 6, cathepsin H+TPI-I+cysteine 4 mM; 7, TPI-I+TPI-II+cysteine 4 mM; 8, cathepsin H+TPI-I+TPI-II+cysteine 4 mM.

However, when it was stored at −20°C for several days, most of the activity showed a pI of 5.0. The highly purified inhibitor also often gave two protein bands on polyacrylamide gel electrophoresis at pH 8.0 without SDS (Fig. 4). The upper band (TPI-I) had a pI of 5.2, and the lower band (TPI-II) a pI of 5.0. TPI-II was converted to TPI-I by the addition of cysteine; the conversion was complete on addition of 1 mM L-cysteine, but only 50% on addition of 0.5 mM cysteine. For the conversion, cysteine could be replaced by reduced glutathione, 2-mercaptoethanol or dithiothreitol. TPI-I was reversibly converted to TPI-II by addition of oxidized glutathione, but the conversion was partial even with a high concentration of oxidized glutathione. The activity of the inhibitors separated by polyacrylamide gel electrophoresis was examined: TPI-I was active and formed a complex with thiol proteinases, but TPI-II was inactive and did not form complexes. For complete formation of complexes of TPI-II with thiol proteinases, it was necessary to add 1 mM L-cysteine to the reaction mixture, which is the concentration required to convert TPI-II to TPI-I. The apparent molecular weights of the two types of inhibitor were the same (Mr=12,500).

The CD profile of TPI-I is different in far ultraviolet region from that of TPI-II and TPI-I is much more unfolded than TPI-II. That is, the binding site to protease might be exposed in TPI-I.

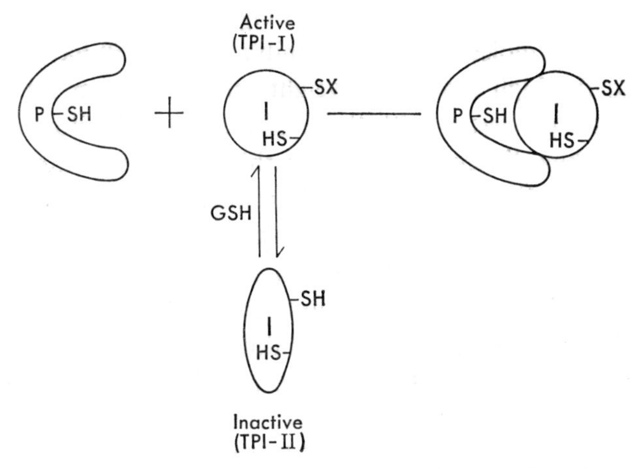

Fig. 5. Regulation of thiol proteinase activities by liver thiol proteinase inhibitor. P, thiol proteinase; I, inhibitor; X, mixed disulfide.

Thus our working hypothesis on regulation of thiol proteinase activities by endogenous inhibitor is shown in Fig. 5. There is a high intracellular level of glutathione (several μmol/g liver (*3*)) and the level changes with dietary conditions (*13*) and on administration of amino acids or inhibitors of enzymes involved in the γ-glutamyl cycle (*20*). Therefore, the ratio of TPI-I to TPI-II may be regulated by the concentration of glutathione *in vivo*.

TISSUE DISTRIBUTION OF THIOL PROTEINASE INHIBITORS AND THEIR COMPARISON WITH THE INHIBITORS IN SERUM

Thiol proteinase inhibitor, assayed for inhibitory activity against cathepsin H, was found in all organs of rats tested (*16*). Only low activity was detected in the serum. Immunological diffusion analysis with anti-liver thiol proteinase inhibitor serum indicated that the inhibitor in rat liver is immunologically identical with the inhibitors in other rat tissues. The presence of the inhibitor in various tissues of mammals, including pigs, hamsters, rabbits, and humans has been reported (*18*, *26*); thus, the inhibitor seems to be ubiquitous in mammals. But species specificity is observed: anti-rat inhibitor serum did not cross-react with the human or rabbit inhibitor (*24*).

In serum, three high molecular weight thiol proteinase inhibitors (*22*) and one low molecular weight inhibitor (*1*, *8*, *19*) have been found. We purified the high molecular weight inhibitors from rat serum and compared their properties with those of the liver inhibitor (*24*). Results showed that they differed from tissue thiol proteinase inhibitor in the following points: 1) all three serum forms of inhibitor had much higher molecular weights than the liver inhibitor. 2) No cross-reactivity was observed between the inhibitors from serum and liver in tests with antiserum inhibitor or anti-liver antiserum. 3) The inhibitors from serum and liver were all specific for thiol proteinases, but their inhibition spectra differed; the inhibitory spectra of the three inhibitors from serum were very similar. The inhibitors from serum and liver had similar inhibitory effects on cathepsins L and B, papain, and ficin. The inhibitor from liver, but not those from serum, also inhibited cathepsins H and C. Conversely, the inhibitors from serum, but not that from liver, inhibited calcium-dependent neutral protease. 4) The inhibitor from liver did not bind to concanavalin A-Sepharose, whereas those from serum did bind and were eluted with α-methyl mannoside.

Recently, a thiol proteinase inhibitor of low molecular weight was discovered in rat serum (*8*), human serum (*19*), and calf and horse sera (*1*). The inhibitor from rat serum had a molecular weight of 16,000 and an isoelectronic point of 9.2. This protein differed from liver thiol proteinase inhibitor in its pI value and immunological properties: it showed no cross-reactivity with the inhibitor from liver in a test with anti-liver inhibitor antiserum (unpublished observation).

SUBCELLULAR LOCALIZATION

On subcellular fractionation of rat liver, most of the thiol proteinase inhibitor was recovered in the cytosol fraction, whereas inhibitor sensitive thiol proteinases are recovered in the lysosomal fraction (*16*). When rat liver was homogenized with cold water and the homogenate was used for assay of cathepsin H, cathepsin B+L (Suc-Tyr-Met-β-naphtylamide hydrolyzing activity) and thiol proteinase inhibitor, significant activities of thiol cathepsins were detected, but no inhibitor was found. These results indicate that cytoplasmic thiol proteinase inhibitor bound to the lysosomal thiol cathepsins with loss of both activities and that there was less inhibitor than lysosomal thiol cathepsins in liver. Since E-64 is known to be an irreversible inhibitor of cathepsins B and L (*5*), the amount of E-64-bound protein in liver was measured using [^{3}H]-E-64 (*6*). Results showed that liver contains about 4–6 nmol/g of proteins binding to E-64. The amount of thiol proteinase inhibitor in liver was calculated as 2–3 nmol/g liver based on the specific activity of the purified inhibitor. Thus if proteins bound to E-64 are considered to be lysosomal thiol cathepsins, these results also show that proteinases are present in greater quantity than inhibitor in the liver.

BIOLOGICAL FUNCTIONS

Two possible physiological functions of the inhibitor may be considered: (1) protection against inappropriate proteolysis, and (2) control of lysosomal thiol proteinase activities. Injection of leupeptin or chloroquine into rats caused increase in the activities of free lysosomal proteinases in the liver cytosol (*17*), and after the injection the level of

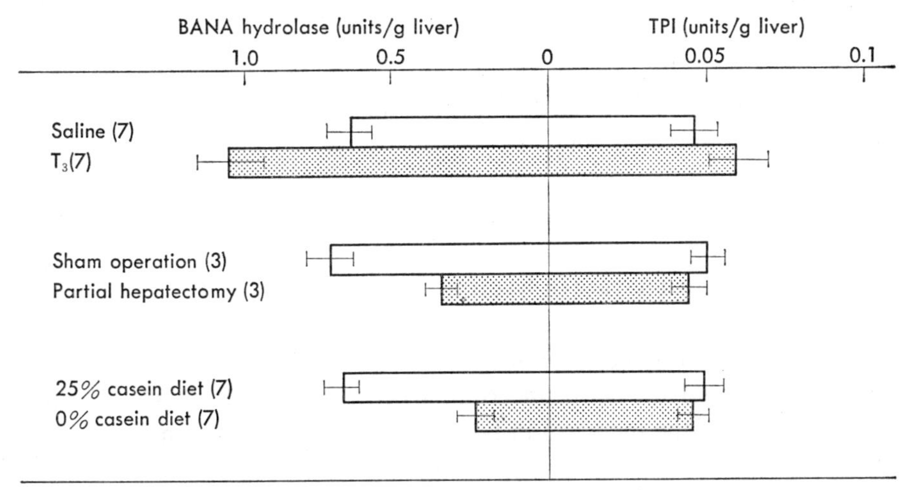

Fig. 6. Changes in levels of Bz-Arg-2-naphthylamide hydrolyzing activities and thiol proteinase inhibitor in the liver under various physiological conditions.
Six rats weighing 150 g were maintained on laboratory diet or special diets before experiments. Triiodothyronine (T_3) was injected subcutaneously at a dose of 0.1 mg/100 g body weight. Thiol proteinase inhibitor (TPI) was assayed as inhibitory activity against cathepsin H. Parenthesis indicates duration (day) of treatment of animals.

thiol proteinase inhibitor decreased markedly (*10*). These results suggest that the inhibitor has a protective effect *in vivo*.

There is no evidence that cytoplasmic thiol proteinase inhibitor regulates lysosomal thiol proteinase activities. To examine this problem, we first tested whether lysosomal thiol proteinases and thiol proteinase inhibitor are regulated coordinately under various physiological and pathological conditions. The effects of hormones and dietary factors on the activities of cathepsins B and H (Bz-Arg-2-naphthylamide hydrolyzing activity) and thiol proteinase inhibitor in the liver are shown in Fig. 6. Administration of T_3 caused 2-fold increases in the activities of the two cathepsins, but did not increase the level of the inhibitor. Moreover, cathepsin B and H activities were decreased in partial hepatectomized liver with no change in inhibitor level. Thus, the levels of lysosomal thiol proteinases were found to change under various physiological conditions, but that of the inhibitor did not. This may be because cathepsins B and H have short half lives (20–24 hr), whereas the thiol proteinase inhibitor has a long half life (more than 72 hr) (unpublished data).

There is evidence that net loss of protein in muscular dystrophy is due to increased protein degradation (*2*). We found that the levels of both lysosomal thiol cathepsins and thiol proteinase inhibitor were increased in skeletal muscle of dystrophic hamsters (*12*). Changes in lysosomal enzyme activities in dystrophic muscle can be classified into two types: some enzymes, including cathepsins B and L, β-glucuronidase and β-galactosidase, were markedly increased whereas others, including acid phosphatase and cathepsin C, were not. Our preliminary data indicated that the increase in lysosomal proteinases in dystrophic muscle was mainly that of macrophages infiltrating local necrotic sites. However, the reason for the increase in level of the thiol proteinase inhibitor is unknown, and it is also uncertain whether the inhibitor participates in regulation of the increases in lysosomal thiol cathepsins in dystrophic muscle.

Cytoplasmic thiol proteinase inhibitor may be sequestered in autophagic vacuole-lysosomes, like other cytoplasmic proteins (*27*), and may regulate lysosomal thiol proteinase activities. But this possibility is unlikely because in lysosomes the amount of sequestered inhibitor is probably much smaller than the amount of thiol proteinases.

REFERENCES

1. Christopher, C. W., Cullinane, P. M., Stebbins, D. R., and Kelley, C. A. *Fed. Proc.*, **41**, 1017 (1982).
2. Goldspink, D. F. *Biochem. J.*, **156**, 71 (1976).
3. Griffith, O. W. and Meister, A. *Proc. Natl. Acad. Sci. U.S.*, **76**, 5605 (1979).
4. Grinde, B. and Seglen, P. O. *Biochim. Biophys. Acta*, **632**, 73 (1980).
5. Hashida, S., Towatari, T., Kominami, E., and Katunuma, N. *J. Biochem.*, **88**, 1805 (1980).
6. Hashida, S., Kominami, E., and Katunuma, N. *J. Biochem.*, **91**, 1373 (1982).
7. Hirado, M., Iwata, D., Niinobe, M., and Fujii, S. *Biochim. Biophys. Acta*, **669**, 21 (1981).
8. Iwata, D., Hirado, M., Niinobe, M., and Fujii, S. *Biochem. Biophys. Res. Commun.*, **104**, 1525 (1982).
9. Järvinen, M. *Acta Chem. Scand. Ser. B.*, **30**, 933 (1976).
10. Katunuma, N. and Kominami, E. *Curr. Top. Cell. Regul.*, **22**, 77 (1983).
11. Katunuma, N., Kominami, E., Hashida, S., and Wakamatsu, N. *Adv. Enzym. Regul.*, **20**, 337 (1982).
12. Katunuma, N., Kominami, E., Noda, T., and Isogai, K. *In* "Muscular Dystrophy—Biomedical Aspects," eds. S. Ebashi and E. Ozawa, p. 237 (1983). Japan Sci. Soc. Press, Tokyo/Springer-Verlag, Berlin.
13. Khairallah, E. A. International Symposium on Medical and Biological Aspects of Proteinase Inhibitors, Tokushima, Japan (Abst. supplement) (1982).
14. Kirschke, H., Langner, J., Riemann, B., Wiederanders, B., Ansorges, S., and Bohley, P. *Ciba Found. Symp.*, **75**, 15 (1980).
15. Kominami, E., Wakamatsu, N., and Katunuma, N. *Biochem. Biophys. Res. Commun.*, **99**, 568 (1981).
16. Kominami, E., Wakamatsu, N., and Katunuma, N. *J. Biol. Chem.*, **257**, 14648 (1982).
17. Kominami, E., Hashida, S., and Katunuma, N. *Biochim. Biophys. Acta*, **659**, 378 (1981).
18. Lenney, J. F., Tolan, J. R., Sugai, W. J., and Lee, A. G. *Eur. J. Biochem.*, **101**, 153 (1979).
19. Lenney, J. F., Liao, J. R., Sugg, S. L., Gopalakrishnan, V., Wong, H. C. H., Ouye, K. H., and Chan, P. W. H. *Biochem. Biophys. Res. Commun.*, **108**, 1581 (1982).
20. Meister, A. *Curr. Top. Cell. Regul.*, **18**, 21 (1981).
21. Neely, A. N., Cox, J. R., Fortney, J. A., Schworer, C. M., and Mortimore, G. E. *J. Biol. Chem.*, **252**, 6948 (1977).
22. Sasaki, M., Taniguchi, K., and Minakata, K. *J. Biochem.*, **89**, 169 (1981).
23. Takio, K., Towatari, T., Kominami, E., Wakamatsu, N., Katunuma, N., and Titani, K. International Symposium on Medical and Biological Aspects of Proteinase Inhibitors, Tokushima, Japan (Abst.), p. 13 (1982).
24. Wakamatsu, N., Kominami, E., and Katunuma, N. *J. Biol. Chem.*, **257**, 14653 (1982).
25. Takio, K., Towatari, T., Katunuma, N., Teller, D. C., and Titani, K. *Proc. Natl. Acad. Sci. U.S.*, **80**, 3666 (1983).
26. Turk, V., Kopitar, M., Brzin, J., Ločnikar, P., Popović, T., Kregar, I., Lah, T., Longer, M., Giraldi, T., and Sava, G. International Symposium on Medical and Biological Aspects of Proteinase Inhibitors, Tokushima, Japan (Abst.), p. 56 (1982).
27. Kominami, E., Hashida, S., Khairallah, E. A., and Katunuma, N. *J. Biol. Chem.*, **258**, 6093 (1983).

Proteinase Inhibitors: Medical and Biological Aspects, eds. N. Katunuma et al., pp. 147–163 (1983).
Japan Sci. Soc. Press, Tokyo/Springer-Verlag, Berlin

Intracellular Proteinases Catalyzing Limited Proteolysis and Their Endogenous Inhibitors

S. PONTREMOLI,*[1] E. MELLONI,*[1] and B. L. HORECKER*[2]

Institute of Biological Chemistry, University of Genova[1]
and Roche Institute of Molecular Biology[2]

The results to be reported here support the existence of a hitherto unrecognized function for lysosomal proteinases, namely, the modification of selected cytosolic proteins by limited proteolysis.

For such processes to occur several conditions must be met. 1) The proteolytic modification must take place in the cytosol, because proteins entering the lysosomes would be expected to be completely degraded. 2) The lysosomal proteinases involved must be active at the physiological pH of the cytosol. 3) There must be a selective release of lysosomal proteinases into the cytosol or, alternatively, access of the lysosomal proteinases to cytosolic proteins without their release into the cytosol.

The hypothesis to be developed derives from studies of two previously unidentified lysosomal proteinases that were characterized using as substrates cytosolic enzymes from rabbit liver, namely, fructose 1,6-bisphosphatase (Fru-P_2ase) and fructose 1,6-bisphosphate aldolase (*8*, *12*, *17*). These cytosolic enzymes catalyze key reactions in the pathway of gluconeogenesis and were shown to be modified *in vivo* in response to changes in the nutritional state of the animals (*15*, *16*) or to the

*[1] 16132 Genova, Italy.
*[2] Nutley, NJ 07110, U.S.A.

administration or deprivation of hormones (*11*). The new lysosomal proteinases that were suspected of catalyzing these modifications were named Fru-P_2ase converting enzyme (CE) (*5, 8*) and cathepsin M (*17*), respectively. These proteinases were shown to be partly membrane-bound, in which form they were active at neutral pH (see below). The membrane-associated activities appear to be located on the external surface of the lysosomes, where they are accessible to proteins in the cytosol. The changes in molecular structure of Fru-P_2ase and aldolase observed *in vivo* (*11, 15, 16*) are similar to those produced by the new lysosomal proteinases *in vitro* (*5, 12, 13*).

CHARACTERIZATION OF THE NEW LYSOSOMAL PROTEINASES

1. Effects on the Catalytic Activity and Molecular Structure of Fru-P_2 ase and Aldolase

Incubation of Fru-P_1ase with lysosomal connecting enzyme (CE) increases by several-fold its catalytic activity measured at pH 9.2 (Fig. 1). This increase in activity, which converts the enzyme into an "alkaline" type of phosphatase, was found to be the result of the hydrolysis of a single peptide bond, located between residues 64 and 65 from the NH_2-terminus (*5*). The four nicked subunits remained associated and no change was detected in the molecular weight of the enzyme (*5*). The primary structure surrounding this proteinase-sensitive bond has been shown to be highly conserved in Fru-P_2ases isolated from tissues of vertebrates, ranging from reptiles through mammals (*7*), suggesting an important function for the proteinase-sensitive structure.

The other new lysosomal proteinase, cathepsin M, acts as a peptidyl peptidase, catalyzing the hydrolysis of one or more peptide bonds near the COOH-terminus of both Fru-P_2ase and aldolase (*12*). In the case of Fru-P_2ase this modification results in loss of catalytic activity measured at pH 7.5. The combined action of CE and cathepsin M on Fru-P_2ase causes a dramatic shift in the pH optimum for this enzyme from the neutral to the alkaline pH range. These effects are also observed, as shown below, with lysosomal membranes and even with undisrupted lysosomes.

The modification of rabbit liver aldolase by cathepsin M that results in loss of its catalytic activity (Fig. 2) was correlated with the

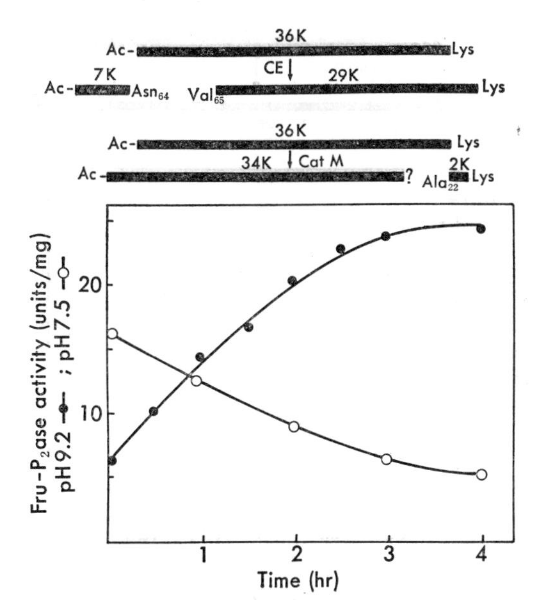

Fig. 1. Changes in catalytic activity and subunit structure of rabbit liver Fru-P_2ase incubated at pH 50 with purified CE or cathepsin M (Cat M).
Fru-P_2ase activity was measured at pH 9.2 (●) or at pH 7.5 (○). For details see refs. *8*, *5*, and *12*. Changes in subunit structure are schematically represented at the top of the figure. It should be noted that the fragments generated during digestion with CE remain associated at neutral pH and dissociation of the 7K fragment is not required for the observed increase in activity at pH 9.2 (*5*).

removal of a small peptide or peptides from the COOH-terminus, without significant change in the subunit molecular weight (*12*, *17*). Aldolase was also inactivated by purified CE, which catalyzes a similar modification of the COOH-terminus (*17*). This peptidyl peptidase activity of CE with aldolase as the substrate is very different from its action on Fru-P_2ase and suggests that the substrate makes a major contribution to the specificity of these lysosomal proteinases. Both CE and cathepsin M contribute to the inactivation of rabbit liver aldolase by lysosomal membranes or intact lysosomes (see below).

2. *Purification and Properties of the New Lysosomal Proteinases, CE and Cathepsin M*

Other proteinases, particularly cathepsins B and L, will also catalyze the limited proteolysis of Fru-P_2ase (*8*, *10*) and liver aldolase (*2*, *17*),

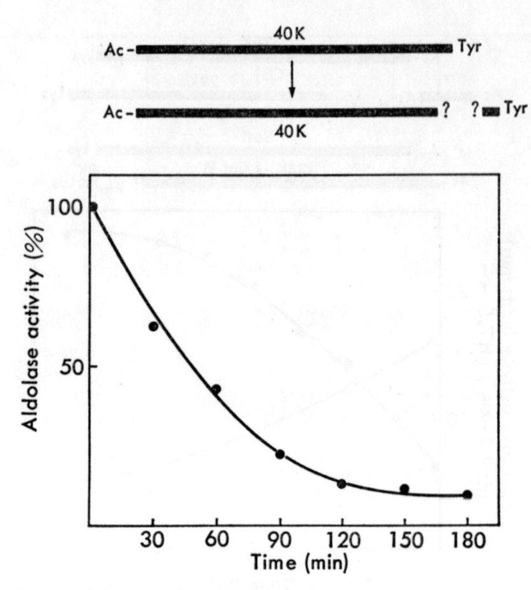

Fig. 2. Changes in catalytic activity and subunit structure of rabbit liver aldolase on digestion with cathepsin M.
The activity of aldolase was measured at pH 7.4. The precise site of cleavage remains unknown. For experimental details see refs. *2* and *3* (reproduced with permission).

but only CE and cathepsin M are present in the lysosomal membranes. These activities can be solubilized with 1 M NaCl, separated from each other by filtration on AcA34, and further purified by chromatography on DEAE-cellulose (Fig. 3). The purified preparations showed little activity with most synthetic substrates, including α-*N*-benzoyl-L-arginine-β-naphtylamide (BANA), the typical substrate for cathepsins B and L (Table I). Cathepsin M was also distinguished from cathepsins B and L by its ablity to hydrolyze unlocked amino acid naphthylamides.

Converting enzyme and cathepsin M were both characterized as thiol proteinases on the basis of their activation by sulfhydryl compounds and inhibition by iodoacetate (Table II). Converting enzyme was resistant to the classical inhibitors of lysosomal proteinases, which further served to distinguish it from these proteinases. Cathepsin M, on the other hand, was highly sensitive to inhibitors of cathepsin B-type proteinases. Neither CE nor cathepsin M was sensitive to the specific inhibitor of angiotensin-converting enzyme, Pyr-Trp-Pro-Arg-Pro-Glu-Ile-Pro-Pro. It may therefore be concluded that CE and cathepsin M

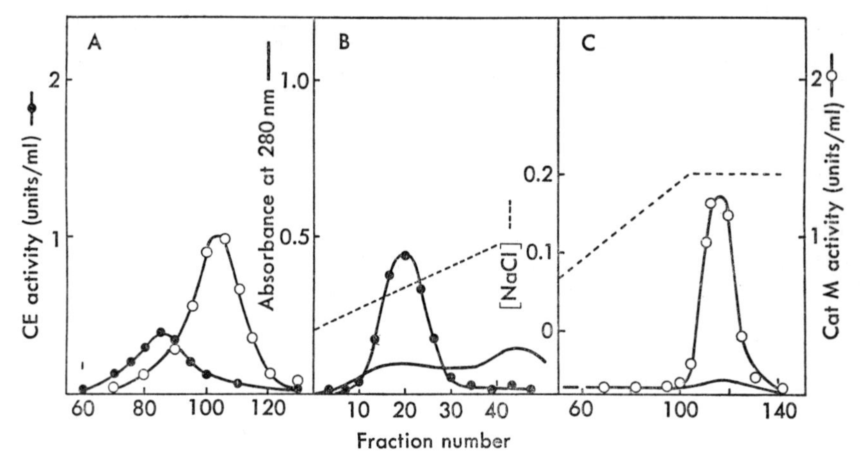

Fig. 3. Purification of CE and cathepsin M from rabbit liver lysosomal membranes. A: membranes were treated with 1 M NaCl to solubilize the proteinase activities and chromatographed on Ultrogel AcA34. CE activity was measured by following the increase in activity of Fru-P_2ase at pH 9.2 (see Fig. 1). Cathepsin M activity was determined with aldolase as the substrate, following the rate of inactivation (see Fig. 2). B: fractions 70–85 from Exp. A were pooled and chromatographed on a DEAE-cellulose column (*12*). C: fractions 106–125 from Exp. A were pooled and chromatographed on CM-cellulose. For details see ref. *2* (reproduced with permission).

TABLE I. Activities of Cathepsin M and CE with Synthetic Substrates

Substrates	Enzyme activity (nmol of naphthylamine/2hr/ml)			
	Cat M	CE	Cat B	Cat L
BANA	5	ND	39.2	13.2
Z-Gly-Gly-Arg-4-methoxy-2-naphthylamide	—	8.3	ND	5.3
Z-Ala-Arg-Arg-4-methoxy-2-naphthylamide	20	n.d.	88	100
Bz-Val-Lys-Lys-Arg-4-methoxy-2-naphthylamide	10	1.7	152	131
Arginine-2-naphthylamide	50	0.4	n.d.	3.0
Leucine-2-naphthylamide	30	n.d.	n.d.	2.0

ND, not detectable; Z,CBz, benzyloxycarbonyl; Bz, benzoyl. For experimental details see ref. *17* (reproduced with permission).

are new proteinases belonging to the cathepsins B and L in substrate specificity, response to inhibitors and stability to alkaline pH (*5*, *12*, *17*). Cathepsin M most closely resembles cathepsin H from human liver, but it is distinguished from cathepsin H by its sensitivity to leupeptin, its stability to alkaline pH, and its insensitivity to puromycin (*17*).

Other properties of the new lysosomal proteinases are summarized

TABLE II. Effect of Activators and Inhibitors on the Activity of Proteolytic Enzymes Purified from the Soluble and Membrane Fractions of Rabbit Liver Lysosomes

Addition	Concentration	Proteolytic activity (% of control) CE	Cat B	Cat L	Soluble Cat M	Membrane-bound Cat M
None		100	100	100	100	100
Cysteine	5 mM	284	386	379	127	132
DTT	1 mM	290	394	402	136	140
Iodoacetate	0.1 mM	44	56	51	45	40
Pepstatin	10 μM	100	100	100	100	100
Leupeptin	1.2 μM	100	12	8	0	0
Chymostatin	7.2 μM	100	15	14	5	7
Antipain	8.2 μM	100	18	20	4	3
Elastinal	10 μM	100	29	31	9	4
TPCK	10 μM	100	54	60	65	80
TLCK	10 μM	100	58	60	71	94
PMSF	0.1 mM	95	86	88	84	87
EDTA	5 mM	94	90	90	95	95
O-Phenanthroline	5 mM	91	90	88	91	92

DTT, dithiothreitol; TPCK, tosylamide phenylethyl chloromethyl ketone; TLCK, tosyllysine chloromethyl ketone; PMSF, phenylmethyl sulfonyl fluoride. For experimental details see ref. *17* (reproduced with permission).

TABLE III. Properties of CE and Cathepsin M

Property	CE Soluble	CE Membrane-bound	Cathepsin M Soluble	Cathepsin M Membrane-bound
Specific activity				
Fed (units/g liver)	1.1	0.09	35.4	31.6
Fasted (units/g liver)	3.6	0.70	49.4	63.0
Stability at alkaline pH	Unstable	Stable	Stable	Stable
pH optimum	5.0	6.5 (80% at pH 7.25)	5.0 (20% at pH 7.25)	5.0 (80% at pH 7.25)
Effect on Fru-P_2ase	Activation		Inactivation	
Effect on aldolase	Inactivation		Inactivation	

The activity of CE was measured by the rate of increase in activity of Fru-P_2ase measured at pH 9.2. The activity of cathepsin M was measured by the loss of aldolase activity, measured at pH 7.25. For details see refs. *8*, *9*, and *17* (reproduced with permission).

in Table III. Important evidence for a physiological role for these proteinases is provided by the fact that their activity is substantially increased during fasting, with the largest increase in the membrane-bound fraction. It is also noteworthy that association with the membranes

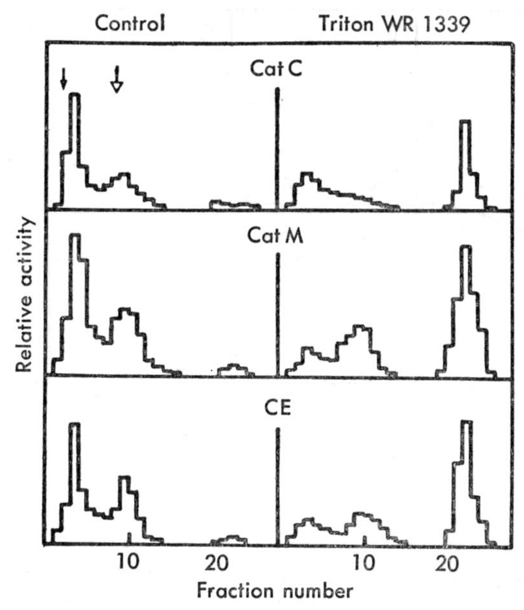

Fig. 4. Sucrose density gradient sedimentation of the heavy particle fraction from livers of control and Triton WR1339-injected rabbits.
The peak for the markers for peroxisomes (catalase) and for mitochondria (NADH oxidase) are indicated by the filled and open arrows, respectively. The sucrose density gradient experiments were carried out as described (*14*).

increases both their stability to alkaline pH and their activity in the neutral pH range. The soluble forms of CE and cathepsin M are typical acid proteinases, with pH optima at 5.0. In the case of CE, binding to the membranes shifts the pH optimum to 6.5 (*8*). Membrane-bound cathepsin M is still most active at pH 5.0, but its activity at pH 7.25 is 75% of the maximum, compared to 25% for the soluble form (*17*).

Based on their elution from the AcA34 columns, the molecular weights of CE and cathepsin M are approximately 75,000 and 25,000, respectively (*8*) (see Fig. 3).

3. *Location of CE and Cathepsin M in the Lysosomes*

The lysosomal localization of the new proteinases was confirmed by analysis of the heavy particle fraction by sucrose density-gradient sedimentation (Fig. 4). With the heavy particle fraction from livers of untreated rabbits both activities were recovered in two peaks that also

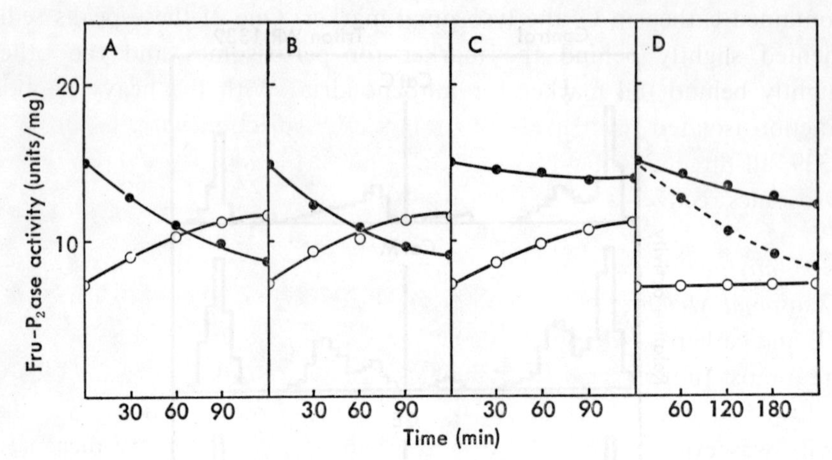

Fig. 5. Limited proteolysis of Fru-P_2ase by lysosomes, lysosomal membranes, and purified cathepsin M.
The digestions were carried out at pH 7.25 except for the dashed curve in the experiment with purified cathepsin M, where the incubation was at pH 5.0. Fru-P_2ase was assayed at pH 9.2 or pH 7.5, as indicated. Leupeptin was 1 μM. A: intact lysosomes. B: intact membranes. C: intact membranes-leupeptin. D: purified cathepsin M. ● pH 7.5; ○ pH 9.2. For details see ref. *17* (reproduced with permission).

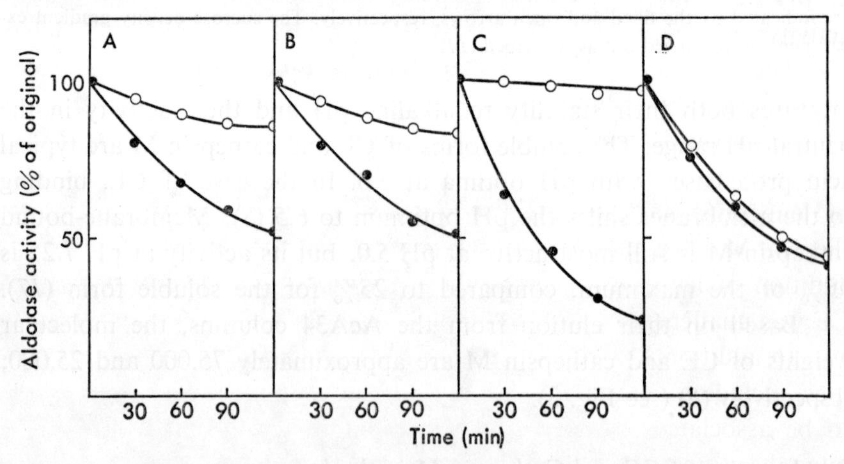

Fig. 6. Limited proteolysis of rabbit liver cellulose by lysosomes (A), lysosomal membranes (B), and by purified cathepsin M (C) or converting enzyme (CE) (D).
The digestions were carried out at pH 7.25 with the lysosomes and lysosomal membranes, at pH 5.0 with purified cathepsin M, and at pH 6.0 with purified CE. Aldolase activity was measured at pH 7.4. ○ +leupeptin; ● control. For details see ref. *17* (reproduced with permission).

contained cathepsin C, the lysosomal marker. One of these peaks sedimented slightly behind the marker for peroxisomes and the other slightly behind the marker for mitochondria. With the heavy peptide fraction isolated from livers of rabbits after injection with Triton WR 1339, all three activities were also present in the less dense Triton-filled lysosomes recovered from the top of the gradient.

4. Activities of Lysosomal Proteinases Expressed by Lysosomes and by Lysosomal Membranes

CE and cathepsin M activities associated with the lysosomal membranes are almost fully expressed by intact lysosomes. Both activities could be followed at pH 7.25 using Fru-P_2ase as the substrate (Fig. 5). CE activity was evaluated from the increase in Fru-P_2ase activity measured at pH 9.2 (alkaline activity) while cathepsin M activity was followed by the loss of Fru-P_2ase activity measured at pH 7.5 (neutral activity). These activities were nearly the same with intact lysosomes or with an equivalent quantity of membranes prepared from these lysosomes. Only the loss of neutral Fru-P_2ase activity due to cathepsin M was inhibited by leupeptin. Purified cathepsin M was more effective at pH 5.0 than at pH 7.25 and did not catalyze the increase in alkaline Fru-P_2ase activity.

The same lysosomes and lysosomal membranes catalyzed the inactivation of rabbit liver aldolase, again at comparable rates (Fig. 6). This inactivation was due mainly to cathepsin M, as indicated by its sensitivity to leupeptin. A small fraction of the activity, not inhibited by leupeptin, was attributed to the presence of CE. With purified cathepsin M inactivation of aldolase was completely prevented by the addition of leupeptin whereas the inactivation of aldolase by purified CE was unaffected by this inhibitor.

In addition to CE and cathepsin M, several other lysosomal activities, including hexosaminidase and leucineaminopeptidase, were found to be associated with lysosomal membranes (*1*, *9*) and expressed by intact lysosomes (*9*). Cathepsins A, B, and C activities were not detected in this compartment.

The evidence that the activity expressed by intact lysosomes is not due to leakage of lysosomal enzymes or to increased permeability of the lysosomes (*9*, *14*) may be summarized as follows: 1) the activation of Fru-P_2ase by intact lysosomes was not inhibited by leupeptin, proving that cathepsin B activity remained fully cryptic in these lysosomes. 2) No activity was detected in the medium after removal of the lysosomes by centrifugation. This was true also for the activity with synthetic substrates such as BANA. 3) The activities of lysosomal proteinases not associated with the membranes, such as cathepsins B and C, remained fully latent in the intact lysosomes and were not detected in the medium during the incubation period.

5. *Localization of Membrane-bound CE and Cathepsin M on the Outer Surface of the Lysosomes*

More direct evidence for the presence of these lysosomal proteinases on the outer surface was obtained by evaluating their susceptibility to exogenous trypsin. Incubation of the intact lysosomes with either soluble or immobilized trypsin resulted in loss of 80% and 60% of the membrane-bound CE and cathepsin M activities, respectively (Table IV). The decrease in the total activity of CE and cathepsin M was fully accounted for by the decreased activity on the membranes, confirming that the soluble activities with the lysosomes were not accessible to trypsin.

The surface orientation of cathepsin M was further established using monoclonal antibodies, prepared by the hybridoma technique, with spleen cells from mice immunized with cathepsin M purified from

TABLE IV. Effect of Exposure of Intact Rabbit Liver Lysosomes to Trypsin on CE and Cathepsin M Activities (units/g liver)

Treatment	CE, assayed with Fru-P_2ase		Cathepsin M, assayed with Fru-P_2ase		Cathepsin M, assayed with Aldolase	
	Total	Membrane-bound	Total	Membrane-bound	Total	Membrane-bound
None	1.28	0.14	68	31	61	26
Soluble trypsin	1.16	0.03	50	12	31	10
Immobilized trypsin	1.17	0.03	48	11	—	—

Incubation of the lysosome preparation with soluble or immobilized (Sepharose-linked) trypsin was carried out in 0.25 M sucrose at pH 7.25. The lysosomes were recovered, lysed by freezing and thawing, and the menbranes separated by centrifugation. Total CE and cathepsin M activities were measured with Fru-P_2ase or aldolase as substrates before centrifugation. Membrane-bound activity was measured after washing the membranes with distilled water, as indicated in the legend to Fig. 6. For details see ref. *14* (reproduced with permission).

lysosomal membranes. The specificity of these monoclonal antibodies was confirmed by demonstrating that cathepsin M was the only fraction adsorbed to the Sepharose-linked antibody from a crude lysate of lysosomes. This monoclonal antibody was specifically adsorbed to intact lysosomes, which did not adsorb non-specific immunoglobulin G (IgG) (unpublished observations). Binding of the anti-cathepsin M IgG to the lysosomes was demonstrated by the subsequent binding of radioiodinated anti-mouse IgG. During these procedures the lysosomes remained intact, as demonstrated by the unaltered latency of cathepsins B and C.

The presence of cathepsin M on the outer lysosomal surface was also demonstrated by the adsorption of lysosomes to plastic dishes that had been coated with the monoclonal antibody. No binding of lysosomes was observed when the dishes were coated with several other non-specific IgG preparations (unpublished observations).

ENDOGENOUS INHIBITORS OF LYSOSOMAL PROTEINASES

The presence of proteinases on the outer surface of lysosomes and accessible to cytosolic proteins raises interesting questions regarding the function of these proteinases and their regulation. We have already shown that these activities are influenced by nutritional conditions. They may also be regulated by specific inhibitors similar to the cytosolic inhibitors of cathepsins B, L, and H described in reports from the laboratories of Lenney (*6*) and Katunuma (*4*). Our search for such endogenous regulators of the lysosomal proteinases CE and cathepsin M has revealed the presence in rabbit liver of at least three distinct inhibitors, each specific for a different lysosomal proteinase or class of proteinases. These inhibitors are located not only in cytosol, as expected from the work of other investigators, but also in the lysosomes, where they are present in both soluble and membrane-bound form. Their distribution in the lysosomal compartments parallels the distribution of the proteinases affected by these inhibitors.

1. Purification of Specific Inhibitors from Rabbit Liver Cytosol

Rabbit liver cytosol contains at least three distinct inhibitors that were separated by chromatography on Tris-acryl DEAE (Fig. 7) and further

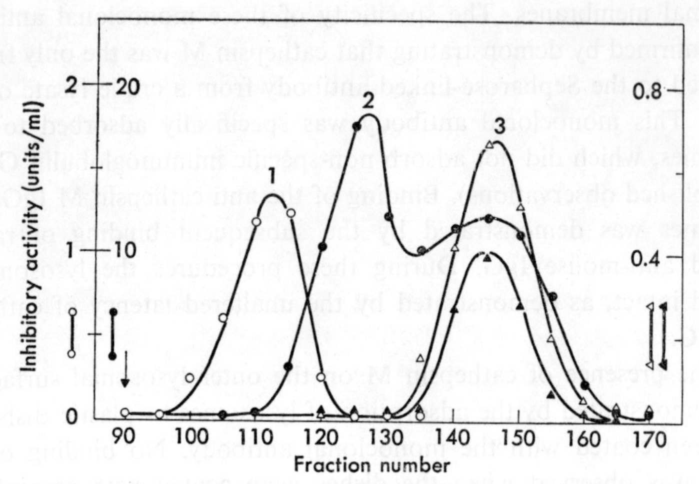

Fig. 7. Separation by ion-exchange chromatography of inhibitors of lysosomal proteinases present in the cytosol.
A post-lysosomal fraction was heated at pH 2.0 for 10 min at 90°C. The clear supernatant solution was concentrated and chromatographed at pH 6.7 on a Tris-acryl DEAE column, eluting with a gradient of NaCl. The eluted fractions were assayed for inhibition of CE (○), cathepsin M (●), cathepsin B (▲), and cathepsin L (△). For details see ref. *18*.

purified by gel filtration on Sephadex G-75 (Fig. 8). Based on their selectivity toward lysosomal proteinases, they were designated as cytosolic I_{CE}, cytosolic I_M, and cytosolic $I_{B/L}$, respectively. Purified cytosolic $I_{B/L}$ also inhibited the activity of cathepsin M; these three inhibitor activities appear to reside in the same polypeptide and attempts to separate them were unsuccessful.

The apparent molecular weights of these inhibitors, based on thier elution from Sephadex G-75, were 40,000 for I_{CE}, 12,500 for I_M, and 12,500 for $I_{B/L}$. Cytosolic $I_{B/L}$ may be identical to the inhibitor of thiol proteinases described by Lenney *et al.* (*6*), Kominami *et al.* (*4*), and Hirado *et al.* (*3*).

2. *Inhibitors of Lysosomal Proteinases Isolated from Rabbit Liver Lysosomes*

The same three classes of inhibitors were present in the soluble fraction isolated from lysosomes disrupted by freezing and thawing (Fig. 9A). Chromatography of this fraction on Sephadex G-75 yielded four peaks with inhibitor activity. Peaks 1 and 3 contained specific inhibitors of

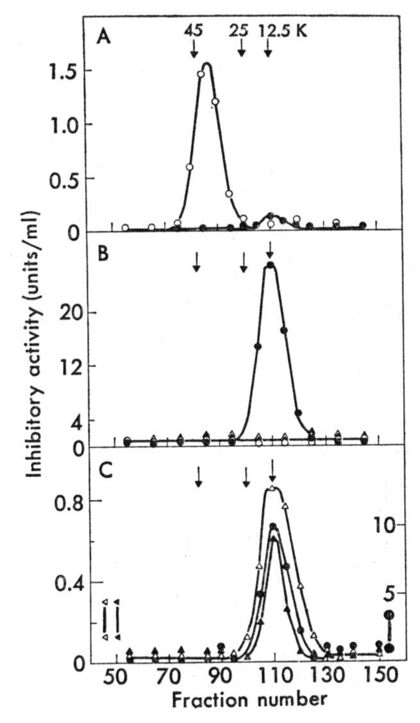

Fig. 8. Gel filtration of inhibitor fractions recovered from ion-exchange chromatography. A: fractions containing cytosolic I_{CE} (fractions 95–115, Fig. 7) were pooled, concentrated and applied at pH 6.0 to a Sephadex G-75 column. B: fractions 120–132 (Fig. 7) containing cytosolic I_M were pooled and gel-filtered as in A. C: fractions 138–160 (Fig. 7) containing cytosolic $I_{B/L}$ were pooled and gel-filtered as in A. Symbols for inhibitor activities were as in Fig. 7.

CE, corresponding to molecular weights of 20,000 and 10,000, respectively. A second peak, corresponding to a molecular weight of 12,500, inhibited cathepsins M, B, and L. This peak could be resolved into two components by treatment with concanavalin A (Con A)-Sepharose. The immobilized lectin adsorbed the $I_{B/L}$ activity but left the I_M activity in the supernatant solution. We therefore conclude that the 12,500-dalton peak contains two distinct inhibitors, one designated lysosomal I_M and the other lysosomal $I_{B/L}$. In addition, the intralysosomal contents contained a second form of $I_{B/L}$ with a molecular weight of approximately 5,000.

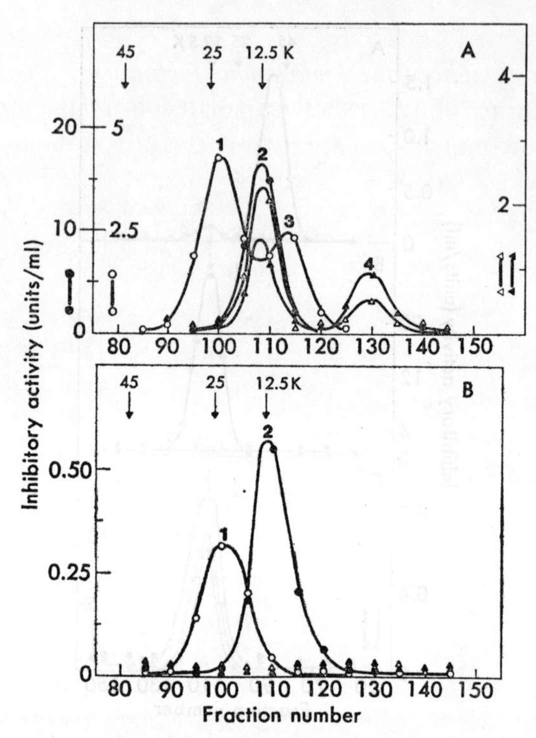

Fig. 9. Separation by gel filtration of inhibitors of lysosomal proteinases present in the lysosomes.
The heavy particle fraction from rabbit liver was disrupted by freezing and thawing and the soluble and membrane-bound fractions collected. A: the soluble lysosomal contents were heated to 90°C at pH 4.0 for 10 min and the clear supernatant solution concentrated and filtered through a column of Sephadex G-75 as described in Fig. 8. B: the washed membranes were solubilized with 1 M NaCl and after heating at pH 2 and 90°C for 10 min adjusted to pH 6.0 and the clear supernatant solution concentrated and chromatographed as in A. Symbols for I_{CE}, I_M, and $I_{B/L}$ are the same as indicated in Fig. 8.

3. *Inhibitors Associated with the Lysosomal Membranes*

The lysosomal membrane fraction, isolated after disruption of the enzymes by freezing and thawing, contained two inhibitors that were solubilized by treating the membranes with 1 M NaCl, as described previously for the membrane-bound proteinases (*8*). These inhibitors were separated by gel filtration into two components, membrane I_{CE}, *Mr* 25,000 and membrane I_M, *Mr* 12,500 (Fig. 9B). No $I_{B/L}$ activity could be detected in association with the lysosomal membranes.

4. *Molecular Weight of the Endogenous Inhibitors*

The several forms of $I_{B/L}$, whether isolated from the cytosol or lysosomes, were of similar size, approximately 12,500 daltons, except that the lysosomes also contained a smaller species of approximately 5,000 daltons. The I_M species also corresponded to polypeptides of approximately 12,500 daltons. The cytosolic and intralysosomal species of I_{CE} were larger, having apparent molecular masses of 40,000 and 20,000 daltons, respectively. However, in the presence of 5 mM cysteine the former dissociated into polypeptides of *Mr* 20,000 and 10,000, respectively and the latter to a polypeptide of *Mr* 10,000. The 10,000 species is probably the active inhibitor.

The polypeptide nature of the inhibitors was established by their inactivation by trypsin.

5. *Lysosomal Localization of the Inhibitor*

To confirm the localization of the inhibitors isolated from the lysosomes-rich heavy particle fraction, the rabbits were injected with Triton WR1339, as described earlier for the lysosomal proteinases (*14*). The inhibitors were found to be present in the light Triton-filled lysosomes that contained CE and cathepsins C and M (*18*).

CONCLUSIONS

The results reported here support a new model for the function of lysosomal proteinases, in which they serve as regulators of cytosolic enzymes by catalyzing the limited proteolysis of these enzymes. The model proposes that certain lysosomal proteinases are located on the outer surface of the lysosomal membrane, in which form they are accessible to the cytosol and are capable of acting at neutral pH on cytosolic proteins.

Regulation of these lysosomal proteinases appears to involve two different mechanism. One is based on changes in the amount of enzyme and/or its activity in response to nutritional conditions, possibly mediated by hormones such as insulin and glucocorticoids. The second mechanism involves the action of specific inhibitors, present in both the cytosolic and lysosomal compartment. The presence of these inhibitors in the cytosol may, as suggested (*6*), serve to protect cytosolic proteins against the action of released lysosomal proteinases. On the other hand, the presence of inhibitors of lysosomal proteinases within the lysosomes was unexpected and the function of these intralysosomal inhibitors remains to be determined. We have obtained preliminary evidence (unpublished) indicating that the level of inhibitors is markedly decreased on fasting.

It is of interest that all three classes of inhibitors are found in the intralysosomal compartment, which contains all of the corresponding proteinases, whereas only I_{CE} and I_M are present on the lysosomal membranes, corresponding to the distribution of the respective proteinases. Further studies should help to elucidate the role of the new proteinases and their inhibitors in the regulation of intracellular metabolism and/or protein turnover.

REFERENCES

1. Crivellaro, O., Lazo, P. S., Tsolas, O., Pontremoli, S., and Horecker, B. L. *Arch. Biochem. Biophys.*, **189**, 490 (1978).
2. Hannappel, E., MacGregor, J. S., Davoust, S., and Horecker, B. L. *Arch. Biochem. Biophys.*, **214**, 293 (1982).
3. Hirado, M., Iwata, D., Niinobe, M., and Fujii, S. *Biochim. Biophys. Acta*, **669**, 21 (1981).
4. Kominami, E., Wakamatsu, N., and Katunuma, N. *Biochem. Biophys. Res. Commun.*, **99**, 568 (1981).
5. Lazo, P. S., Tsolas, O., Sun, S. C., Pontremoli, S., and Horecker, B. L. *Arch. Biochem. Biophys.*, **188**, 308 (1978).
6. Lenney, J. F., Tolan, J. R., Sugai, W. J., and Lee, A. G. *Eur. J. Biochem.*, **101**, 153 (1979).
7. MacGregor, J. S., Hannappel, E., Xu, G-J., Pontremoli, S., and Horecker, B. L. *Arch. Biochem. Biophys.*, **217**, 652 (1982).
8. Melloni, E., Pontremoli, S., Salamino, F., Sparatore, B., Michetti, M., and Horecker, B. L. *Arch. Biochem. Biophys.*, **208**, 175 (1981).
9. Melloni, E., Pontremoli, S., Salamino, F., Sparatore, B., Michetti, M., and Horecker, B. L. *Proc. Natl. Acad. Sci. U.S.*, **78**, 1499 (1981).
10. Nakashima, K. and Ogino, K. *J. Biochem.*, **75**, 355 (1974).
11. Pontremoli, S., DeFlora, A., Salamino, F., Melloni, E., and Horecker, B. L. *Proc. Natl. Acad. Sci. U.S.*, **72**, 2969 (1975).
12. Pontremoli, S., Melloni, E., Michetti, M., Salamino, F., Sparatore, B., and Horecker, B. L. *Proc. Natl. Acad. Sci. U.S.*, **79**, 2451 (1982).
13. Pontremoli, S., Melloni, E., Michetti, M., Salamino, F., Sparatore, B., and Horecker, B. L. *Proc. Natl. Acad. Sci. U.S.*, **79**, 5194 (1982).
14. Pontremoli, S., Melloni, E., Michetti, M., Salamino, F., Sparatore, B., and Horecker, B. L. *Biochem. Biophys. Res. Commun.*, **106**, 903 (1982).

15. Pontremoli, S., Melloni, E., Salamino, F., de Flora, A., and Horecker, B. L. *Proc. Natl. Acad. Sci. U.S.*, **71**, 1776 (1974).
16. Pontremoli, S., Melloni, E., Salamino, F., Sparatore, B., Michetti, M., and Horecker, B. L. *Proc. Natl. Acad. Sci. U.S.*, **76**, 6323 (1979).
17. Pontremoli, S., Melloni, E., Salamino, F., Sparatore, B., Michetti, M., and Horecker, B. L. *Arch. Biochem. Biophys.*, **214**, 376 (1982).
18. Pontremoli, S., Melloni, E., Salamino, F., Sparatore, B., Michetti, M., and Horecker, B. L. *Proc. Natl. Acad. Sci. U.S.*, **80**, 1261 (1983).

15. Pontremoli, S., Melloni, E., Salamino, F., de Flora, A. and Horecker, B. L. Proc. Natl. Acad. Sci. U.S., 71, 1776 (1974).
16. Pontremoli, S., Melloni, E., Salamino, F., Sparatore, B., Michetti, M., and Horecker, B. L. Proc. Natl. Acad. Sci. U.S., 76, 6323 (1979).
17. Pontremoli, S., Melloni, E., Salamino, F., Sparatore, B., Michetti, M. and Horecker, B. L. Arch. Biochem. Biophys. 214, 376 (1982).
18. Pontremoli, S., Melloni, E., Salamino, F., Sparatore, B., Michetti, M., and Horecker, B. L. Proc. Natl. Acad. Sci. U.S., 80, 1261 (1983).

Proteinase Inhibitors: Medical and Biological Aspects, eds. N. Katunuma et al., pp. 165–172 (1983).
Japan Sci. Soc. Press, Tokyo/Springer-Verlag, Berlin

Calpastatin, an Endogenous Inhibitor Protein Acting Specifically on Calpain

Takashi MURACHI,[*1] Emiko TAKANO,[*1,*2] and Kazuyoshi TANAKA[*1]

Department of Clinical Science, Kyoto University Faculty of Medicine[*1]

Calpastatin is a collective name given to a family of endogenous inhibitor proteins which act specifically on Ca^{2+}-dependent cysteine proteinases [EC 3.4.22.17] now also collectively called calpain (*9*, *13*).

An inhibitory factor against the activation of phosphorylase *b* kinase reported earlier to be present in bovine cardiac muscle (*3*) was identified as a protease inhibitor in 1978 (*18*). Almost concurrently, the homogenate of rat liver (*12*) and brain (*11*), which showed no Ca^{2+}-dependent proteolysis as such, was found upon DEAE-cellulose chromatography to contain both Ca^{2+}-protease and its specific inhibitor protein. Soon after this, the very wide distribution of a set of Ca^{2+}-protease and its inhibitor protein among mammalian and avian cells was revealed (see for review, ref. *9*).

An intracellular protease which required both Ca^{2+} and cysteine for the activity was first discovered in rat brain in 1964 (*2*). Since then, similar Ca^{2+}-proteases have been reported to be involved in degradation of myofibrillar, neurofilament and other cytoskeletal proteins, activation of phosphorylase *b* kinase and protein kinase C, and transformation

*1 Kawahara-cho, Shogoin, Sakyo-ku, Kyoto 606, Japan.
*2 Present address: Central Clinical Laboratory, National Kyoto Hospital, Kyoto 612, Japan.

of steroid hormone-binding proteins (*9*). In spite of their diverse functions, these proteases were shown to have remarkable similarity in molecular size, Ca^{2+}-requirement and catalytic properties so that a grouping under the name *calpain* was proposed (*9*). *Cal* stands for Ca^{2+} and *-ain* implies its belonging to cysteine proteinases. The specific inhibitor protein was accordingly named *calpastatin* (*9*).

DISTRIBUTION

A simple method has been devised which permitted quantifying calpain and calpastatin in a given tissue by a single run of Ultrogel AcA-34 chromatography (*8*). Calpastatin as an aggregate emerged from the gel column well ahead of calpain. Differential determination of two sub-

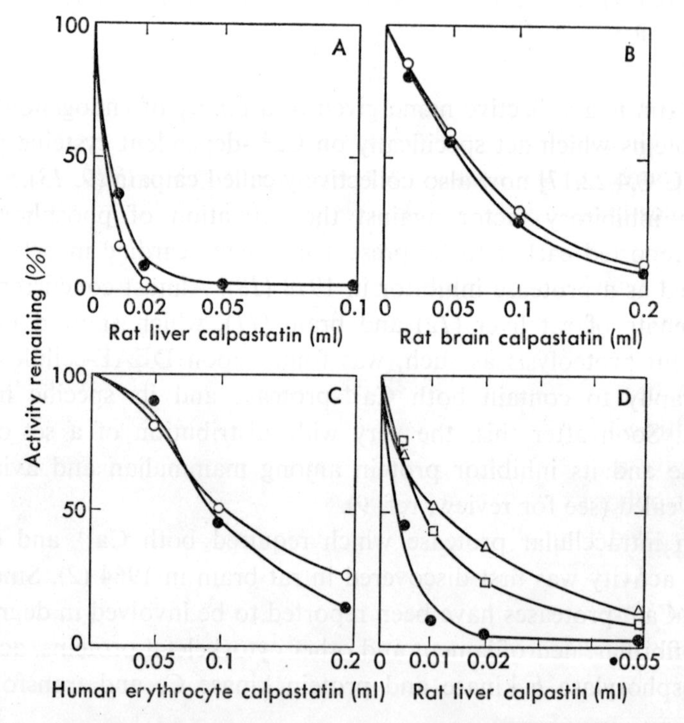

Fig. 1. Cross-reactivity of calpain and calpastatin from different sources. Proteolytic activity was determined using casein as the substrate. ○ rat liver calpain II; ● rat brain calpain II; □ rat liver calpain I; △ human erythrocyte calpain.

classes of calpain, I and II*, was made by carrying out protease assays in both 0.1 mM and 4.5 mM Ca^{2+}. By utilizing this method, the distribution of calpastatin among various tissues and cells was found to be almost ubiquitous but remarkably diverse (*8*).

Rat liver, lung, skeletal, and cardiac muscle, and human liver and erythrocytes contained larger amounts of calpastatin than calpain, whereas the reverse was true in rat spleen cells and erythrocytes, mouse liver, and AH 130 ascites tumor cells. Rat kidney contained no detectable amount of calpastatin. In these and other distribution studies, only the soluble fraction of tissue homogenates was analyzed, because most, if not all, calpastatin was found to be localized in the cytosol together with calpains I and II (*9*).

As shown in Fig. 1, the inhibition of calpain by calpastatin shows little organ- and species-specificity (*9*).

PURIFICATION AND PROPERTIES

Calpastatin proteins have recently been purified from several different sources. The hemolysate of human erythrocytes has proven to be a unique source of calpastatin and calpain I, a low Ca^{2+}-requiring protease, since it does not contain calpain II, a high Ca^{2+}-requiring species (*10*).

After removing the major part of hemoglobin from the hemolysate, the material was chromatographed on a DEAE-cellulose column at pH 7.5, from which calpastatin was eluted at a NaCl concentration of 0.1 M. A heat treatment and further chromatographies on DEAE-cellulose, Ultrogel AcA-34 and Sephacryl S-200 yielded the final product which gave a single protein band of 70K daltons upon polyacrylamide gel electrophoresis in the presence of SDS (*16*). When the 70K protein was extracted from the gel, concentrated and chromatographed on a Sephadex G-200 column, the protein peak with calpastatin activity appeared only at the position corresponding to 280K daltons, a molecular size found for the final product before SDS-gel electrophoresis (Fig. 2). These results indicate reversible dissociation of 280K calpa-

* Calpain I requires 10–50 μM Ca^{2+} for half-maximal activity, whereas calpain II requires 1–5 mM Ca^{2+} for maximal activity (*5*, *9*). In an alternative terminology, μCANP and m-CANP may correspond to calpains I and II, respectively (*14*).

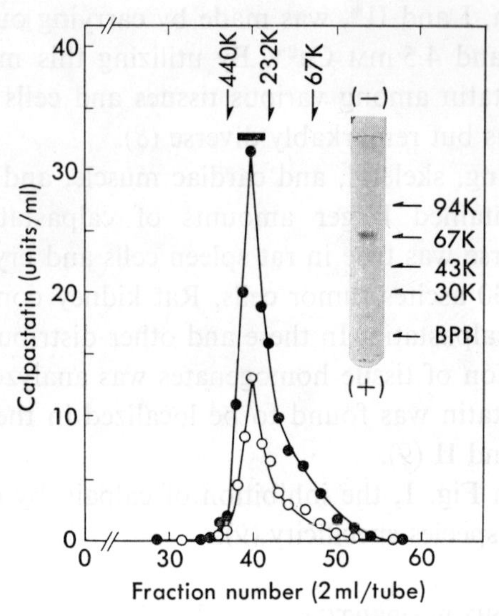

Fig. 2. Sephacryl S-200 column chromatography of human erythrocyte calpastatin before (●) and after (○) SDS-polyacrylamide gel electrophoresis.
The original material collected from the gel column (horizontal bar) gave a 70K protein band on the electrophoresis (see inset), which was recovered by extraction from the gel and then rechromatographed on an identical Sephacryl gel column.

TABLE I. Molecular Size of Calpastatins

Source (ref.)	Molecular mass (K daltons)	
	Oligomer	Monomer
Rat liver[a] (*12*)	280–300	
Rat brain[a] (*11*)	300	
Bovine cardiac muscle[a] (*17*, *18*)	260–270	60–70
Rabbit skeletal muscle (*15*)	68	34
Rabbit skeletal muscle (*1*)		70+13
Chicken skeletal muscle (*4*)		68
Human erythrocytes (*16*)	280	70
Human erythrocytes (*6*)	240–250	60

[a] Data on crude preparations.

statin into 70K monomers. Table I summarizes the molecular sizes reported so far on calpastatins from various sources, suggesting a similar unit size but with different self-associabilities.

The amino acid compositions of calpastatins from human erythrocytes (*16*) and chicken skeletal muscle (*4*) are compared in Table II. The data indicate non-identity of these two proteins, but with such

TABLE II. Amino Acid Composition of Calpastatin from Human Erythrocytes as Compared with That from Chicken Skeletal Muscle

Amino acid	Human erythrocytes[a]	Chicken skeletal muscle[b]
Aspartic acid	81	53
Threonine	32	38
Serine	62	52
Glutamic acid	96	96
Proline	65	43
Glycine	34	70
Alanine	61	47
Half-cystine	0.8	ND[c]
Valine	21	44
Methionine	7	7
Isoleucine	14	19
Leucine	51	41
Tyrosine	6	18
Phenylalanine	8	8
Lysine	78	36
Histidine	6	18
Arginine	20	22
Tryptophan	0[d]	ND[c]
Total	643	612

[a] Taken from ref. *16*. All values were based upon 21 valine residue per molecule. [b] Taken from ref. *4*; calculated for *Mr* 68,000. [c] Not determined. [d] By hydrolysis in the presence of 0.5% thioglycolic acid.

common features as scarceness in aromatic amino acids and richness in proline and glutamate. Human erythrocyte calpastatin has no carbohydrate, no tryptophan, and only less than one, possibly zero, residue of half-cystine per 643 residues corresponding to a 70K monomer. Preliminary data on circular dichroism have shown very low α-helix content of the molecule. All the known calpastatins are so extremely heat-stable that they do not lose inhibitory potency even after heating at 100°C for 5–20 min (*9*). The mechanism of such heat-stability remains to be elucidated in its relation to the structure of a calpastatin molecule.

FUNCTION

The first two reports, which independently described the discovery of calpastatin (*12*, *18*), agreed in demonstrating that the inhibition of

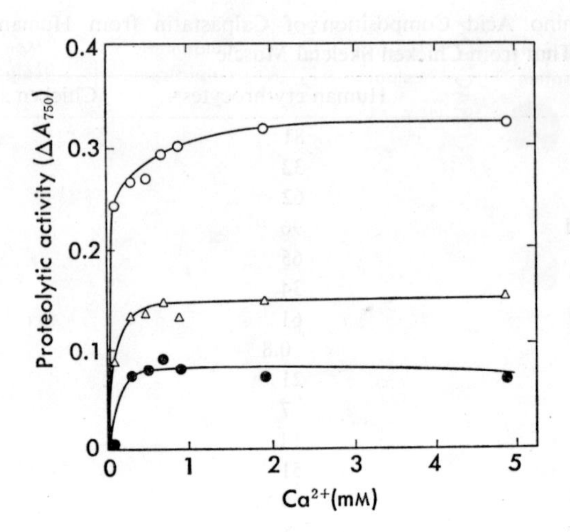

Fig. 3. Effect of Ca^{2+} concentration on the inhibition of calpain by calpastatin, both from human erythrocytes.
Each tube contained 0.33 units of calpain. ○ without calpastatin; △ with 0.21 μg of calpastatin; ● with 0.35 μg of calpastatin.

calpain by calpastatin could be made observable only in the presence of Ca^{2+}, while the inhibition was not due to sequestering of Ca^{2+} from the medium by calpastatin. Figure 3 illustrates the data obtained with highly purified human erythrocyte calpastatin when incubated with calpain I preparation of the same origin (*16*). The nature of the Ca^{2+}-dependence of calpain activity remains unaltered whether calpastatin is present or absent.

Both calpains I and II were found to be inhibited by calpastatin, though to different extents (*9*). None of several other proteases, including trypsin, chymotrypsin, and papain, were inhibited at all (*9*, *12*, *18*). Specific inhibition by calpastatin occurred equally against different substrate proteins such as casein, protein kinase C, and troponins (*9*). These results indicate the primary interaction of calpastatin to Ca^{2+}-activated calpain molecule rather than to the substrate protein.

Several lines of evidence have indicated the formation from Ca^{2+}-activated calpain and calpastatin molecules of a complex which no longer shows enzymatic or inhibitory activity (*6*, *9*, *16*). The dissociation of such complex into component enzyme and inhibitor could be achieved by removing Ca^{2+} from the complex (*6*, *9*).

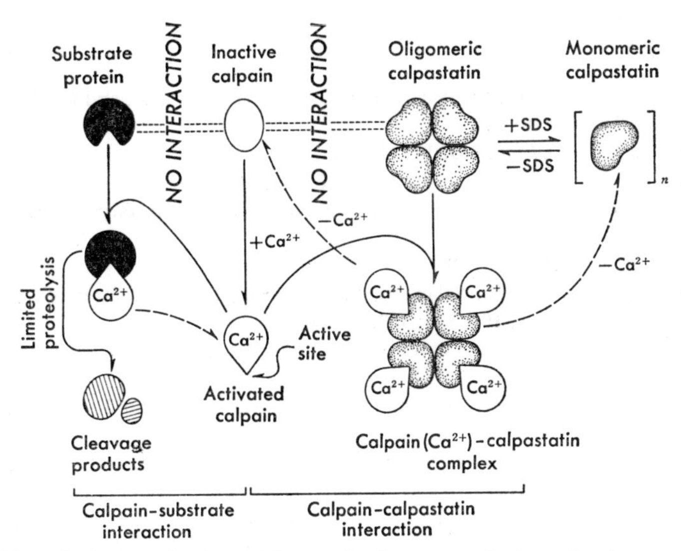

Fig. 4. Hypothetical mechanism of interaction between calpain and calpastatin, and the substrate molecules in the presence and absence of Ca^{2+}.

Figure 4 is a schematic representation of the modes of interaction between calpain, calpastatin, and substrate molecules in the presence or absence of Ca^{2+}. Calpain interacts neither with the substrate nor with calpastatin unless the Ca^{2+} concentration in the medium becomes sufficiently high to activate it. Ca^{2+} induces conformational changes of a calpastatin molecule so that its active site is exposed, and such Ca^{2+}-activated calpain is now capable of interacting either with the substrate leading to peptide bond cleavages or with calpastatin resulting in specific inhibition. The situation postulated in Fig. 4 for calpain resembles that already known for calmodulin, wherein calmodulin can interact only when it has been activated by Ca^{2+}, either with its target enzyme or with one of the other calmodulin-binding proteins (*7, 9*).

Acknowledgment

This work was supported in part by grants for Scientific and Cancer Research from the Ministry of Education, Science and Culture and by a grant for New Drug Development Research from the Ministry of Health and Welfare, Japan.

REFERENCES

1. Cottin, P., Azanza, J. L., Vidalenc, P., Ducastaing, A., Valin, C., and Ouali, A. *Reprod. Nutr. Dev.*, **21**, 309 (1981).
2. Guroff, G. *J. Biol. Chem.*, **239**, 149 (1964).
3. Drummond, G. I. and Duncan, L. *J. Biol. Chem.*, **241**, 3097 (1966).
4. Ishiura, S., Tsuji, S., Murofushi, H., and Suzuki, K. *Biochim. Biophys. Acta*, **701**, 216 (1982).
5. Mellgren, R. L. *FEBS Lett.*, **109**, 129 (1980).
6. Melloni, E., Sparatore, B., Salamino, F., Michetti, M., and Pontremoli, S. *Biochem. Biophys. Res. Commun.*, **106**, 731; **107**, 1053 (1982).
7. Murachi, T. *In* "Calcium and Cell Function," Vol. 4, ed. W. Y. Cheung, (1983), in press. Academic Press, New York.
8. Murachi, T., Hatanaka, M., Yasumoto, Y., Nakayama, N., and Tanaka, K. *Biochem. Intern.*, **2**, 651 (1981).
9. Murachi, T., Tanaka, K., Hatanaka, M., and Murakami, T. *In* "Advances in Enzyme Regulation," Vol. 19, ed. G. Weber, p. 407 (1981). Pergamon Press, New York.
10. Murakami, T., Hatanaka, M., and Murachi, T. *J. Biochem.*, **90**, 1809 (1981).
11. Nishiura, I., Tanaka, K., and Murachi, T. *Experientia*, **35**, 1006 (1979).
12. Nishiura, I., Tanaka, K., Yamato, S., and Murachi, T. *J. Biochem.*, **84**, 1657 (1978).
13. Nomenclature Committee of the International Union of Biochemistry. *Eur. J. Biochem.*, **116**, 423 (1981).
14. Suzuki, K., Tsuji, S., Kubota, S., Kimura, Y., and Imahori, K. *J. Biochem.*, **90**, 275 (1981).
15. Takahashi-Nakamura, M., Tsuji, S., Suzuki, K., and Imahori, K. *J. Biochem.*, **90**, 1583 (1981).
16. Takano, E. and Murachi, T. *J. Biochem.*, **92**, 2021 (1982).
17. Waxman, L. *In* "Protein Turnover and Lysosome Function," eds. H. L. Segal and D. J. Doyle, p. 363 (1978). Academic Press, New York.
18. Waxman, L. and Krebs, E. G. *J. Biol. Chem.*, **253**, 5888 (1978).

Proteinase Inhibitors: Medical and Biological Aspects, eds. N. Katunuma et al., pp. 173–179 (1983).
Japan Sci. Soc. Press, Tokyo/Springer-Verlag, Berlin

Calcium Activated Neutral Protease (CANP) and Its Exogenous and Endogenous Inhibitors

Kazutomo IMAHORI,[*1] Koichi SUZUKI,[*2] and Seiichi KAWASHIMA[*1]

Tokyo Metropolitan Institute of Gerontology[*1] *and Faculty of Medicine, The University of Tokyo*[*2]

The proteases which are responsible for the break down of intracellular proteins have not been clarified. Although cathepsins are good candidates they need reservation since they cannot cleave the proteins unless the latter are incorporated into the lysosomal particle. Another possible candidate would be calcium activated neutral protease (CANP). It exists in the cytosol and its activity can be finely regulated by the influx or efflux of Ca^{2+} ion. Moreover, unfavorable break down of the proteins by CANP can be prevented by its specific endogenous inhibitor. CANP was first identified in rabbit skeletal muscle by Huston and Krebs (*2*). They called it kinase activating factor (KAF) since it activated phosphorylase *b* kinase by limited proteolysis. Afterwards, the enzyme was partially purified from several organs and called by different names: Ca^{2+}-activated sarcoplasmic facotr (CASF) (*1*), Ca^{2+}-activated factor (CAF) (*8*), and receptor transforming factor (RTF) (*7*).

In 1978, we succeeded in preparation of this enzyme in homogeneity and characterized its property (*4*). Accordingly we called it CANP, instead of using the term of the factor; later, Murachi *et al.* proposed the name of calpain (*6*). Notwithstanding the different names,

*1 35-2 Sakae-cho, Itabashi-ku, Tokyo 173, Japan.
*2 7-3-1 Hongo, Bunkyo-ku, Tokyo 113, Japan.

the preparations obtained from different sources showed very similar enzymatic properties. All of them were thiol proteases and required millimolar Ca^{2+} for activity. However, recently Mellgren reported the existence of another type of CANP which requires micromolar Ca^{2+} for activity (*5*). In order to discriminate between the two types of CANPs we will call them m-CANP and μ-CANP, designating their requirement of millimolar and micromolar Ca^{2+}, respectively. The properties of these CANPs as well as the relation between them will be described below. Reading the endogenous inhibitor, several authors have reported its preparation and characterization. However the results are controversial. This may be partly due to the difference in the sources but also to their preparations not being in homogeneity. Since we have succeeded in purification of the inhibitor (*13*) we will describe its property, especially stoichiometry, in its inhibition of CANP.

PROPERTIES OF m-CANP

We have prepared m-CANP from chicken (*4*), human (*9*), and rabbit (*14*) skeletal muscles, all in homogeneity. All of them have very similar properties. The optimum pH exists at around 7.5 and the isoelectric point around 4.5. They require several millimolar Ca^{2+} for full activity. As an example, the Ca^{2+} dependency of m-CANP obtained from rabbit

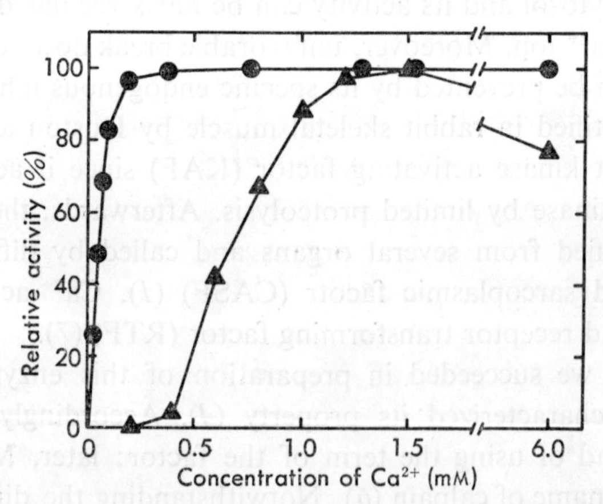

Fig. 1. Effect of Ca^{2+} on the activity of μ-CANP (●) and m-CANP (▲). Purified μ-CANP (0.68 unit) or m-CANP (0.53 unit) was assayed for the protease activity using casein as the substrate.

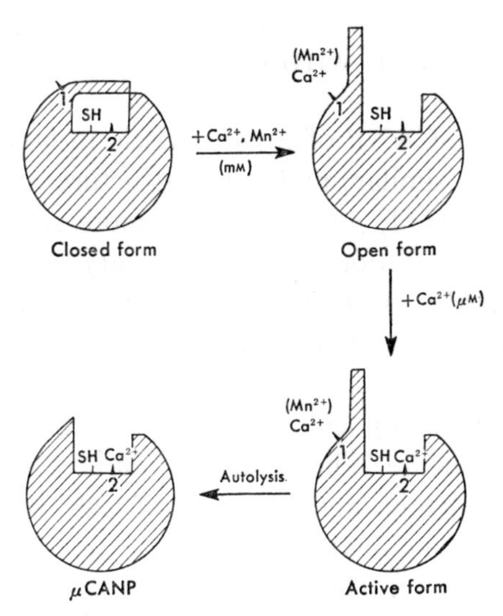

Fig. 2. Schematic models for the activation of m-CANP.
The numbers in the figure (1 and 2) represent the first and second Ca^{2+}-binding sites, respectively.

muscle is shown in Fig. 1 (*10*) m-CANPs obtained from other sources gave similar curves. As will be discussed later, the curve showed a typical sigmoidal nature.

m-CANP is classified as thiol protease, however the treatment of m-CANP with iodoacetic acid but without Ca^{2+} scarcely inactivated the enzyme. On the contrary, in the presence of Ca^{2+}, the reagent inactivated the enzyme very rapidly and complete inactivation was attained when the incorporation of carboxy methyl groups into m-CANP became one mol/mol enzyme (*11*). This suggests that the role of Ca^{2+} is to open up the active site and to expose the essential thiol group (Fig. 2). This idea was supported by the following facts. m-CANP contained three classes of thiol groups: those originally exposed (class I), those that became exposed after the addition of Ca^{2+} (class II), and those that remained tightly buried unless 5 M urea was added (class III). When various amounts of Ca^{2+} were added to the enzyme the activity appeared in proportion to the exposure of the class II thiol group.

μ-CANP AND ITS RELATION WITH m-CANP

We have prepared μ-CANP of rabbit muscle in homogeneity (*3*); thus prepared it had a similar molecular structure to m-CANP. However, as shown in Fig. 1, μ-CANP is far more sensitive to Ca^{2+} than m-CANP. The concentration of Ca^{2+} necessary for half maximal activity (K_a) is 40 μM. Moreover, it should be noted that the curve of μ-CANP is hyperbolic while that of m-CANP is sigmoidal. This sigmoidal nature can be explained schematically by Fig. 2. In this model we assumed two Ca^{2+}-binding sites, site 1 and site 2. Site 2 has far larger affinity for Ca^{2+} than site 1, however it is not accessible for Ca^{2+} unless it is unmasked. Probably, upon binding of Ca^{2+} to site 1, the conformational change of m-CANP takes place, unmasking both site 2 and the essential thiol group. This indicates that two Ca^{2+} ions binding to site 1 and site 2 play different roles. It is also suggested that the role of Ca^{2+} for site 1 can be substituted with other cations or factors. Actually, in the presence of 2 mM Mn^{2+} the sensitivity of m-CANP to Ca^{2+} gave almost the same curve as that of μ-CANP of Fig. 1 (*12*). If the chemical modification of site 1 would give the same effect as binding of Ca^{2+} or Mn^{2+} the modified enzyme should behave as μ-CANP. μ-CANP is somewhat larger than m-CANP (*3*). We have reported that the limited autolysis of m-CANP converted it to μ-CANP (*10*). μ-CANP thus obtained is called derived μ-CANP in order to discriminate from native μ-CANP, which can be obtained directly from the cell. Derived μ-CANP has the same Ca^{2+} sensitivity as native μ-CANP. This conversion is also illustrated in Fig. 2. If the cap part, which masks the active site of m-CANP, is removed by autolysis the enzyme will be converted into μ-CANP. The model of Fig. 2 can be supported from the following experimental results. The presence of 2 mM Ca^{2+} or Mn^{2+} shifted the isoelectric point of m-CANP from 4.5 to 5, while both native and derived μ-CANPs had their isoelectric points at pH 5. In accordance with this difference of isoelectric point both μ-CANPs can be eluted from DEAE-cellulose column by lower salt concentration than that required for the elution of m-CANP (*3*, *10*).

ENDOGENOUS INHIBITOR

Although the activity of CANP is regulated by the concentration of Ca^{2+} another regulatory mechanism is necessary to prevent the unfavorable break down of the intracellular proteins. The endogenous inhibitor would meet this necessity. There are several proteins which inhibited the activity of CANP but we purified one of them from the skeletal muscle of rabbit (*13*). The inhibitor was identified as a protein since, among several hydrolases tested, only proteases inactivated it. The molecular weight of the inhibitor was estimated to be 70,000 and 35,000 by gel filtration and SDS-gel electrophoresis, respectively; thus it appears to be the dimeric protein. When the activity of m-CANP was assayed in the presence of various amounts of the inhibitor the results obtained are as shown in Fig. 3A. Apparently the activity of m-CANP decreased in proportion to the amount of inhibitor added. This suggests that the inhibitor forms a tight, stoichiometric and inert complex with m-CANP. When the results of Fig. 3A are replotted as shown in Fig. 3B it is evident that each subunit of the inhibitor forms a complex with one molecule of m-CANP, with inhibition of the enzyme activity. When the complex was examined by disc electrophoresis the complex formation took place only when a sufficient amount of Ca^{2+} was supplemented. The inhibitor of rabbit muscle was effective

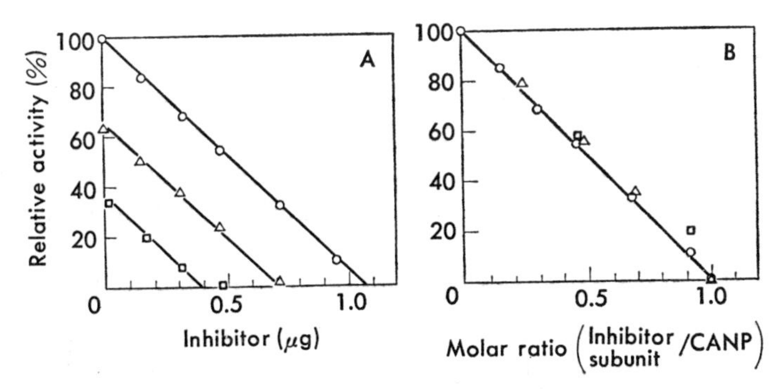

Fig. 3. The stoichiometry in the reaction between m-CANP and its inhibitor.
A: titration of CANP activity with purified inhibitor. To 0.8 (□), 1.6 (△) or 2.4 (○) μg of CANP was added the inhibitor as indicated in the abscissa and the remaining activity was assayed. B: the remaining activity obtained in A was plotted against the molar ratio of the inhibitor subunit to CANP.

not only against m-CANP of rabbit muscle but also that of chicken muscle as well as against the native or derived μ-CANP. However the inhibitor was ineffective for any other protease so far tested, including trypsin, chymotrypsin, papain, plasmin, bromelain, and cathepsin B. As indicated above, the physiological role of the endogenous inhibitor would be to protect physiologically important proteins from unfavorable breakdown. However, reservation is necessary for any conclusion since even the location of the inhibitor in the cell has not been clarified.

EXOGENOUS INHIBITOR

Since CANPs belong to the thiol enzyme they are easily inactivated by such a thiol reagent as iodoacetic acid, when enough Ca^{2+} is supplemented. In fact, when m-CANP was incubated with 100-fold molar excess of iodoacetic acid in the presence of 2 mM Ca^{2+}, complete inactivation took place within 3 min, however, in the absence of Ca^{2+}, no appreciable inactivation was observed even after 60 min (*11*). This can be explained by Fig. 2 if we assume that the inactivation is due to the modification of the essential thiol group.

Next, we explored exogenous inhibitors which specifically inhibit thiol protease but not other metabolically important thiol enzymes. After examining a number of compounds we selected leupeptin, antipain, and E-64, which inhibited m-CANP with K_i values of 0.46, 2.0, and 0.90 μM, respectively (*11*). They also inhibited other thiol proteases such as papain, cathepsins B and L, but were ineffective against the other thiol enzymes so far tested, including several dehydrogenases. Leupeptin and antipain inhibited CANP non-competitively and reversibly. Even after incubation with Ca^{2+} and 50-fold molar excess of either inhibitor for 60 min, m-CANP regained full activity by dialyzing out the inhibitor.

On the other hand, the incubation of m-CANP with E-64 in the presence of Ca^{2+} resulted in complete and irreversible inactivation of the enzyme. However, in the absence of Ca^{2+}, E-64 did not inactivate the enzyme. This can be easily understood from Fig. 2 if we assume that E-64 would react with the essential thiol group at the active site. Using [^{3}H]-E-64 we confirmed that the incorporation of one mol of E-64 per mol of m-CANP was just enough to abolish the latter activity.

As will be expected from Fig. 2, E-64 inactivated native and derived μ-CANPs irreversibly. These exogenous inhibitors, especially E-64, would be useful to clarify the physiological role of CANP since the introduction of E-64 would result in a specific inactivation of the enzyme. On the other hand, if we assume that muscle dystrophy would be due to CANP-dependent breakdown of muscle protein, the progress of the disease might be stopped by administration of the inhibitor to the patient.

REFERENCES

1. Busch, W. A., Stromer, M. H., Goll, D. E., and Suzuki, A. *J. Cell Biol.*, **52**, 367 (1972).
2. Huston, R. B. and Krebs, E. G. *Biochemistry*, **7**, 2116 (1968).
3. Inomata, M., Hayashi, M., Nakamura, M., Imahori, K., and Kawashima, S. *J. Biochem.*, **93** (1983), in press.
4. Ishiura, S., Murofushi, H., Suzuki, K., and Imahori, K. *J. Biochem.*, **84**, 225 (1978).
5. Mellgren, R. L. *FEBS Lett.*, **109**, 129 (1980).
6. Murachi, T., Tanaka, K., Hatanaka, M., and Murakami, T. *Adv. Enzymol. Regul.*, **19**, 407 (1981).
7. Pucca, G. A., Nola, E., Succa, V., and Brescinai, M. *J. Biol. Chem.*, **252**, 1358 (1977).
8. Reddy, M. K., Estringer, J. D., Robinowitz, M., Fischman, D. A., and Zak, R. *J. Biol. Chem.*, **250**, 4278 (1975).
9. Suzuki, K., Ishiura, S., Tsuji, S., Katamoto, T., Sugita, H., and Imahori, K. *FEBS Lett.*, **102**, 355 (1979).
10. Suzuki, K., Tsuji, S., Kubota, S., Kimura, Y., and Imahori, K. *J. Biochem.*, **90**, 275 (1981).
11. Suzuki, K., Tsuji, S., and Ishiura, S. *FEBS Lett.*, **130**, 119 (1981).
12. Suzuki, K. and Tsuji, S. *FEBS Lett.*, **140**, 16 (1982).
13. Takahashi-Nakamura, M., Tsuji, S., Suzuki, K., and Imahori, K. *J. Biochem.*, **90**, 1583 (1981).
14. Tsuji, S. and Imahori, K. *J. Biochem.*, **90**, 233 (1981).

Proteinase Inhibitors: Medical and Biological Aspects, eds. N. Katunuma et al., pp. 181–190 (1983).
Japan Sci. Soc. Press, Tokyo/Springer-Verlag, Berlin

Regulation of Proteinases in *Saccharomyces cerevisiae*

Helmut HOLZER

Biochemisches Institut, Universität Freiburg[*1] *and GSF-Abteilung Enzymchemie*[*2]

Only during the last few years first insights into the control mechanisms of intracellular proteolysis have been obtained (for a summary see ref. *12*). Based on work in the author's laboratory two aspects concerning the control of proteinases in *Saccharomyces cerevisiae* will be discussed: 1) the role of endogenous proteinase inhibitors and 2) the function of covalent modification of the target proteins in the initiation of proteolysis.

ROLE OF ENDOGENOUS PROTEINASE INHIBITORS IN THE CONTROL OF PROTEOLYTIC PROCESSES IN YEAST

In Table I, the macromolecular proteinase inhibitors from yeast are listed (*24*). These inhibitors are localized in the cytosol. They are highly specific and possess high affinities for their respective yeast proteinases. The association constants (*6*, *7*, *19*, *20*, *28*) are shown in Table II. The main portion of the proteolytic activity in yeast is localized in the vacuoles (*21*). This intracellular compartmentation led to the assumption that the main function of the inhibitors is the protection of pro-

[*1] Hermann-Herder-Straße 7, D-7800 Freiburg, FRG.
[*2] D-8042 Neuherberg bei München, FRG.

TABLE I. Proteinase Inhibiting Polypeptides from Yeast (*24*)

Inhibitor	Isoelectric point	Specific inhibition of	Hydrolyzed by	Molecular weight	Stability
I^A^2	5.7	Proteinase A	Proteinase B	7,700	Heat- and acid-resistant
I^A^3	6.3	Proteinase A	Proteinase B	7,700	
I^B^1	8.0	Proteinase B	Proteinase A	8,500	Heat- and acid-resistant
I^B^2	7.0	Proteinase B	Proteinase A	8,500	
I^B^3	4.6	Proteinase B	Proteinase A	11,500	
I^C^	6.6	Carboxy-peptidase Y	Proteinases A and B	23,800	Heat- and acid-labile

TABLE II. Association Constants of Endogenous Proteinase Inhibitors

Inhibitor	Proteinase	$K^{app}_{assoc.}$ (M^{-1})	References
$I^{A}3$	A	1.8×10^{7}	*28*
$I^{B}1$	B	3.7×10^{9}	*7*
$I^{B}2$	B	3.9×10^{9}	*7*
3-NitroTyr$^{41}I^{B}2$	B	3.2×10^{8}	*7*
$I^{B}3$	B	3.3×10^{9}	*19*
I^{C}	CPY	4.0×10^{8}	*20*

CPY, carboxypeptidase Y.

teins of the cytosol against degradation by proteinases liberated from the vacuoles by leakage or lysis (*5, 12, 14*). However, an additional function of the inhibitors may consist in the control of physiological proteolytic processes during the life cycle of the yeast cell. During the transition from vegetative growth to spore formation, proteinase B activity increases whereas proteinase B inhibitor activity decreases as shown in Fig. 1 (*4*). This behaviour suggests a control of proteinase B activity by the proteinase B inhibitor. In experiments with mutants deficient of proteinase A inhibitor which have been isolated in Dr. Wolf's laboratory (*1*), evidence was obtained that the concentration of the proteinase B inhibitor is under control of proteinase A and of proteinase A inhibitor. *In vitro* experiments by Saheki and Holzer (*31*) have demonstrated that the proteinase B inhibitor is rapidly degraded (and thereby inactivated) by proteinase A. In the proteinase A inhibitor deficient mutant, the rate of degradation of proteinase B inhibitor is about twice as fast as it is in the wild type strain (Fig. 2) (*34*). The most plausible explanation for this observation is that the proteinase A in-

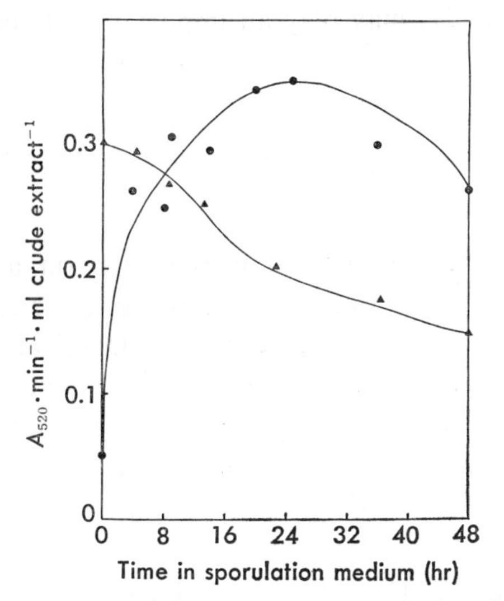

Fig. 1. Activities of proteinase B (●) and proteinase B inhibitor I^B (▲) at the transition from vegetative growth to sporulation (*4*).

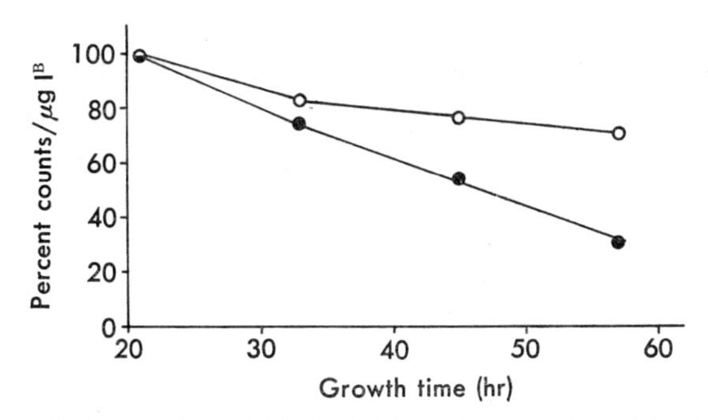

Fig. 2. Turnover of the proteinase inhibitor I^B in wild type cells and in the proteinase A inhibitor deficient pai mutant (*34*). ○ wild type; ● pai 1-mutant.

hibitor controls proteinase A activity and that in turn proteinase A controls the proteinases B inhibitor level by degradation. However, at present it is not known how the proteinase A and B inhibitors, localized in the cytosol, come into contact with the proteinases A and B, which are compartmentalized in the vacuoles. It might be that under certain physiological conditions, vacuoles are lysed or become leaky to fulfill proteolytic functions outside of their compartment. Local changes in the pH which strongly affect the extent of inhibition of the proteinases by their inhibitors (*24*) may play an important role in these regulatory mechanisms.

INITIATION OF SELECTIVE PROTEOLYSIS BY COVALENT MODIFICATION OF THE SUBSTRATE

The addition of glucose to yeast cells grown on acetate or ethanol causes a rapid inactivation of certain gluconeogenic enzymes which are not required in the presence of glucose. The inactivation of cytoplasmic malate dehydrogenase, fructose-1,6-bisphosphatase, and phosphoenolpyruvate carboxykinase (*cf.* Fig. 3) has been demonstrated to be the result of proteolytic degradation (*8*, *26*, *27*). In the author's laboratory, the inactivation of fructose-1,6-bisphosphatase has been studied in more detail. The inactivation occurs in two characteristic phases (*15*). In a first and very rapid phase the disappearance of about 60% of the enzyme activity is observed followed by a second and slower phase in which the residual enzyme activity disappears (*cf.* Fig. 4). As shown in Table III, during the rapid inactivation phase ^{32}phosphate is incorporated into fructose-1,6-bisphosphatase (*22*, *25*). By thin layer chromatography of an acid hydrolysate of the labeled fructose-1,6-bisphosphatase it was shown that serine residues of the enzyme are phosphorylated

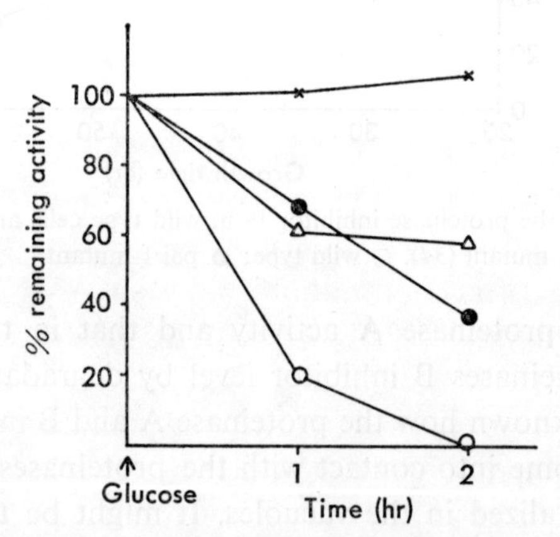

Fig. 3. Catabolite inactivation of cytoplasmic malate dehydrogenase (MDH, △), phosphoenolpyruvate carboxykinase (PEPCK, ●), and fructose-1,6-bisphosphatase (FBPase, ○) in *S. cerevisiae* M 1. Reference enzyme: glucose-6-phosphate dehydrogenase (G6PDH, ×).

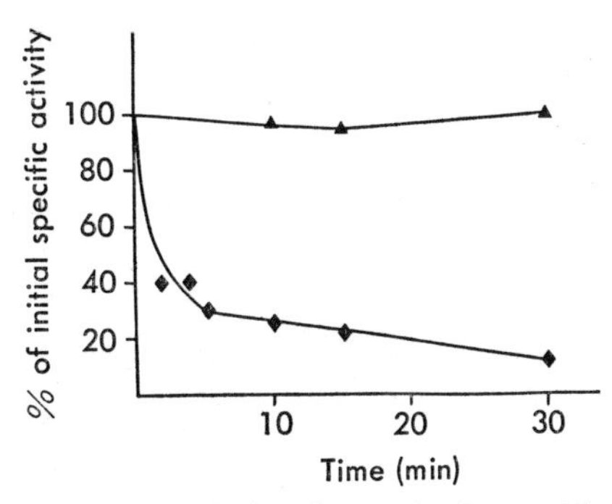

Fig. 4. Inactivation of fructose-1,6-bisphosphatase in *S. cerevisiae* M 1 after addition of glucose (*15*).

TABLE III. Specific Catalytic Activity and ^{32}P-label of Yeast Fructose-1,6-bisphosphatase (*25*)

Pretreatment of yeast	Specific catalytic activity (units/mg protein)	Radioactivity (^{32}P) in immunochemically cross-reacting material (cpm in FBPase peak)
Derepressed (glucose-starved)	0.073	29
3 min glucose-inactivated	0.028	1,897
3 min glucose-inactivated, thereafter 2 hr glucose-free reactivated	0.073	194

(*22, 25*). In cell free extracts, a cyclic AMP-dependent protein kinase, which partially inactivates fructose-1,6-bisphosphatase could be demonstrated (*18, 30*), as well as a Mg^{2+} (or Mn^{2+}) dependent enzyme which reactivates the phosphorylated fructose-1,6-bisphosphatase and which is inhibited by inorganic phosphate (D. Müller and M. Birtel, unpublished experiments, *cf.* also ref. *11*). Based on these results the phosphorylation/dephosphorylation cycle shown in Fig. 5 was postulated (*30*). A rapid increase in the concentration of cyclic AMP after addition of glucose to starved yeast cells, observed by van der Plaat in studies on the regulation of trehalase (*33*) and by Purwin *et al.* (*30*) in studies on the inactivation of fructose-1,6-bisphosphatase supports the scheme shown in this figure. It postulates that cyclic AMP is a mediator of the inactivating effects of glucose on fructose-1,6-bisphosphatase. Since not only glucose, but also uncouplers and ionophores such as 2,4-dinitrophenol, carbonyl-cyanide m-chlorophenyl-hydrazone (CCCP) and nystatin cause inactivation of fructose-1,6-bisphosphatase by phos-

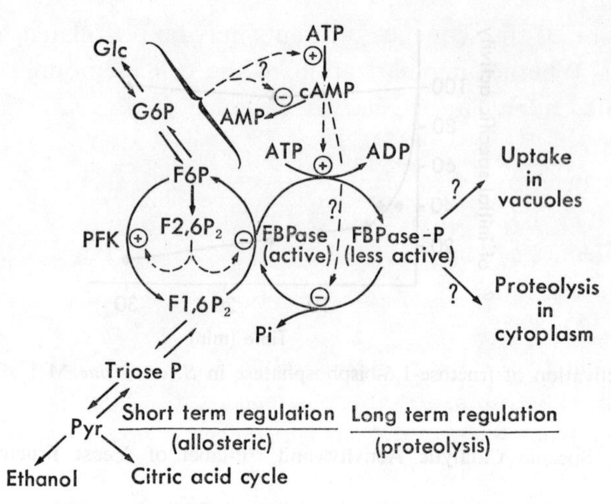

Fig. 5. Regulation of yeast fructose-1,6-bisphosphatase by cyclic AMP-dependent phosphorylation/dephosphorylation (*30*).

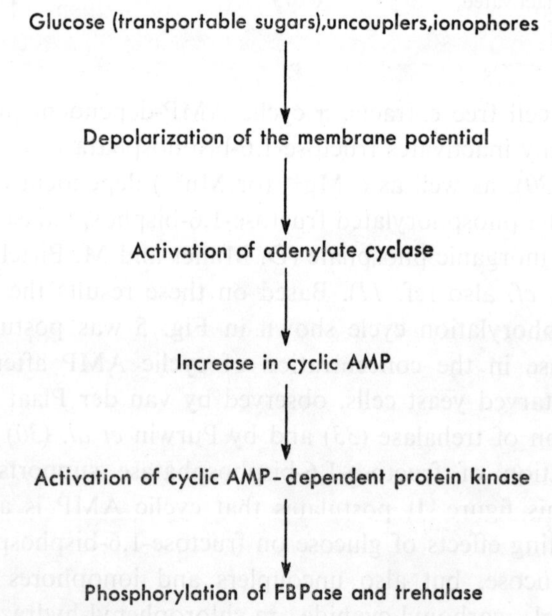

Fig. 6. Hypothesis for the sequence of events leading to phosphorylation of fructose-1,6-bisphosphatase in intact cells (*11*).

phorylation (*23*) the sequence of events may be postulated as shown in Fig. 6 (*11*). Whether depolarization of the cell membrane potential in fact activates adenylate cyclase as assumed in the scheme has to be established.

It has been assumed that the partial inactivation of fructose-1,6-bisphosphatase by a glucose induced phosphorylation prevents a "futile cycle" which would hydrolyze ATP to ADP and inorganic phosphate when both fructose-1,6-bisphosphatase and 6-phosphofructokinase are active (*15*, *22*, *25*). Since only 60% of the fructose-1,6-bisphosphatase activity is inactivated there were doubts about the efficiency of such a mechanism. The discovery of fructose-2,6-bisphosphate as a potent inhibitor of fructose-1,6-bisphosphatase in yeast (*13*) strongly suggests that the "futile cycle" is effectively prevented by this product of glucose metabolism (*cf.* Fig. 5). Most likely, the biological function of the phosphorylation is not to prevent the "futile cycle" but to initiate the proteolytic degradation of fructose-1,6-bisphosphatase. Proteolysis of the enzyme as a consequence of the addition of glucose to starved or ethanol grown yeast cells has been demonstrated by Funayama *et al.* (*8*). At present, two possibilities for the mechanism of initiation of proteolysis by phosphorylation of the target protein are being tested in the author's laboratory:

1) The phosphorylated enzyme may be more susceptible to proteinases than the non-phosphorylated form.
2) In contrast to the non-phosphorylated enzyme, the phosphorylated enzyme may rapidly be taken up into the vacuoles where proteolysis occurs.

Several other cases in which evidence has been obtained that a covalent modification initiates proteolysis are listed in Table IV. In addition to fructose-1,6-bisphosphatase, yeast NAD-glutamate dehydrogenase is probably phosphorylated prior to proteolysis (*9*). In rat liver pyruvate kinase is selected for proteolysis by phosphorylation (*2*) and a similar situation has been found for degradation of troponins in bovine cardiac muscle (*32*). Oxidation of an amino acid residue in glutamine synthetase in *Escherichia coli* (*16*) and oxidation of the Fe-S-center in glutamine phosphoribosylpyrophosphate amidotransferase in *Bacillus subtilis* (*3*) may initiate proteolysis of those enzymes under certain conditions. There is good evidence that fructose-1,6-bisphosphatase in

TABLE IV. Control of Intracellular Proteolysis by Covalent Modification

Type of covalent modification		References
Limited proteolysis	Rabbit liver: aldolase fructose-1,6-bisphosphatase	*29*
Conjugation with ubiquitin	Reticulocytes: (all eucariotes?): hemoglobin and other proteins	*10*
Lack of glycosylation	African claw frog: ACTH-β-lipotropin precursor	*17*
Oxidation of Fe-S center	*B. subtilis:* Gln-PRPP-amido transferase	*3*
Oxidation of an amino acid-residue	*E. coli:* Gln-synthetase	*16*
Phosphorylation/dephosphorylation	Yeast: NAD-glutamate dehydrogenase	*9*
	Yeast: fructose-1,6-bisphosphatase	*22, 25*
	Rat liver: pyruvate kinase	*2*
	Bovine cardiac muscle: troponins	*32*

rabbit liver becomes susceptible to proteolytic degradation after limited proteolysis with lysosomal proteinases (*29*). From experiments with reticulocytes, it was concluded that proteolysis of hemoglobin and other proteins is initiated by conjugation with ubiquitin (*10*). Therefore, it appears that in various organisms the selection of enzymes for proteolytic degradation may be initiated by an enzyme catalyzed covalent modification of the target protein.

Acknowledgments

Part of this article was written when the author was a Scholar-in-Residence at the Fogarty International Center, NIH, Bethesda, Md. (U.S.A.) and was reported at the International Symposium on Medical and Biological Aspects of Proteinase Inhibitors on August 5th–8th, 1982 at Tokushima City, Japan 770. The author acknowledges support for attendance at the Symposium in Tokushima to Prof. Hamo Umezawa, Tokyo, and to the Japan Society for the Promotion of Sciences. The experimental work from the author's laboratory included in this paper was supported by the Deutsche Forschungsgemeinschaft, Sonderforschungsbereich 46, and the Verband der Chemischen Industrie e.V. (Fonds der Chemische Industrie). The author thanks Dr. Peter Bünning for discussions and critical reading of the manuscript. Thanks are also due to W. Fritz and Ulrike Eitel for help with the figures and for typing the manuscript.

REFERENCES

1. Beck, I., Fink, G. R., and Wolf, D. H. *J. Biol. Chem.*, **255**, 4821 (1980).
2. Bergström, G., Ekman, P., Humble, E., and Engström, L. *Biochim. Biophys. Acta*, **532**, 259 (1978).
3. Bernlohr, D. A. and Switzer, R. L. *Biochemistry*, **20**, 5675 (1981).
4. Betz, H. and Weiser, U. *Eur. J. Biochem.*, **62**, 65 (1976).
5. Betz, H., Hinze, H., and Holzer, H. *J. Biol. Chem.*, **249**, 4515 (1974).
6. Bünning, P. and Holzer, H. *In* "Limited Proteolysis in Microorganisms," eds. G. N. Cohen and H. Holzer, p. 81 (1979). U.S. Government Printing Office, Washington, D.C.
7. Bünning, P. and Holzer, H. *J. Biol. Chem.*, **252**, 5316 (1977).
8. Funayama, S., Gancedo, J. M., and Gancedo, C. *Eur. J. Biochem.*, **109**, 61 (1980).
9. Hemmings, B. A. *Biochem. Soc. Trans.*, **10**, 328 (1982).
10. Hershko, A. and Ciechanover, A. *Annu. Rev. Biochem.*, **51**, 335 (1982).
11. Holzer, H. *In* "Molecular Aspects of Cellular Regulation," Vol. 3, ed. Ph. Cohen, Elsevier, Amsterdam, in press.
12. Holzer, H. and Heinrich, P. C. *Annu. Rev. Biochem.*, **49**, 63 (1980).
13. Lederer, B., Vissens, S., v. Schaftingen, E., and Hers, H. G. *Biochem. Biophys. Res. Commun.*, **103**, 1281 (1981).
14. Lenney, J. F., Matile, Ph., Wiemken, A., Schellenberg, M., and Meyer, J. *Biochem. Biophys. Res. Commun.*, **60**, 1378 (1974).
15. Lenz, A. G. and Holzer, H. *FEBS Lett.*, **109**, 271 (1980).
16. Levine, R. L., Oliver, C. N., Fulns, R. M., and Stadtman, E. R. *Proc. Natl. Acad. Sci. U.S.*, **78**, 2120 (1981).
17. Loh, Y. P. and Gainer, H. *FEBS Lett.*, **96**, 269 (1978).
18. Londesborough, J. *FEBS Lett.*, **144**, 269 (1982).
19. Matern, H., Weiser, U., and Holzer, H. *Eur. J. Biochem.*, **101**, 325 (1979).
20. Matern, H., Barth, R., and Holzer, H. *Biochim. Biophys. Acta*, **567**, 503 (1979).
21. Matile, Ph. and Wiemken, A. *Arch. Microbiol.*, **56**, 148 (1967).
22. Mazón, M. J., Gancedo, J. M., and Gancedo, C. *J. Biol. Chem.*, **257**, 1128 (1982).
23. Mazón, M. J., Gancedo, J. M., and Gancedo, C. *Eur. J. Biochem.*, **127**, 605 (1982).
24. Müller, M. and Holzer, H. *In* "Enzyme Regulation and Mechanisms of Action," Vol. 60, eds. P. Mildner and B. Ries, p. 339 (1980). Pergamon Press, Oxford.
25. Müller, D. and Holzer, H. *Biochem. Biophys. Res. Commun.*, **103**, 926 (1981).
26. Müller, M., Müller, H., and Holzer, H. *J. Biol. Chem.*, **256**, 723 (1980).
27. Neeff, J., Hägele, E., Neuhaus, J., Heer, U., and Mecke, D. *Eur. J. Biochem.*, **87**, 489 (1978).
28. Núnez de Castro, I. and Holzer, H. *Hoppe-Seyler's Z. Physiol. Chem.*, **357**, 727 (1976).
29. Pontremoli, S., Melloni, E., and Horecker, B. L. *In* "Metabolic Interconversion of Enzymes 1980," ed. H. Holzer, p. 186 (1981). Springer-Verlag, Berlin.
30. Purwin, C., Leidig, F., and Holzer, H. *Biochem. Biophys. Res. Commun.*, **107**, 1482 (1982).

31. Saheki, T. and Holzer, H. *Biochim. Biophys. Acta*, **384**, 203 (1975).
32. Toyo-Oka, T. *Biochem. Biophys. Res. Commun.*, **107**, 44 (1982).
33. van der Plaat, J. B. *Biochem. Biophys. Res. Commun.*, **56**, 580 (1974).
34. Wolf, D. H., Ehmann, C., and Beck, I. *In* "Biological Function of Proteinases," eds. H. Holzer and H. Tschesche, p. 55 (1979). Springer-Verlag, Berlin.

Proteinase Inhibitors: Medical and Biological Aspects, eds. N. Katunuma et al., pp. 191–200 (1983).
Japan Sci. Soc. Press, Tokyo/Springer-Verlag, Berlin

Biochemical Aspects of an Inhibitor Protein from Uterine Myometrium

Ernst-Günter AFTING

*Institut für Biochemie, Abteilung Proteinbiochemie, Georg-August-Universität**

The uterus is unique in undergoing extreme changes in size and function during the various phases of its physiological behavior. The enlargement of this organ during pregnancy by tremendeous increase in protein content and the rapid involution in the postpartal period by controlled protein degradation are striking. These processes are also mirrored in a smaller scale during every estrus cycle (Fig. 1). By measuring the DNA content in the rat myometrium we have observed only a 2-fold increase in DNA content in contrast to a 10-fold increase in tissue weight (Fig. 1). Therefore, about 80% of the weight change during pregnancy and involution can be calculated to be due to hyper- and hypotrophy. The uterine involution therefore requires a specific *intracellular* proteolytic system (*10*).

Intracellular enzyme levels within growing, involuting or steady state tissues are controlled by changes in the rate of both enzyme synthesis and degradation (*22*, *23*). The detailed mechanism of protein synthesis and the factors influencing its rate are well documented, but the knowledge about the protein degradation and the proteolytic enzymes and organelles involved is rather fragmentary (*4*, *12*). De Duve and Wattiaux (*8*) have shown that intracellular organelles, the lyso-

* Humboldtallee 7 D-3400 Göttingen, FRG.

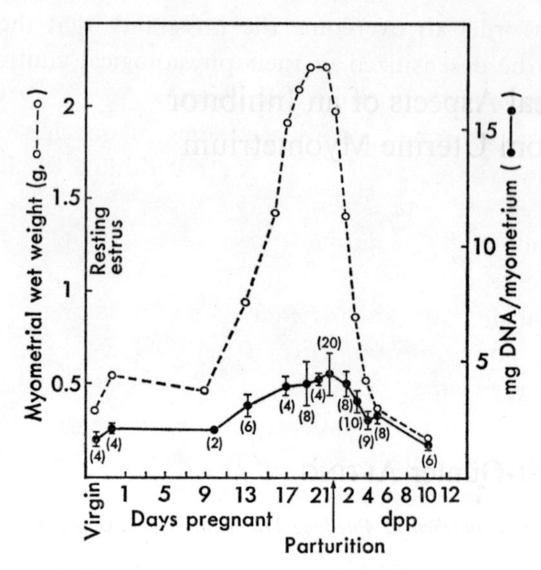

Fig. 1. Wet weight and DNA content of the rat uterine myometrium during pregnancy and involution.
dpp, days postpartum.

somes, could play a predominant role during intra- and extracellular protein degradation. Lysosomes contain a range of protein- and peptide hydrolases with a wide spectrum of endo- and exopeptidase activities to degrade most, if not all, cellular proteins. One might therefore be tempted to assume that protein breakdown is solely mediated by lysosomes. But the specificity of protein turnover still needs to be elucidated, *i.e.*, the mechanism by which different uterine proteins are broken down at different rates (*11*). Lysosomes do not seem to possess either the specificity of protein uptake or the sensitivity to hormonal and other controls which are required to explain the observed selective and precisely regulated protein catabolism.

Another possibility is that besides an intralysosomal pathway of protein degradation an extralysosomal protein breakdown may also take place. The latter mechanism requires the existence of extralysosomal proteolytic enzymes which should show a neutral to alkaline pH-optimum of activity in contrast to the lysosomal enzymes which are tailored to the acidic pH range of the lysosome (*18*). These extralysosomal proteolytic enzymes should at first inactivate their substrate enzyme(s) in order to overcome the possibility that the substrate enzymes could be desensitized to their physiological control by one proteolytic cleavage of their amino acid chain.

We have performed extensive studies in our lab on these intracellular proteolytic enzymes which are responsible for the control and degradation of the major intracellular protein of the uterine cell, the contractile actomyosin complex. Two peaks of activity could be detected for the degradation of uterine actomyosin in the rat myometrium, one in the acidic pH range, typical for a lysosomal enzyme and the other with a optimum of activity at pH 8 which is characteristic for a non-lysosomal enzyme.

The lysosomal enzyme degrading actomyosin turned out to be identical to cathepsin D isolated from various organs of mammals (*1*). In contrast to the report by Sapolsky and Woessner (*21*), only one form of the enzyme could be detected in the rat uterus. In addition to the other lysosomal proteinases cathepsins B and L (*17*), this enzyme splits several components of the actomyosin complex of muscle cells, preferentially myosin light and heavy chains as was already described by Schwarz and Bird (*24*). In intact muscle cells the actomyosin complex is not readily available to the proteolytic system of the lysosomal enzymes. In addition, to explain the different degradation and turnover rates of the components of the actomyosin complex (*28*), the non-lysosomal proteolytic enzymes characterized by a neutral to a slightly alkaline pH-optimum of activity seems to play the major role in the initiation of actomyosin degradation. Only a few proteinases capable of degrading myofibrillar proteins at neutral or slightly alkaline pH have been identified in muscular tissue (*2, 3, 5, 7, 15, 27*).

We have characterized an alkaline proteinase from uterine smooth muscle which immediately inactivates the target enzyme as having to be postulated for an extralysosomal, intracellular proteolytic enzyme. The uterine enzyme degrades the actomyosin complex with a concomitant decrease in the myosin ATPase activity (*2*). The enzyme is not specific for the degradation of myosin itself but also attacks some other high molecular weight proteins of the actomyosin complex by limited proteolysis. This is documented by the fact that after inactivation of the myosin ATPase the production of low molecular weight splitting products ceases (*2*). The enzyme is stable in the pH range between 6 and 9 which is in good agreement with its characterization as a non-lysosomal enzyme (*20*), but is very sensitive to inactivation by increasing temperatures. Temperatures above 40°C rapidly inactivate the alkaline proteinase. This enzyme is sensitive to SH- and OH-modifying agents, to trypsin inhibitor and, in contrast to cathepsin D, completely insensitive to pepstatin (*19*). However, the hormones estrogen and progesterone which trigger the physiological status of the uterus are without any direct effect on the enzyme.

The lysosomal enzymes are *in vivo* inactivated after a possible leakage from the lysosomal compartment simply by the pH shift occuring within the cytosol, whereas alkaline proteinases are always very harmful to every cell as they are not affected by a physiological pH change. They can act at the physiological pH of the cytosol and must be carefully controlled to exclude their activities towards susceptible substrate enzymes. Therefore special safety mechanisms have to be developed to overcome the uncontrolled intra- and extracellular effects of the alkaline proteinases which can leak out of cells during toxic effects on membranes, cell lysis, *etc.* A broad spectrum of proteinase inhibitors has been shown to be present in the serum of a variety of species (*14*), in order to prevent the interference of alkaline proteinases leaking out from cells with the very sensitive cascades of limited proteolysis of the serum, *e.g.*, the blood clotting system and the complement system (*6, 16*).

It is therefore not surprising that the extraction of tissues with buffered media always results in the inactivation of the alkaline proteolytic enzymes by serum inhibitors which can be found in the tissues even after prolonged perfusion of the organs with buffeı solutions. On the other hand we have shown by careful determinations of the serum content of several tissues including the uterine myometrium that, besides the serum inhibitors, a tissue specific intracellular inhibitor also exists against proteolytic enzymes with alkaline pH optima (Table I). In addition to the alkaline proteinase from the uterine myometrium we have used yeast proteinase B as a test tool because the yeast enzyme was also very sensitive to inhibition by the uterine myometrial inhibitor. In contrast to the uterine alkaline proteinase the yeast enzyme was available in large amounts and in a homogeneous form. The inhibitor was not active against the acid proteinase cathepsin D from all tissues.

TABLE I. Proteinase B Inhibiting Activities in Different Tissues of the Rat

Tissue	Units/g wet weight[a]
Serum	55.7[b]
Uterus	
4 days prepartal	7.6
3 days postpartal	4.8
4 days postpartal	3.8
Heart	2.9
Brain	1.8
Gastrocnemius	1.6
Spleen	0.9
Lungs	3.8
Liver	3.1

[a] Units of inhibition are expressed as units of proteinase B inhibited and corrected for the serum content in every tissue. [b] Units/ml serum.

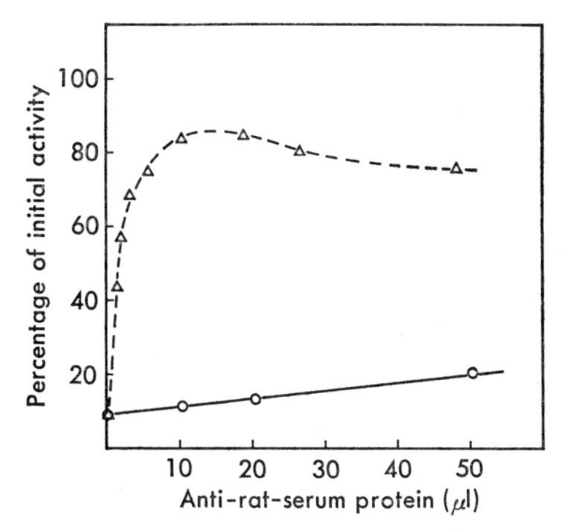

Fig. 2. Effect of anti-rat-serum immunoglobulins on the inhibitory activity of rat uterine and serum inhibitors.
Yeast proteinase B was 90% inhibited by a sufficient amount of uterine inhibitor (○) or serum inhibitor (△). This amount of inhibitor was preincubated with increasing amounts of anti-rat-serum immunoglobulins, followed by addition of proteinase B and substrate and the proteolytic activity was measured.

The existence of an intracellular inhibitor in uterine tissues does not rule out that this inhibitor was taken up by the uterine cells and is identical with the serum inhibitor. This has been recently described for α_1-proteinase inhibitor in the mouse uterus (*9*). To rule out this possibility, antibodies against rat serum proteinase inhibitors were raised in rabbits and the immunoglobulin fraction was purified until it was free of serum inhibitors. If all the inhibitor activity of the rat uterus was due to the serum content, this inhibition could be expected to be relieved by the addition of sufficient amounts of antibody. As shown in Fig. 2, if yeast proteinase B was 90% inhibited by a sufficient amount of the uterine inhibitor, the inhibition was not affected by the addition of increasing amounts of rabbit immunoglobulins directed against serum proteinase inhibitors. As a control 90% inhibition of yeast proteinase B by the rat serum inhibitor was nearly totally overcome by addition of the rabbit-anti-rat immunoglobulins (Fig. 2). Both experiments therefore independently point out that the myometrium contains an endogeneous proteinase inhibitor which differs from the known serum inhibitors. The inhibitor is much more stable to heat than the alkaline proteinase. This property has been used for the partial purification because brief heat treatment of the uterine extract stabilizes the inhibitor. Without prior heating the uterine extract loses its inhibitory activity within a few days presumably as a result of the proteolysis of the inhibitor protein itself.

A possible physiological importance of the inhibitor protein can be suggested on the basis of variations in its activities during the pre- and

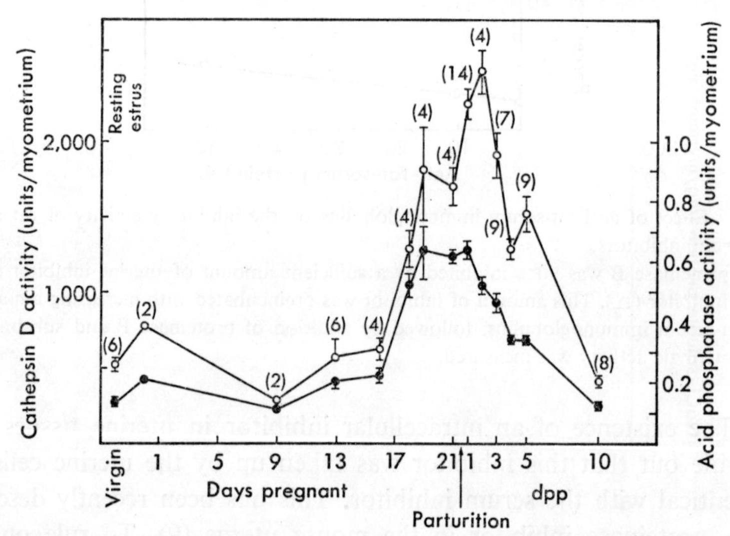

Fig. 3. Activity of the cathepsin D (○) and lysosomal acid phosphatase (●) throughout pregnancy and involution ±S.E.M.

postpartal period. A most interesting result was that the inhibitor protein as well as the proteinase cathepsin D and the alkaline proteinase are not induced after parturition by the abrupt fall of the steroid hormones estrogen and progesterone. This can be shown by measuring the increasing activity during gestation. The activity profile of lysosomal cathepsin D and another lysosomal marker enzyme, acid phosphatase, is shown in Fig. 3. Both enzyme activities increase dramatically during the last third of the trimenon and highest activities are found just before parturition. Cathepsin D activity as compared to the acid phosphatase activity shows a burst at the first and second day postpartum. This could be due to both enzymes being derived at least partly from different populations of lysosomes. Compared to the overall involution of the uterine proteins (Fig. 1) the decrease of the proteinases was significantly delayed. This may be of physiological importance because the degradation of cellular proteins within the lysosomes as the main proteolytic cellular organelle must always be guaranteed.

The alkaline proteinase activity in marked contrast to the other proteins was measured to be essentially constant throughout pregnancy and involution (Fig. 4). This rather surprising result may arise from the fact that the uterine and serum inhibitors interfere with our test

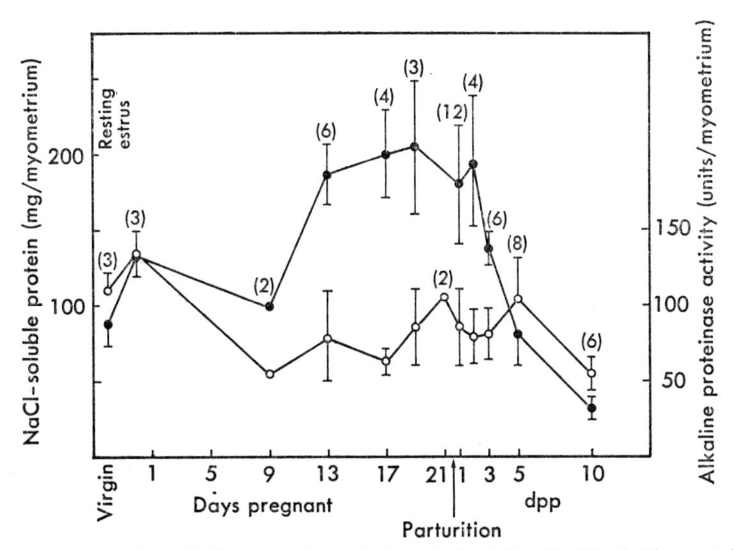

Fig. 4. Activity of the alkaline proteinase (○) and the 0.6 M NaCl-soluble protein (●) during pregnancy and involution ±S.E.M.

system and may not have been completely separated during the extraction of the tissue prior to the assay, which may also account for the rather wide fluctuations of the alkaline proteinase activity which were measured. On the other hand, the NaCl-soluble protein as a measure of the actomyosin complex increased and decreased in parallel to the uterine wet weight. The increase of inhibitor activity was the highest as compared to all other variables tested during pregnancy and involution (Fig. 5). Its activity rose 15-fold during pregnancy and reached its highest value shortly before delivery. After parturition the activity fell very abruptly so that even in uteri taken during the process of parturition the activity had already fallen to 60% of its maximum value. Both the rise and the fall of the inhibitory activity occured 1–2 days earlier than the change in uterine wet weight and in the activity of several other proteinases and enzymes which were measured (*11*).

The variation in the activity of the inhibitor, its properties and its presence in other tissues than the uterus all combine to suggest that it could possess important physiological functions in different organs.

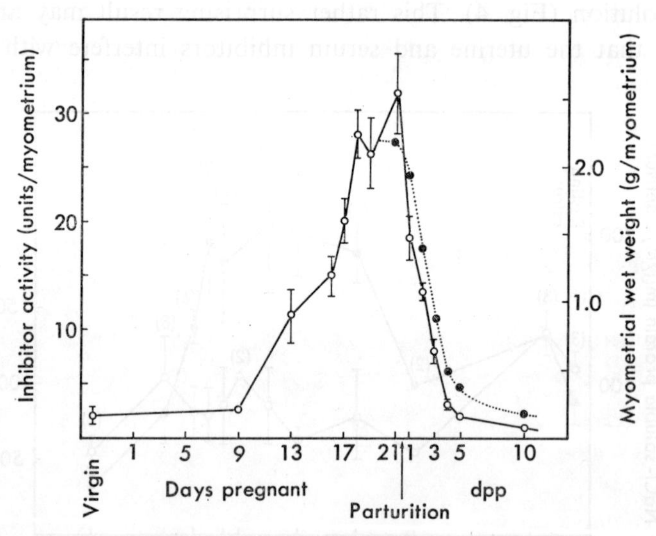

Fig. 5. Variation of the proteinase inhibitor activity (○) throughout pregnancy and involution ±S.E.M.
For comparison the postpartal decrease in myometrial wet weight is also indicated (●). The number of experimental animals averaged four per time point before parturition and nine per time point afterwards.

The growth of the uterine tissue during pregnancy is only possible if protein synthesis prevails over protein degradation. The rise in inhibitor activity, occurring earlier than the increase in proteolytic activity and increasing to a much greater degree, could block the initial steps of intracellular extralysosomal proteolysis by alkaline proteinases and permit organ growth. The fall in inhibitory activity at parturition which precedes the involution of the tissue could release the initial steps of the proteolytic apparatus and start the controlled protein degradation, and these later steps may possibly occur within the lysosomes. The protein degrading cathepsins are sequestered in the specialized lysosomal organelle which shows an acidic pH value (*18*) and is tailored for the degradation of nearly all cellular biomolecules. Besides this compartmentalisation there seems to be no need for an additional control of lysosomal enzymes by extralysosomal and/or extracellular inhibitors in mammals. The lysosomal enzymes are inhibited by the occurring pH shift if they escape from the lysosomal compartment. In contrast, the extralysosomal alkaline proteinases seem not to be sequestered by specific organelles and therefore require a very sensitive and strict control in order to prevent their action on susceptible substrates and to exclude uncontrolled damage to the cells.

The abrupt fall of the inhibitor even during the process of delivery may be the predominant reason why the involution cannot be inhibited by postpartal application of the hormones estrogen and progesterone even if they are injected within 3 hr after parturition (*3*, *13*, *26*). At first, the inhibitor seems not to be sensitive to the hormones estrogen and progesterone and, second, if the inhibitor has been inactivated presumably by proteolysis itself, then the activation of the proteolytic system and the involution of the tissue has passed a point of no return.

Further experiments are necessary to elucidate these complex and carefully regulated mechanisms of uterine involution and the intracellular degradation of the actomyosin complex.

Acknowledgments

I am very grateful to my colleagues Prof. J. S. Elce, Dr. M. Roth, and Miss M. L. Becker for the contributions that they made to this research. The research was supported in part by the Deutsche Forschungsgemeinschaft (SFB 89, Göttingen).

REFERENCES

1. Afting, E. G. and Becker, M. L. *Biochem. J.*, **197**, 519 (1981).
2. Afting, E. G. and Roth, M. *Hoppe-Seyler's Z. Physiol. Chem.*, **362**, 453 (1981).
3. Azanza, J. L., Raymond, J., Robin, J. M., Cottin, P., and Ducastaing, A. *Biochem. J.*, **183**, 339 (1979).
4. Ballard, F. J. *Essays Biochem.*, **13**, 1 (1977).
5. Bird, J. W. C. and Carter, J. H. *In* "Degradative Processes in Heart and Skeletal Muscle," ed. K. Wildenthal, p. 51 (1980). North-Holland Biomedical Press, Amsterdam.
6. Davie, E. W., Fujikawa, K., and Kurachi, K. (1979). *In* "Biological Functions of Proteinases," eds. H. Holzer and H. Tschesche, p. 238 (1979). Springer-Verlag, Berlin.
7. Dayton, W. R., Goll, D. E., Stromer, M. H., Reville, W. J., Zeece, M. G., and Robson, R. M. *In* "Proteinases and Biological Control," eds. E. Reich, D. B. Rifkin, and E. Shaw, p. 551 (1975). Cold Spring Harbor Laboratory, New York.
8. De Duve, Ch. and Wattiaux, R. *Annu. Rev. Physiol.*, **28**, 435 (1966).
9. Finlay, T. H., Katz, J., Rasums, A., Seiler, S., and Levitz, M. *Endocrinology*, **108**, 2129 (1981).
10. Geyer, H., Afting, E. G., and Müller, U. *Arch. Gynäk.*, **222**, 231 (1977).
11. Geyer, H., Müller, U., and Afting, E. G. *Eur. J. Biochem.*, **79**, 483 (1977).
12. Goldberg, A. L. and Dice, J. F. *Annu. Rev. Biochem.*, **43**, 835 (1974).
13. Goodall, F. R. *Science*, **152**, 356 (1966).
14. Heimburger, N. *In* "Protease and Biological Control," eds. E. Reich, D. B. Rifkin, and E. Shaw, p. 367 (1975). Cold Spring Harbor Laboratory, New York.
15. Ishiura, S., Sugita, H., Suzuki, K., and Imahori, K. *J. Biochem.*, **86**, 579 (1979).
16. Laurell, A. B. *In* "Biological Functions of Proteinases," eds. H. Holzer and H. Tschesche, p. 223 (1979). Springer-Verlag, Berlin.
17. Matsukawa, U., Okitani, A., Nishimuro, T., and Kato, H. *Biochem. Biophys. Acta*, **662**, 41 (1981).
18. Ohkuma, S. and Poole, B. *Proc. Natl. Acad. Sci. U.S.*, **75**, 3327 (1978).
19. Roth, M. and Afting, E. G. *Hoppe-Seyler's Z. Physiol. Chem.*, **360**, 357 (1979).
20. Roth, M., Hoechst, M., and Afting, E. G. *Acta Biol. Med. Germ.*, **40**, 1357 (1981).
21. Sapolsky, A. I. and Woessner, J. F. *J. Biol. Chem.*, **247**, 2069 (1972).
22. Schimke, R. T. *Curr. Top. Cell Regul.* **I**, 77 (1969).
23. Schimke, R. T. and Doyle, D. *Annu. Rev. Biochem.*, **39**, 929 (1970).
24. Schwarz, W. N. and Bird, J. W. C. *Biochem. J.*, **167**, 811 (1977).
25. Strecker, F. Thesis (1980). Medizinische Fakultät, Universität Freiburg.
26. Wenner, G. F. Thesis (1980). Medizinische Fakultät, Universität Freiburg.
27. Yasogawa, N., Sanada, Y., and Katunuma, N. *J. Biochem.*, **83**, 1355 (1978).
28. Zack, R., Martin, A. F., Reddy, M. K., and Rabinowitz, M. *Circ. Res.*, **38**, 145 (1976).

Proteinase Inhibitors: Medical and Biological Aspects, eds. N. Katunuma et al., pp. 201–210 (1983).
Japan Sci. Soc. Press, Tokyo/Springer-Verlag, Berlin

Naturally-Occurring Inhibitors of Aspartic Proteinases

John KAY,[*1] Martin J. VALLER,[*1] and Ben M. DUNN[*2]

Department of Biochemistry, University College[*1] *and Department of Biochemistry and Molecular Biology, J. Hillis Miller Health Centre, University of Florida*[*2]

Proteolytic enzymes are classified on the basis of their catalytic mechanism as belonging to one of four groups—the serine, cysteine, metallo, and aspartic proteinases (*16*). In contrast to the detailed sequence information and 3-dimensional structures that have been produced for many enzymes belonging to the first three groups, relatively little is known about the aspartic proteinases.

These enzymes have been identified from four major sources:

1) Microorganisms (but not bacteria) *e.g.*, the proteinases from *Endothia parasitica, Penicillium janthinellum, Mucor pusillus*, and *Scytalidium lignicolum*

2) The intracellular hydrolases in the lysosomes of many cell types (cathepsins D and E).

3) From the stomachs of a number of species. Three different types of gastric enzyme have been resolved—the pepsins, chymosins, and gastricsins (*15*, *20*). Gastricsin is also produced by the prostate which is the source of the aspartic proteinase of seminal fluid.

4) Renins from tissues such as kidney and sub-maxillary gland.

[*1] Cardiff. CF1 1XL., Wales, U.K.
[*2] Gainesville, Fl.32610, U.S.A.

These aspartic proteinases participate in a variety of physiological processes and alterations in the levels of activity expressed may be associated with the onset of pathological conditions such as hypertension, gastric ulcers, muscular dystrophy, and neoplastic diseases. In addition to their biological roles, aspartic proteinases have been used for centuries in two major areas of food processing—milk coagulation and the production of fermented foods from soya beans, rice, and cereals. They are thus of enormous economic importance and the relative scarcity of the traditional milk-clotting enzyme, calf chymosin, has led to the widespread introduction of rennet substitutes for cheese-making (*19*, *20*). Microbial enzymes, pig, cow, and chicken pepsins have all been used industrially. However, the recent success in cloning and sequencing the cDNA for chymosin (*13*, *24*) should soon result in chymosin grown in microorganisms being available for commercial purposes.

STRUCTURE AND MECHANISM OF ACTION

The primary structures of a number of the aspartic proteinases have been determined, *e.g.*, several microbial enzymes, pig and chicken pepsins, calf chymosin, and most recently, mouse renin (*23*, *25*), but to date only the crystal structures of the microbial enzymes (*e.g.*, penicillopepsin and *Endothia* proteinase) have been solved (*14*, *36*). The microbial enzymes have almost identical three-dimensional structures and the sequence homologies that have been demonstrated between these and the mammalian proteinases suggest that all of the aspartic proteinases must have very similar three-dimensional conformations overall. The microbial enzymes, however, do not appear to be synthesised in precursor form, whereas pepsinogen, prochymosin, progastricsin, *etc.* are well-documented zymogens whose activation mechanisms have been described in detail (*9*, *15*). A precursor of cathepsin D has been reported recently (*27*).

All of these enzymes operate maximally in an acidic pH environment and the use of active site directed irreversible inhibitors has indicated that the carboxyl side-chains of two aspartic acid residues are responsible for operating the catalytic mechanism in all of them (except the *Scytalidium* proteinases). It is generally accepted that no covalent intermediates are formed during the catalytic cycle of these enzymes.

Thus, while all of the aspartic proteinases possess these fundamental similarities, each type of enzyme must have evolved subtle distinctions in structure and activity from the others in order to carry out its specific physiological function in its own environment. For example, cathepsin D is unstable at pH 2, the pH at which the pepsins are maximally active. At this pH, the pepsins readily digest most proteins (including immunoglobulins) whereas chymosin has very little general proteolytic activity towards proteins other than caseins. This may be important for the development of immunity in the new-born. Similarly, renin has a very restricted specificity, directed to only one peptide bond in the circulating angiotensinogen (*35*). All of the enzymes appear to have an extended active site cleft (*36*) which can accommodate about 7 amino acid residues of the substrate in the P_4-P_3' sub-sites with a distinct preference for cleaving the bond between hydrophobic residues in the P_1-P_1' sub-sites (*26*). The differences in specificity and activity may be explained by discrete alterations in some or all of the sub-sites in the various enzymes.

INHIBITORS

In contrast to the ubiquitous distribution of inhibitors of serine, cysteine, and metalloproteinases throughout the biological world, naturally-occurring inhibitors of aspartic proteinases are relatively uncommon (*16*). Apart from multifunctional proteins (*e.g.*, from the seeds of the *Bauhinia* tree (*11*) and from *Scopolia japonica* cells (*32*)) which inhibit several classes of proteinase, inhibitors that are specific for aspartic proteinases have been described in only a few locations:

a) Proteins (*Mr* approx. 17,000) from *Ascaris lumbricoides*

b) Acylated pentapeptides (pepstatins) from various species of Actinomycetes

c) The inhibitor peptide (containing 16/17 residues) released on activation of (pig/cow) pepsinogen(s).

d) Renin-binding proteins. It is not certain whether the interaction of these proteins with renin actually blocks the activity of the enzyme (*38*).

However, the susceptibility of individual aspartic proteinases to

TABLE I. The Interaction of Various Inhibitors with Aspartic Proteinases

	Isovaleryl -pepstatin	Acetyl -pepstatin	Lactyl -pepstatin	Ascaris inhibitors
		K_i ($\times 10^8$ M)		
Pig pepsin	0.1	0.1	0.1	0.05[a]
Pig gastricsin	1.9	1.8	7	0.17[a]
Human gastricsin	10	12	580	Weak[a]
Chicken pepsin	0.1	0.1	5	Strong[b]
Calf chymosin	7	6.3	190	V. weak[b]
(Cow) cathepsin D	0.12	0.08	3	V. weak[b]

All assays were done at pH 3.1, 37°C.
[a] Data from ref *1*. [b] Data from ref *22*.

inhibition by these compounds differs considerably. The *Ascaris* proteins are effective inhibitors of pig and chicken pepsins, pig gastricsin (Table I), human pepsin (*1*), and cathepsin E (*22*) but have little or no activity towards human gastricsin, chymosin, cathepsin D or penicillopepsin.

PEPSTATINS

In the series of naturally-occurring pepstatins produced by Actinomycetes, two forms of the Acyl-Val-Val-Sta-Ala-Sta pentapeptide structure, differing only in the nature of the acyl substituent are found most commonly. These are the isovaleryl (*37*) and acetyl (*33*) derivatives. Both are very poorly soluble in aqueous solution but are very effective inhibitors (Table I) of most of the aspartic proteinases except human gastricsin, chymosin, and renin (*29*). The original proposal that the tight-binding was due to the first statine residue acting as a transition state analogue has been confirmed by recent NMR and ESR studies (*28*, *30*, *34*), and data from X-ray crystallographic analyses of complexes between *Rhizopus* proteinase and penicillopepsin (*14*) with pepstatin (fragments) is now being obtained. A series of synthetic analogues of pepstatin has been prepared (*29*, *31*) and it has been deduced from this work that the C-terminal statine residue is relatively unimportant and that the Val-Sta-Ala combination together with the 3S, 4S configuration at the first statine residue are the vital factors for inhibitor potency. However, these synthetic analogues are just as poorly soluble in water as the naturally-occurring pentapeptides. By introduc-

	1 5 10
Mouse prorenin	-Thr-Thr-Phe-Glu-Arg-Ile -Pro-Leu———— Lys -Lys—— Met-Pro
Human progastricsin	Ala-Val-Val-Lys -Val-Pro-Leu-Lys -Lys -Phe-Lys-Ser-Leu-Arg
Calf prochymosin	Ala-Glu-Ile -Thr-Arg-Ile -Pro-Leu-Tyr-Lys -Gly-Lys-Ser-Leu-Arg
Chicken pepsinogen	Ser-Ile -His-Arg-Val-Pro-Leu-Lys-Lys -Gly-Lys-Ser-Leu-Arg
Human pepsinogen	Ile -Met-Tyr-Lys -Val-Pro-Leu-Ile -Arg-Lys -Lys-Ser-Leu-Arg
Cow pepsinogen	Ser-Val-Val-Lys -Ile -Pro-Leu-Val-Arg-Lys -Lys-Ser-Leu-Arg
Pig pepsinogen	Leu-Val-Lys -Val-Pro-Leu-Val-Arg-Lys -Lys-Ser-Leu-Arg

Fig. 1. Amino terminal sequences of the zymogens of aspartic proteinases (numbering based on the pig pepsinogen sequence).

ing a hydrophilic lactyl residue as the acylating group, the resultant Lactyl-Val-Sta-Ala-Sta=Lactyl-pepstatin (*17*) is rendered totally water-soluble. Lactyl-pepstatin is as effective as isovaleryl and acetyl-pepstatins in inhibiting pig pepsin and gastricsin (Table I) but it is a considerably weaker inhibitor of the other enzymes tested than the two other pepstatins. This suggests that the residue occupying the P_3 sub-site of the enzymes is just as important as the Val-Sta-Ala sequence in determining the tightness of inhibitor binding.

INHIBITION BY AN ACTIVATION PEPTIDE

This effect is seen also with the pepsin inhibitor peptide obtained upon activation of pepsinogen (*12*). It has long been known that one of the peptides released on activation of (pig) pepsinogen binds to pepsin above pH 4.5 to stabilise the enzyme at higher pH values and to inhibit it (usually as measured in a milk-clotting assay at pH 5.3). We have shown that this inhibitor is the peptide released in the first step in the sequential activation of pig (and cow) pepsinogen(s) (*8*) and is derived from the first 16/17 residues in the zymogens (*4*, *12*, *18*). It can be seen (Fig. 1) that there is considerable homology in the sequences of this region of calf prochymosin, human progastricsin, human, chicken, cow, and pig pepsinogens, and even mouse prorenin (*25*). Despite these similarities, however, the naturally-occurring 1–16 peptide from pig pepsinogen, although a good inhibitor of pig and cow

TABLE II. Inhibition of Aspartic Proteinases by Synthetic (1–12) Activation Peptide

Enzyme	$K_i(\times 10^8$ M)
Pig pepsin	10
Pig gastricsin	40
Human gastricsin	300
Chicken pepsin	17,000
Calf chymosin	∞
Pig cathepsin D	∞

All assays were carried out at pH 5.3, 37°C with pre-incubation of enzyme and inhibitor for 15 min before the addition of substrate.

TABLE III. Apparent Dissociation Constants for the Interaction of Synthetic Peptides with Pig Pepsin

Peptide	$K_i(\times 10^8$ M)
1–16	13
1–12	10
Gly 4,5; 1–12	530
Gly 6,7; 1–12	100
Gly 4–7; 1–12	3,000

pepsins, is without effect on calf chymosin and chicken pepsin (*18*). Similarly, the first peptide released on activation of chicken pepsinogen (containing 26 residues) had virtually no inhibitory effect on the activity of any of the aspartic proteinases tested (*21*).

In order to investigate this further, a series of peptides was prepared by solid phase synthesis. Previous work had shown that residues 1, 13, 14, 15, and 16 were not essential for inhibitory potency and so the peptide corresponding to residues 1–12 was tested against various enzymes. It was found to inhibit pig (and cow) pepsin(s) strongly (Table II), pig gastricsin weakly and to have virtually no effect on the other mammalian enzymes. In addition, the peptide was degraded by the microbial enzymes, *Endothia, Rhizopus*, and penicillopepsins.

Systematic replacement or modification of certain residues indicated those that are involved in inhibitor binding. Substitution of Arg 8 by Thr and conversion of the lysine residues at positions 3, 9, and 10 into homocitrulline by carbamylation did not decrease the inhibitory potency towards pig pepsin (*5–7*). However, replacement of the hydrophobic residues in positions 4–7 with glycine lowered the K_i values for pig pepsin dramatically (Table III). The proline residue seems to be the key feature for effective inhibition, probably by occupying the P_1 sub-site (*7*). Thus, it would appear that this inhibitor is bound in the active site cleft by hydrophobic interactions and that electrostative interactions between the positively charged peptide and the negatively-charged enzyme are not influential in determining the final equilibrium position. Nevertheless, it has been demonstrated that the first kinetic step of the pepsin-peptide interaction does depend on an ionic interaction although this does not cause inhibition and may be a surface binding event (*2*, *3*). This is followed by a second, much slower step that leads to the stable, inhibited complex. By making use of a spin-labelled zymogen, we have also shown that this "remote" binding can be observed during activation of the precursor and liberation of the peptide (*10*).

SYNTHETIC PEPTIDE SUBSTRATES

The determination of these interactions between inhibitors and the aspartic proteinases has been greatly facilitated by the use of a synthetic peptide substrate containing the chromophoric (NO_2)-phenylalanine residue in the P_1' position of the scissile bond so that hydrolyses can be followed spectrophotometrically. This peptide substrate was designed according to a survey of the pepsin-susceptible bonds in proteins (*26*). Hydrolysis is followed at 310 nm and the substrate can be used at any pH value (usually between 2–6) at which the enzyme is active. Kinetic parameters for the hydrolysis of this substrate by a

TABLE IV. Kinetic Parameters for the Hydrolysis of Pro-Thr-Glu-Phe-(NO_2)Phe-Arg-Leu by Several Aspartic Proteinases

Enzyme	k_{cat} (sec^{-1})	K_m (mM)	k_{cat}/K_M (sec^{-1} mM^{-1})
Pig pepsin	90	0.077	1,170
Pig gastricsin	54	0.095	568
Human gastricsin	23	0.370	62
Chicken pepsin	17	0.091	184
Calf chymosin	38	0.200	190
Cow cathepsin D	60	0.240	250

Cleavage occurred at the Phe-(NO_2)Phe bond. All reactions were carried out at pH 3.1, 37°C.

P_4	–	P_3	–	P_2	–	P_1	–	P'_1	–	P'_2
Iva	–	Val	–	Val	–	Sta	–	Ala	–	Sta
		Lactyl	–	Val	–	Sta	–	Ala	–	Sta
Val	–	Lys	–	Val	–	Pro	–	Leu	–	Val
Pro	–	Thr	–	Glu	–	Phe	–	(NO_2)Phe	–	Arg

Fig. 2. The interaction of inhibitors and substrates with the sub-sites of an aspartic proteinase.

number of the different types of mammalian aspartic proteinases are shown in Table IV. This peptide would appear to be an excellent substrate for pig pepsin (and to a lesser extent pig gastricsin) but it is hydrolysed rather less readily by the other aspartic proteinases tested.

CONCLUSIONS

These results reflect the situation that was described above for the susceptibility of the individual types of enzyme to inhibition. It would thus appear (Fig. 2) that the nature of the amino acid residue (in the substrate or inhibitor) occupying the P_3 sub-site of the enzyme may have particular significance. Pig pepsin (and possibly pig gastricsin) can apparently tolerate this P_3 sub-site being occupied by a hydrophobic or hydrophilic residue since lactyl-pepstatin is as effective an inhibitor of these enzymes as isovaleryl and acetyl-pepstatins; the 1–12 peptide is an effective inhibitor and if the proline residue does fill the P_1 sub-site, then the hydrophilic lysine (number 3) would occupy P_3; and the synthetic peptide with Thr in P_3 is a good substrate. By contrast, the other enzymes seem to have a distinct preference for filling the P_3 sub-site with a hydrophobic residue—thus, lactyl-pepstatin is less effective than isovaleryl or acetyl-pepstatins; the 1–12 peptide does not inhibit at all and the synthetic peptide is a much poorer substrate that it is for the pepsins. A similar conclusion about the importance of the P_3 sub-site has been reached by others using synthetic pepstatin analogues (*31*).

None of the naturally-occurring inhibitors can be considered to be of physiological importance to the host cell as regulators of the activities of the mammalian aspartic proteinases (*16*). It is likely that modulation of this type of proteinase has to be achieved through regulation of the pH and of the rate of supply of substrate proteins. Nevertheless, the naturally-occurring inhibitors and their synthetic analogues have proved to be of inestimable value biochemically in facilitating distinction among the different types of aspartic proteinase. An understanding of their selectivity may be of importance for the design and synthesis of specific inhibitors for use therapeutically in controlling individual aspartic proteinases. An excellent example is the recent preparation of specific renin inhibitors for the control of hypertension (*35*).

Acknowledgments
The research in Florida was supported by a grant from NIH (No. AM-18865-08) while the work in the U.K. was supported by a grant from the Agricultural Research Council (No. AG 72/31). Our international collaboration was sponsored by awards from the Wellcome Trust and the Burroughs-Wellcome Fund. We are very grateful to our many colleagues throughout the world who participated indirectly in this work by generously supplying us with samples of their purified enzymes. It is also a pleasure to acknowledge the superb secretarial and administrative assistance provided by Mrs. Barbara Power.

REFERENCES

1. Abu-Erreish, G. M. and Peanasky, R. J. *J. Biol. Chem.*, **249**, 1566 (1974).
2. Dunn, B. M., Pham, C., Raney, L., Abayasekara, D., Gillespie, W., and Hsu, A. *Biochemistry*, **20**, 7206 (1981).
3. Dunn, B. M. *Arch. Biochem. Biophys.*, **214**, 763 (1982).
4. Dunn, B. M., Deyrup, C., Moesching, W. G., Gilbert, W. A., Nolan, R. J., and Trach, M. L. *J. Biol. Chem.*, **253**, 7269 (1978).
5. Dunn, B. M. and Deyrup, C. *In* "Peptides," eds. E. Gross and J. Meienhofer, p. 161 (1979). Pierce Chemical Co., Rockford.
6. Dunn, B. M., Deyrup, C., and Merdinger, E. *Farmacia*, **28**, 197 (1980).
7. Dunn, B. M., Lewitt, M., and Pham, C. *Biochem. J.*, **209**, 355 (1983).
8. Dykes, C. W. and Kay, J. *Biochem. J.*, **153**, 141 (1976).
9. Foltmann, B. and Jensen, A. L. *Eur. J. Biochem.*, **128**, 63 (1982).
10. Glick, D. M., Valler, M. J., Rowlands, C. C., Evans, J. C., and Kay, J. *Biochemistry*, **21**, 3746 (1982).
11. Goldstein, Z., Trop, M., and Birk, Y. *Nature New Biol.*, **246**, 29 (1973).
12. Harboe, M., Andersen, P. M., Foltmann, B., Kay, J., and Kassell, B. *J. Biol. Chem.*, **249**, 4487 (1974).
13. Harris, T. J. R., Lowe, P. A., Lyons, A., Thomas, P. G., Eaton, M. A. W., Millican,

T. A., Patel, T. P., Bose, C. C., Carey, N. H., and Doel, M. T. *Nucleic Acid. Res.*, **10**, 2177 (1982).
14. James, M. N. G., Sielecki, A., Salituro, F., Rich, D. H., and Hofmann, T. *Proc. Natl. Acad. Sci. U.S.*, **79**, 4868 (1982).
15. Kay, J. *In* "The Enzymology of Post-Translational Modification of Proteins," eds. R. B. Freedman and H. Hawkins, p. 424 (1980). Academic Press, London.
16. Kay, J. *Biochem. Soc. Trans.*, **10**, 277 (1982).
17. Kay, J. Afting, E-G., Aoyagi, T., and Dunn, B. M. *Biochem. J.*, **203**, 795 (1982).
18. Kay, J. and Dykes, C. W. *In* "Acid Proteases, Structure, Function & Biology," ed. J. Tang, p. 103 (1977). Plenum Press, New York.
19. Kay, J. and Valler, M. J. *Meth. Milk Dairy J.*, **35**, 281 (1981).
20. Kay, J., Valler, M. J., Keilova, H., and Kostka, V. *In* "Proteinases and Their Inhibitors, Structure, Function and Applied Aspects," eds. V. Turk and Lj. Vitale, p. 269 (1981). Pergamon Press, Oxford.
21. Keilova, H., Kostka, V., and Kay, J. *Biochem. J.*, **167**, 855 (1977).
22. Keilova, H. and Tomasek, V. *Biochim. Biophys. Acta*, **284**, 461 (1972).
23. Misono, K. S., Chang, J.-Y., and Inagami, T. *Proc. Natl. Acad. Sci. U.S.*, **79**, 4883 (1982).
24. Nishimori, K., Kawaguchi, Y., Hidaka, M., Uozumi, T., and Beppu, T. *J. Biochem.*, **90**, 901 (1981).
25. Panthier, J-J., Foote, S., Chambraud, B., Strosberg, A. D., Corvol, P., and Rougeon, F. *Nature*, **298**, 90 (1982).
26. Powers, J. C., Harley, A. D., and Myers, D. V. *In* "Acid Proteases, Structure, Function and Biology," ed. J. Tang, p. 141 (1977). Plenum Press, New York.
27. Puizdar, V. and Turk, V. *FEBS Lett.*, **132**, 299 (1981).
28. Rich, D. H., Bernatowicz, M. S., and Schmidt, P. G. *J. Am. Chem. Soc.*, **104**, 3535 (1982).
29. Rich, D. H. and Sun, E. T. O. *J. Med. Chem.*, **23**, 27 (1980).
30. Rich, D. H., Boparai, A. S., and Bernatowicz, M. S. *Biochem. Biophys. Res. Commun.*, **104**, 1127 (1982).
31. Rich, D. H. & Bernatowicz, M. S. *J. Med. Chem.*, **25**, 791 (1982).
32. Sakato, K., Tanaka, H., and Misawa, M. *Eur. J. Biochem.*, **55**, 211 (1975).
33. Satoi, S. and Murao, S. *Agric. Biol. Chem.*, **37**, 2579 (1973).
34. Schmidt, P. G., Bernatowicz, M. S., and Rich, D. H. *Biochemistry*, **21**, 1830 (1982).
35. Szelke, M., Leckie, B., Hallett, A., Jones, D. M., Suieras, J., Atrash, B., and Lever, A. F. *Nature*, **299**, 555 (1982).
36. Tang, J., James, M. N. G., Hsu, I-N., Jenkins, J. A., and Blundell, T. L. *Nature*, **271**, 618 (1978).
37. Umezawa, H. and Aoyagi, T. *In* "Proteinases in Mammalian Cells and Tissues," ed. A. J. Barrett, p. 637 (1977). Elsevier/North-Holland, Amsterdam.
38. Ueno, N., Miyazaki, H., Hirose, S., and Murakami, K. *J. Biol. Chem.*, **256**, 12023 (1981).

Proteinase Inhibitors: Medical and Biological Aspects, eds. N. Katunuma et al., pp. 211–227 (1983).
Japan Sci. Soc. Press, Tokyo/Springer-Verlag, Berlin

Human Leukocyte Collagenase and Regulation of Activity

H. TSCHESCHE and H. W. MACARTNEY

*Biochemistry Department, Faculty of Chemistry, University of Bielefeld**

Human leukocyte collagenase is contained in the secondary or specific granules of the granulocyte cell and has been the subject of interest of several groups who have described its activity, partial purification and specificity. The enzyme is similar to other collagenases isolated from human skin, rheumatoid synovium, periodontium, macrophages, platelets, and other vertebrate tissues (*18*). The true vertebrate collagenase exhibits a unique splitting specificity as it degrades the native collagen triple helix into two fragments of 3/4 and 1/4 size. The exact site of cleavage has been determined to be between the amino acid residues 772 and 773 of the tropocollagen molecule, a Gly-Ile bond in the α_1-chain and a Gly-Leu bond in the α_2-chain, respectively (*2*, *3*, *5*).

The breakdown of the collagen fibre is initiated by the collagenase and the two 3/4 and 1/4 triple helical fragments formed are generally not susceptible to other endoproteinases, though type I collagen would also appear to be degraded by leukocyte elastase at the terminal telopeptides. At body temperature both fragments are thermally instable; the 3/4 triple helical fragment denatures at 32°C and the 1/4 fragment at 28°C (*14*) to yield separate α-chain pieces that form gelatin. The

* D-4800 Bielefeld 1, FRG.

gelatin random coil structure may then be subject to degradation by the leukocyte gelatinase, an enzyme found in another compartment of the granulocyte, tertiary or secretory granules (*1*).

ISOLATION OF HUMAN LEUKOCYTE COLLAGENASE

Though several reports on the existence of human neutrophil (polymorpho-nuclear (PMN)) collagenase have already appeared (*18*), only moderate purification has so far been reported. A purification to apparent homogeneity was achieved in our laboratory using a rapid and reproducible method involving affinity chromatography on concanavalin A (Con A)-Sepharose, collagen-Sepharose, and activated thiol-Sepharose 4 B, gel filtration on Sephacryl S-300 and ion exchange chromatography on DEAE-Sephacel (*9*). The overall procedure is schematically represented in Fig. 1. It yields both a latent and an active enzyme.

The latent and active collagenase were shown by gel filtration on Sephadex G-75 superfine (*9*) and SDS-polyacrylamide gel electrophoresis without mercaptoethanol (*9*) to be single entities with *Mr* 91,000 and 64,000, respectively. It is interesting to note that the latent collagenase passes through the collagen Sepharose column unretarded, while the active enzyme is bound to the affinity resin and is eluted only at 1 M NaCl (Fig. 2). Further purification can then be achieved by passage of the latent and active enzymes over Sephacryl S-300 (Fig. 3) and DEAE-Sephacel (see *9*).

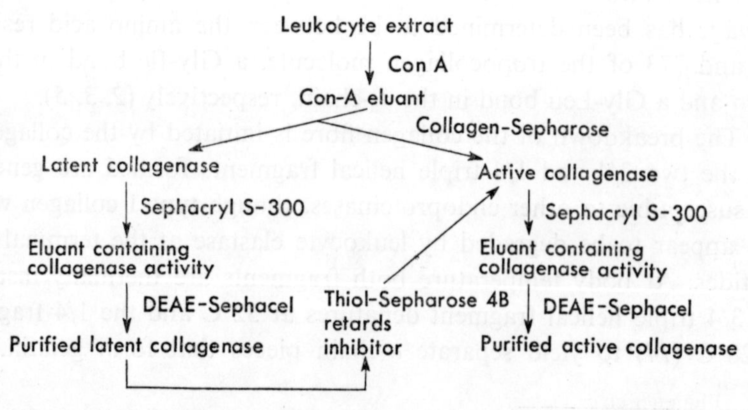

Fig. 1. Isolation and purification of the leukocyte collagenase inhibitor (*9*).

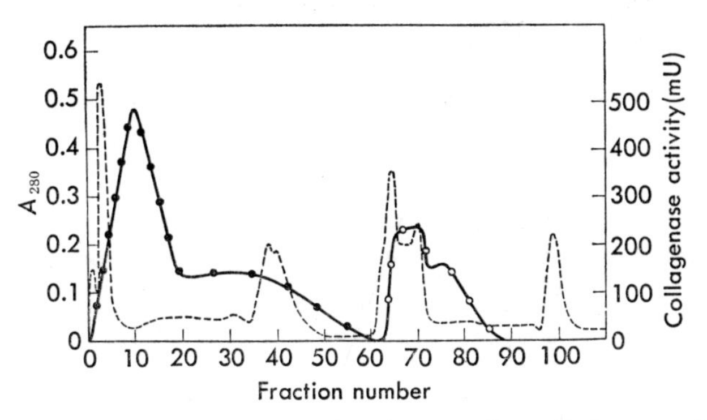

Fig. 2. Elution profile of eluant from Con A on a collagen-Sepharose column (*9*).

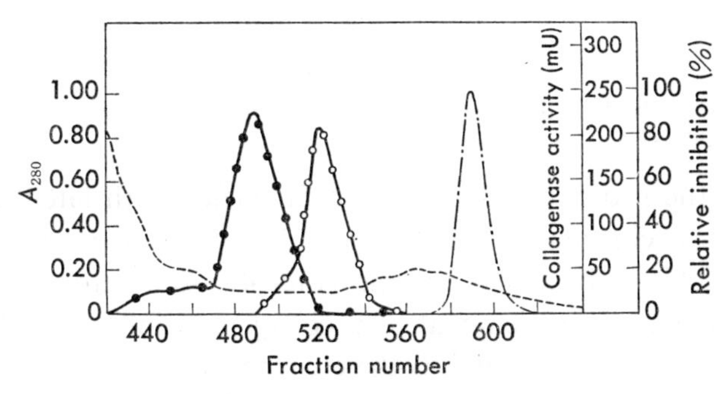

Fig. 3. Elution profile of latent and active collagenases and the leukocyte collagenase inhibitor from a Sephacryl S-300 column (*9*).

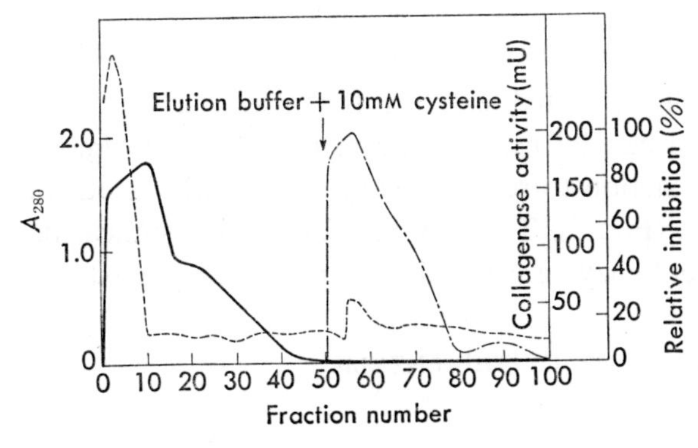

Fig. 4. Elution profile of latent collagenase applied to a column of activated thiol-Sepharose 4 B (*9*).

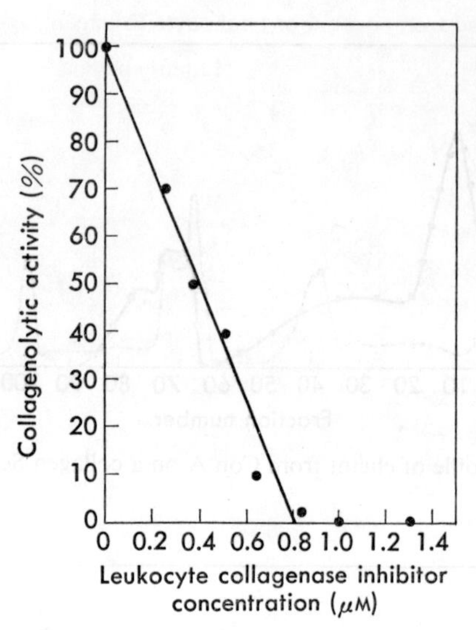

Fig. 5. Titration of active polymorphonuclear leukocyte collagenase with the leukocyte collagenase inhibitor.

While the procedure described so far yields both latent and active collagenase, a chromatographic step of latent enzyme on thiol-Sepharose 4 B yields only active enzyme (Fig. 4). Activation of the latent enzyme is achieved by removal of a collagenase inhibitor (*7, 9*) which seems to be covalently bound to the enzyme by a disulfide bridge, since it is not removed on SDS-gel electrophoresis in the absence of mercaptoethanol.

The activation process on thiol-Sepharose or by disulfide- or mercury-containing activators (see below) is accompanied by a decrease in molecular weight of the latent enzyme by about *Mr* 25,000. The *Mr* 25,000 protein liberated exhibits the characteristics of a collagenase inhibitor and can be used to titrate the active enzyme in a 1 : 1 stoichiometric reaction to reform the proteolytically inactive enzyme-inhibitor complex (Fig. 5). The complex formed is identical in its properties with the latent enzyme isolated from the leukocyte extract.

CHARACTERISATION OF THE ENZYME

All three components of the system have been purified to apparent homogeneity as described (*7, 9, 10*). The latent and the active col-

TABLE I. Amino Acid Composition of Leukocyte Collagenases

	Latent enzyme	Active enzyme
Aspartic acid	82	55
Threonine	42	32
Serine	55	47
Glutamic acid	86	57
Proline	42	33
Glycine	84	55
Alanine	79	50
Half-cystine	31	22
Valine	48	34
Methionine	9/10	6
Isoleucine	31	25
Leucine	66	52
Tyrosine	24	17
Phenylalanine	31	25
Lysine	38	26
Histidine	17	15
Arginine	58	50
Total	823/824	601
Fucose	1	1
N-acetylglucosamine	5	5
N-acetylgalactosamine	6	1
Mannose	3/4	2
Galactose	5/6	5/6
Glucose	5/6	5/6

lagenase revealed the amino acid compositions given in Table I. The active enzyme is comprised of about 600 residues while the latent form contains about 825 residues. Isoelectric points were determined by isoelectric focussing and were estimated to be in the acidic range around pH 5.5 and 6.4 for the latent and active enzymes, respectively. This result seems to be in agreement with the amino acid analyses of the two enzymes in which the acidic amino acid residues amount to approximately 1.3–1.4 times the basic residues.

The optimum pH for collagenolytic activity against type I collagen was observed near neutrality at pH 7.4. The enzyme exhibits proteolytic activity in the pH range from pH 5.7 to about pH 8.0 and is highly specific in splitting the tropocollagen molecule into 3/4 and 1/4 fragments (*9*). The enzymic activity of the collagenase reduces the intrinsic viscosity of a type I collagen solution by about 40%.

Both the latent and the active enzyme are glycoproteins that contain fucose, *N*-acetylglucosamin, *N*-acetylgalactosamine, mannose, galactose, and glucose.

CHARACTERISATION OF THE INHIBITOR

The leukocyte collagenase inhibitor could be isolated from the crude cell extract (*10*) from the latent enzyme after activation by several activators (see below) (*7*, *10*) or by passage of the latent enzyme over thiol-Sepharose 4 B. In the latter procedure the inhibitor is removed from the latent enzyme and is bound to the matrix of the column. It could be eluted by including 10 mM cysteine in the elution buffer (Fig. 4) (*10*). Final purification was achieved by gel filtration on Sephacryl S-300 and two further steps of ion equilibrium chromatography on DEAE-Sephacel as described (*10*). The inhibitor showed a molecular weight of about 24,500 in the gel filtration on Sephacryl S-300 and

TABLE II. Amino Acid Compositions of Collagenase Inhibitor

	Leukocyte inhibitor	β_1-Anticollagenase
Aspartic acid	24	29
Threonine	10	17
Serine	15	22
Glutamic acid	29	36
Proline	17	15
Glycine	24/26	32
Alanine	23/24	31
Half-cystine	7	4
Valine	19	19
Methionine	5	6
Isoleucine	10/12	13
Leucine	13	24
Tyrosine	5	7
Phenylalanine	6	11
Lysine	9/10	17
Histidine	7/8	6
Arginine	16	12
Total	239/246	301
N-acetylglucosamine	1	0
Mannose	1	1
Galactose	1	1
Glucose	1	2

displayed a single band with the same apparent molecular weight in the SDS-gel electrophoresis (*9*). The protein had one free sulfhydryl group per mol as revealed by Ellman's titration method (*10*). Alkylation of the single sulfhydryl by iodoacetamide abolished the inhibitory activity.

The amino acid composition of the inhibitor indicates three disulfide bridges to be present per mol and a free sulfhydryl group. The composition indicates a high content of acidic residues. The ratio of basic to acidic amino acids is approximately 1 : 1.7 (Table II). This would possibly indicate an acidic isoelectric point, which was revealed by isoelectric focussing to be around pH 5.5. The number of residues accounts for about 240 amino acids which is in agreement with the apparent molecular weight. The inhibitor is a glycoprotein containing a carbohydrate moiety of about one residue each of *N*-acetylglucosamine, mannose, galactose, and glucose (*10*). The amino acid composition of the β_1-anticollagenase from human plasma (*11*) is included in the table. It differs with respect to composition and number of residues.

ACTIVATION OF THE LATENT ENZYME

The latent enzyme was shown to be activatable by numerous agents (Table III) (*7*, *9*). Activators include mercurial compounds and active proteinases such as commercial trypsins, cathepsin G and inactive proteinases such as diisopropyl phosphorofluoridate (DFP)- and phenylmethyl sulfonyl fluoride (PMSF)-cathepsin G, tosyl-lysine chloromethyl ketone (TLCK)-trypsin or trypsinogen. Purified β-trypsin was not capable of activating the latent enzyme; thus proteolytic activity was not necessary for activation. The latent enzyme is not a zymogen but rather an enzyme-inhibitor complex that is dissociable into active enzyme and an inhibitor by various disulfides including cystine and oxidized glutathione. The inability to activate the latent enzyme by the selectively reduced and carboxymethylated 179, 203-di(S-carboxymethyl)trypsinogen indicates that this particular disulfide bond in trypsinogen is responsible for the activation (*7*). Similarly, the disulfides in insulin, relaxin or immunoglobulin G (IgG) seem to be responsible for the activation, as are the disulfides in partially degraded commercial trypsin. Thus the activation process seems to be coupled to a disulfide interchange reac-

TABLE III. Activation of Latent Human Polymorphonuclear Leukocyte Collagenase by Disulfide-containing Activators

Assay contents	Activation
1. Oxidized glutathione	100.0
2. Cystine	100.0
3. Trypsinogen	77.0
4. Commercial trypsin	102.6
5. TLCK-commercial trypsin	98.0
6. Cathepsin G	100.9
7. DFP-cathepsin G	98.1
8. PMSF-cathepsin G	100.4
9. *p*-Chloromercuribenzoate	99.7
10. 4-Aminophenylmercuric acetate	99.7
11. Mersalyl	102.0
12. Insulin	80.2
13. Relaxin	50.6
14. IgG	87.0
15. Pancreas kallikrein	48.0
16. β-Trypsin	28.8
17. 179,203-di(S-carboxymethyl)trypsinogen	0
18. 179,203-di(S-carboxymethyl)trypsin	0
19. Unactivated control	1.8

tion, and indeed, excellent activation could be achieved with cystine or oxidized glutathione.

THE LATENT ENZYME IS AN ENZYME-INHIBITOR-COMPLEX

The disulfide-dependent dissociation of the latent collagenase into the active enzyme and an inhibitor and the presence of a free sulfhydryl group in the latent enzyme and another one in the free inhibitor which is essential for inhibitory activity, led us to propose a disulfide-linked structure for the latent enzyme, *i.e.*, the enzyme-inhibitor complex. The hypothetical structure of the latent enzyme being a mixed disulfide-linked enzyme-inhibitor complex is supported by further experimental evidence:

1) The reformed complex has properties identical to the latent enzyme (*7, 9*).

2) The amino acid composition of the reformed complex reproduces that of the latent enzyme (*10*) (Table III).

3) SDS-polyacrylamide gel electrophoresis without mercaptoethanol

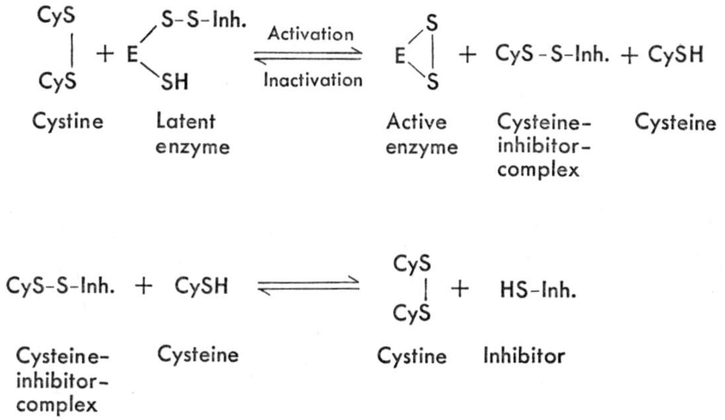

CyS–CyS (Cystine) + E(S-S-Inh.)(SH) (Latent enzyme) ⇌ (Activation / Inactivation) E(S–S) (Active enzyme) + CyS-S-Inh. (Cysteine-inhibitor-complex) + CySH (Cysteine)

CyS-S-Inh. (Cysteine-inhibitor-complex) + CySH (Cysteine) ⇌ CyS–CyS (Cystine) + HS-Inh. (Inhibitor)

Fig. 6. Schematic representation of the mixed disulfide complex of enzyme and inhibitor (latent enzyme) and its activation bei thiol/disulfide interchange reaction with cystine (*8*, *9*).

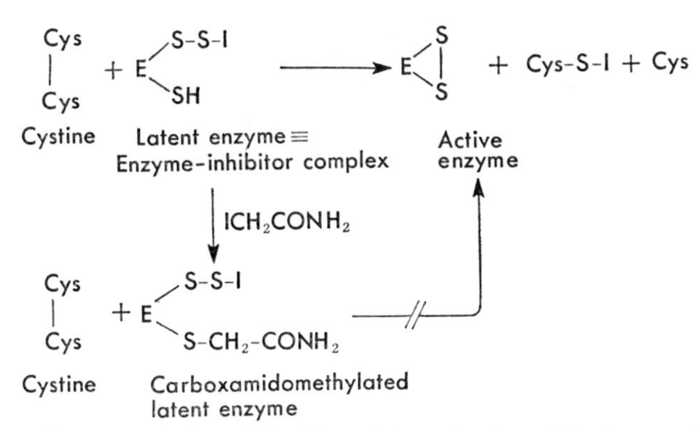

Fig. 7. Schematic representation of blocking of the activation of the latent collagenase by alkylation with iodoacetamide (*9*).

revealed a *Mr* 91,000 for the latent enzyme but *Mr* 61,000 and 25,000 in the presence of mercaptoethanol (*9*).

4) Carboxamidomethylation with iodoacetamide abolished the property of the latent enzyme being activatable by disulfides (*9*) (Fig. 7).

The mechanism of activation *via* an intramolecular disulfide rearrangement resulting in complex dissociation and reformation of an active enzyme with intact peripheral disulfide bond is not yet fully clear. Obviously the association-dissociation equilibrium can be shifted by disulfides in an interchange reaction depending on the overall redox state in the medium. Indeed, disulfides are a prerequisite for the activation, which cannot be achieved if, as mentioned above, disulfides are absent or cannot be formed in the incubation medium.

ACTIVATION COUPLED TO THE GLUTATHIONE CYCLE

Optimal activation of the latent leukocyte collagenase could be achieved by oxidized glutathione. A straightforward activation of the latent enzyme by addition of increasing amounts of oxidized glutathione could be obtained. The oxidized form could be generated *in vitro* from the reduced one if oxidizing equivalents such as hydrogen peroxide and any of the peroxidases, such as glutathione peroxidase, NADH peroxidase, horse radish peroxidase, lactoperoxidase or myeloperoxidase, were added. Without the peroxidase enzyme no activation *via* formation of oxidized glutathione took place even if hydrogen peroxide and/or reduced glutathione were added to the mixture containing the latent enzyme. However, in the presence of a peroxidase the addition of hydrogen peroxide led to a straightforward activation again depending on the amount added (*16*). The activation by hydrogen peroxide could be blocked completely if the peroxidase was poisoned with the inhibitor sodium azide or if catalase or glutathione reductase were added to the system (*16*) (Fig. 8).

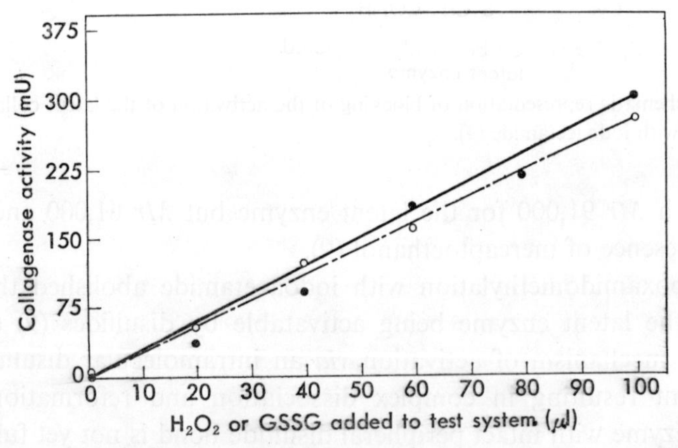

Fig. 8. Activation of latent collagenase with either oxidized glutathione or the hydrogen peroxide/myeloperoxidase activation system (*16*).
GSSG, oxidized glutathione.

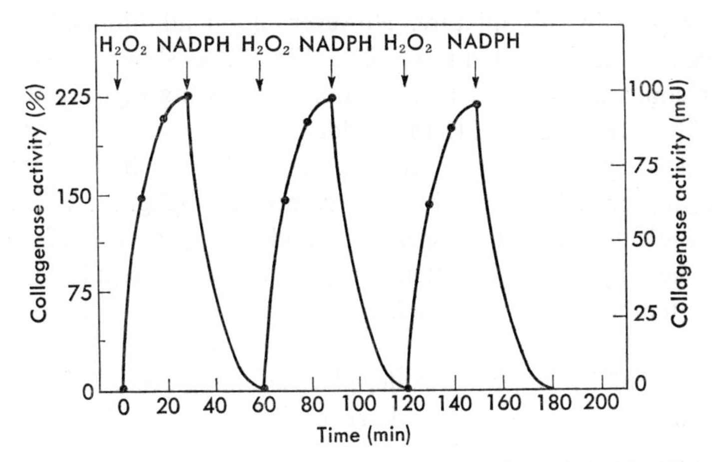

Fig. 9. Activation and inactivation of latent collagenase *via* the H_2O_2/NADPH/glutathione peroxidase/glutathione reductase system employing *in vivo* concentrations of oxidized and reduced glutathione (*16*).

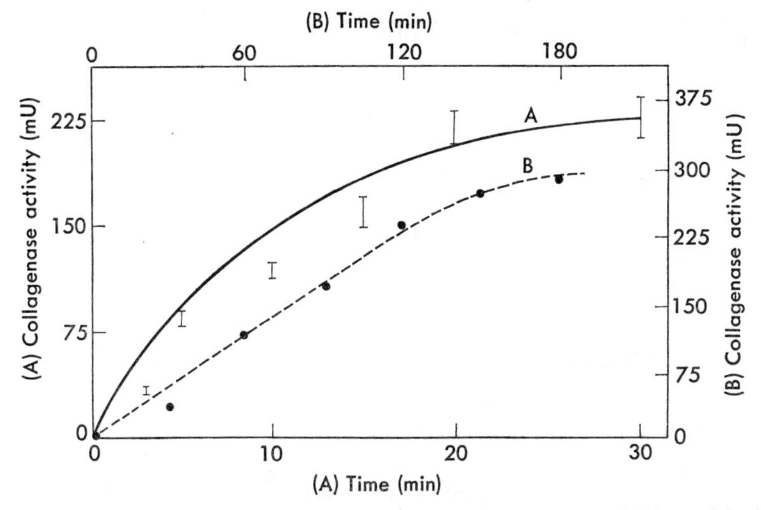

Fig. 10. Time dependence of activation of latent collagenase *via* addition of hydrogen peroxide (A) and *via* the peroxidase/glucose oxidase system (B) (*16*).

It is highly likely that activation of the latent enzyme takes place *in vivo via* the glutathione cycle. An *in vitro* system with latent enzyme was established containing the physiological concentrations of reduced and oxidized glutathione of the resting cell and glutathione peroxidase and reductase (*16*). No activation was observed upon incubation of the latent enzyme in this mixture. However, upon addition of hydrogen peroxide a rapid activation took place (Fig. 9). As shown the activation process could be reversed to inhibition if reducing equivalents in the form of NADH or NADPH were added to the system. The activation/inactivation process could be repeated several times by new additions of oxidizing or reducing equivalents, respectively (Fig. 9). This *in vitro* system indicates that the appearance and disappearance of the proteolytic activity from the latent enzyme could be regulated *via* the redox potential in the system, *i.e.*, the redox state of the glutathione cycle.

Instead of hydrogen peroxide, glucose and glucose oxidase could be used as the hydrogen peroxide-generating system, which could also be coupled to the glutathione/peroxidase activation system (*16*). All types of mammalian phagocytic cells have been shown to exhibit increased respiration during phagocytosis. The changes in oxygen metabolism are coupled with an increase in generation of superoxide anions (O^{2-}), hydrogen peroxide, and other oxygen radicals which serve as

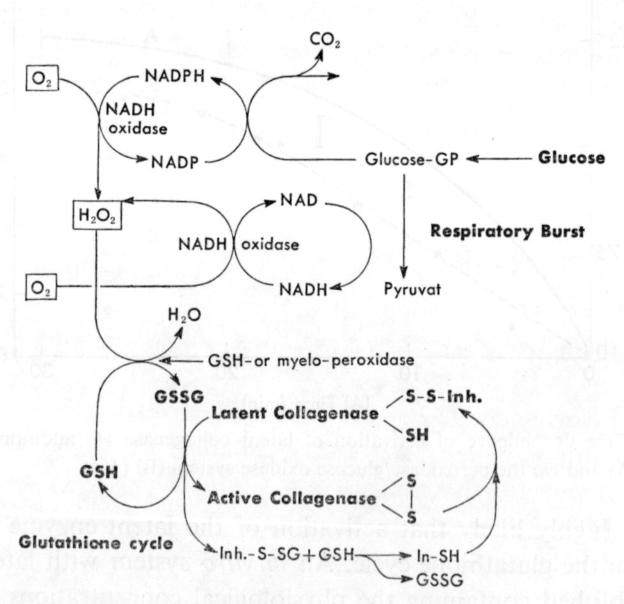

Fig. 11. Scheme of the hypothetical regulation of human leukocyte collagenase *via* the redox state of the glutathione cycle linked to the glucose and oxygen metabolism of the cell (*6*).

microbiocidal agents. The changes in oxygen metabolism are linked to an increase in glucose catabolism through the hexose monophosphate (HMP) pathway. The ratio of NADPH/NADP shifts from 9 to approximately 3 in phagocitizing leukocytes (*19*). Altogether these metabolic changes are referred to as the respiratory burst. During this process the level of oxidized glutathione increases significantly and the amount of reduced glutathione decreases by at least 25–30% in 10–15 min (*17*). Thus, it is feasible that *in vivo* the change in the redox state of the glutathione cycle either in the phagosome or the extracellular space, *e.g.*, in frustrated phagocytosis, might lead to activation of the latent enzyme. The glutathione cycle in leukocytes is an integral link between the phagocytosis-associated respiratory burst and the increase in HMP shunt activity. It becomes evident then that the activation of the latent collagenase *in vivo* might be linked to the phagocitizing activity of the cell and finally to its glucose metabolism (Fig. 11).

THE LEUKOCYTE COLLAGENASE IS A METALLO ENZYME

The human leukocyte collagenase required both calcium and zinc ions for stability and proper catalytic function. In this respect the enzyme exhibited behavior similar to specific collagenases from other tissues. Calcium was probably required as an extrinsic stabilizing cofactor ion for stabilization of the enzyme's tertiary structure, possibly in a similar way as has been elucidated for the calcium-binding loop of trypsin.

The transition metal ion zinc seemed to be an extrinsic and intrinsic requirement, participating in the enzyme's catalytic mechanism and in stabilizing the proper conformation that allowed the conversion of the latent to the active enzyme by the thiol-disulfide interchange activation reaction.

The involvement of zinc in the proper structure of the latent enzyme-inhibitor complex was readily demonstrated. Upon addition of chelator or on dialysis of the latent enzyme against a chelating agent such as EDTA, cysteine or others the activatable property of the latent apoenzyme is abolished. However, the zinc-free apoenzyme regained its activation properties upon readdition of zinc or other double positive metal ions such as Cu^{2+}, Mn^{2+}, Mg^{2+} or Co^{2+} in concentrations equimolar to the amount of chelator. Employing oxidized glutathione ac-

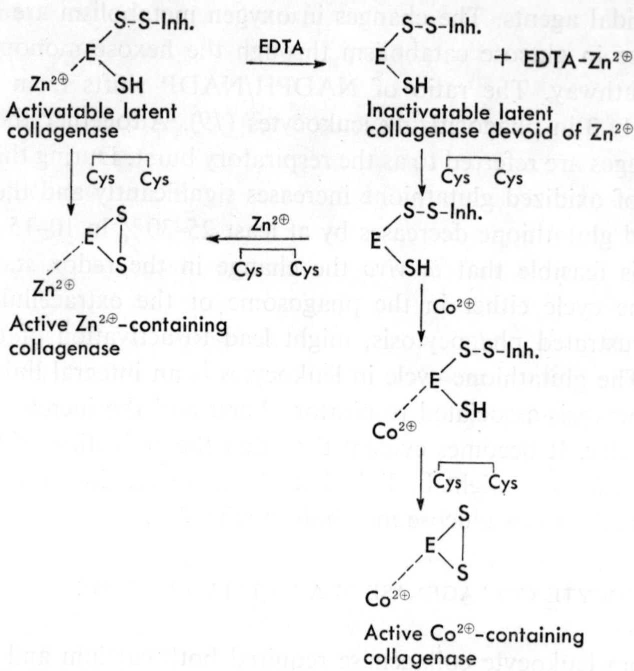

Fig. 12. Schematic representation of intrinsic and extrinsic requirement of latent collagenase for zinc (*8*).

tivator, it could be demonstrated that any of the divalent metal ions tested could substitute for zinc (*8*). The reconstituted metal-enzyme-inhibitor complex could be dissociated into active enzyme and inhibitor. However, addition of more than three times the concentration of zinc required to reconstitute the metal-containing latent enzyme-inhibitor complex from the apoenzyme again led to loss of the activatable properties of the complex.

The double positive metal ion, Zn^{2+}, is also a necessary constituent of the active collagenase. Chelators such as EDTA or cysteine inactivate the enzyme. The activity could be reconstituted with varying degrees of effectiveness upon readdition of a one or up to 3 M concentration of either zinc, copper, manganese or cobalt ion. Zinc ions, however, had the best ability of all metals tested to restore the catalytic activity or to mediate in the activation to the latent enzyme.

The latent enzyme and the active collagenase obviously contain a second binding site for double positive metal ions that functions as an inhibitory site and which, when occupied, inhibits the active enzyme and prevents the latent enzyme from being activatable. A conformational change induced by a second metal ion binding to the free and also to the latent enzyme could explain this inhibitory effect. This perhaps provides a concerted regulatory role of zinc ions in actual and potential collagenase activity. High concentrations would inhibit the active enzyme and concomitantly block further activation of latent enzyme *via* removal of the leukocyte collagenase inhibitor in a peroxidase-coupled oxidative activation process, *e.g.*, linked to the glutathione cycle (*16*).

SIMILAR SYSTEMS

The thiol-disulfide interchange system elucidated for human leukocyte collagenase and its inhibitor was also demonstrated for the interaction between the leukocyte collagenase and β_1-anticollagenase from human plasma (*11*). The plasma collagenase inhibitor was also isolated to apparent homogeneity and characterized as a protein of *Mr* 30,000 exhibiting a different amino acid composition than the leukocyte collagenase inhibitor. It also contained a free sulfhydryl group essential for inhibitory activity. The leukocyte collagenase was inhibited by the plasma inhibitor under formation of an activatable enzyme-inhibitor complex, *i.e.*, a latent enzyme. Activation by dissociation of the complex could be achieved by the same disulfides or the same H_2O_2/myeloperoxidase/glutathione-system as was described for the latent leukocyte collagenase (*16*).

The leukocyte inhibitor and the β_1-anticollagenase both inhibit other collagenases from other tissues, such as fibroblasts and synovial cells, with formation of latent enzymes reactivatable by various disulfide activators (*12*). Thus, the thiol-disulfide interchange system seems to be of more general importance. It was indeed demonstrated that tumor tissue produces an inhibitor for pancreatic trypsin which operates by the same thiol/disulfide interchange mechanism (*15*).

We have found evidence for another system of enzyme and inhibitor interaction which operates by the thiol/disulfide interchange mechanism. This is the gelatinase inhibitor system of human leukocytes. The gelatinase is located in another compartment of the cell, a type of secretory organelle or tertiary granule, while the collagenase is found in the specific or secondary granule.

PROBABLE SIGNIFICANCE IN PATHOPHYSIOLOGICAL PROCESSES

The thiol/disulfide mediated reversible association-dissociation reaction of the human leukocyte collagenase and its inhibitor provides the first example for a proteolytic system in which the activity could be regulated. Regulation could be achieved by coupling this system with the thiol-disulfide redox system of the glutathione cycle.

Indeed, *in vivo* the glutathione cycle is an integral link between the phagocytosis associated respiratory burst and the increase in HMP shunt activity. During inflammatory reactions the phagocytosis stimulated increase in hydrogen peroxide has been shown to be accompanied by a perturbation of the redox state. Changes in the glutathione content in arthritic erythrocytes have been reported (*13*). Many investigators have used the finding of decreased serum sulfhydryl groups as a parameter of increased disease activity in rheumatoid arthritis (*4*, *6*).

Local collagenase activation might well be the result of redox state perturbations either in the phagosome and/or perhaps in the extracellular space of highly inflamed pathologic tissue. Activation could proceed *via* the increase in oxidized glutathione but also by other activating disulfides, *e.g.*, increased IgG levels or various other activators of high molecular weight such as the biological fluid and tissue activators.

The inflammatory activation response might well be one of the deleterious effects of collagenase activation in the disease states of rheumatoid arthritis, osteoarthritis, pulmonary emphysema, cirrhosis, pemphiges, and tumor invasion.

Acknowledgments

We wish to thank the Stiftung Volkswagenwerk, Hannover, FRG and the Fonds der Chemischen Industrie, Frankfurt, FRG, for financial support. The authors are indebted to the Deutsches Rotes Kreuz, Blutspendezentrale Hagen, for generous supply of human leukocytes. We gratefully acknowledge the skillful technical assistance of Mrs. E.

Pieper and Mrs. K. Meyer in the processing of the human cell material, and that of Mrs. V. Süwer in the preparation of latent collagenase.

REFERENCES

1. Dewald, B., Bretz, U., and Baggiolini, M. *J. Clin. Invest.*, **70**, 518 (1982).
2. Fietzek, P. P., Rexrodt, F. W., Hopper, K. E., and Kühn, K. *Eur. J. Biochem.*, **38**, 396 (1973).
3. Gross, J., Harper, E., Harris, E. D., Jr., McCroskry, P. A., Highberger, J. H., Corbott, C., and Kang, A. H. *Biochem. Biophys. Res. Commun.*, **48**, 1147 (1974).
4. Haataja, M. *Scand. J. Rheumatol.*, **4**, 7 (1975).
5. Highberger, J. H., Corbett, C., Kang, A. H., and Gross, J. *Biochemistry*, **14**, 2872 (1975).
6. Lorber, A., Bovy, R., and Chang, C. *Metabolism*, **20**, 446 (1971).
7. Macartney, H. W. and Tschesche, H. *FEBS Lett.*, **119**, 327 (1980).
8. Macartney, H. W. and Tschesche, H. *Hoppe-Seyler's Z. Physiol. Chem.*, **362**, 1523 (1981).
9. Macartney, H. W. and Tschesche, H. *Eur. J. Biochem.*, **130**, 71 (1983).
10. Macartney, H. W. and Tschesche, H. *Eur. J. Biochem.*, **130**, 79 (1983).
11. Macartney, H. W. and Tschesche, H. *Eur. J. Biochem.*, **130**, 85 (1983).
12. Macartney, H. W. and Tschesche, H. *Eur. J. Biochem.*, **130**, 93 (1983).
13. Munthe, E., Kaas, E., and Jellum, E. *In* "Proc. 4th Int. Meet. Future Trends Inflammation," eds. D. A. Willoughby and J. Girond, p. 439 (1980). MTP Press, Lancaster.
14. Sakai, T. and Gross, J. *Biochemistry*, **6**, 518 (1967).
15. Steven, F. S., Podrazky, V., and Itzhaki, V. *Biochim. Biophys. Acta*, **524**, 170 (1978).
16. Tschesche, H. and Macartney, H. W. *Eur. J. Biochem.*, **120**, 183 (1981).
17. Voetmann, A. A., Loos, J. A., and Roos, D. *Blood*, **15**, 741 (1980).
18. Wooley, D. E. and Evanson, J. M. eds. "Collagenases in Normal and Pathological Connective Tissue" (1980). John Wiley & Sons, Cichester-New York-Brisbane-Toronto.
19. Zatti, M. and Rossi, F. *Biochim. Biophys. Acta*, **99**, 557 (1965).

Proteinase Inhibitors: Medical and Biological Aspects, eds. N. Katunuma et al., pp. 229–241 (1983).
Japan Sci. Soc. Press, Tokyo/Springer-Verlag, Berlin

Intracellular Proteinases and Inhibitors Associated with the Hemolymph Coagulation System of Horseshoe Crabs (*Tachypleus tridentatus* and *Limulus polyphemus*)

Takashi MORITA, Takanori NAKAMURA, Sadami OHTSUBO, Shigenori TANAKA, Toshiyuki MIYATA, Masuyo HIRANAGA, and Sadaaki IWANAGA

*Department of Biology, Faculty of Science, Kyushu University**

Limulus hemolymph contains only one type of cells, called amebocytes or granular hemocytes (*1*, *11*, *13*). The amebocytes contain a number of dense granules and exposure of the cells to bacterial endotoxin results in aggregation associated with striking degranulation and clot formation. This endotoxin-induced clotting phenomenon has been known as a possible defense mechanism, serving to immobilize invading gram-negative bacteria (*3*, *6*). Because of this unique property and its extreme sensitivity to endotoxins, the overall molecular events of the clotting system in *Limulus* amebocytes are of great interest, in comparison with those of mammalia.

This paper concerns mainly the biochemical studies on the intracellular clotting system closely related to the function of these hemocytes. First of all, we would like to mention the primary structure of a clottable protein called coagulogen and its structural change during gelation (*4*, *7*, *8*, *15*). Secondly, we will describe the purification and properties of a proclotting enzyme (*9*, *10*, *12*). Thirdly, we shall show a new glucan-mediated pathway associated with the *Limulus* clotting system (*5*). Finally, we shall deal with an anticoagulant, which specifi-

* Hakozaki 6-10-1, Higashi-ku, Fukuoka 812, Japan.

cally inhibits the endotoxin-mediated activation of the clotting enzyme (*16*).

COMPLETE AMINO ACID SEQUENCES OF *TACHYPLEUS* AND *LIMULUS* COAGULOGENS

The hemolymph was collected by inserting a needle into a joint between the cephalothorax and the abdominal region. Fifty to 150 ml of the hemolymph per individual was drawn and amebocytes were obtained by centrifugation. *Limulus* hemolymph thus obtained is shown in Fig. 1. The amebocytes can be seen at the bottom in the first tube. A light blue plasma fraction contained mainly hemocyanin and lectins. The second tube shows a clot formed in the presence of the endotoxins, so called lipopolysaccharides, and this clot seemed to be stabilized on prolonged incubation, as shown in the third tube. We have recently determined the amino acid sequence of this clottable protein.

Figure 2 shows the complete amino acid sequence of coagulogen isolated from the amebocyte lysate of Japanese horseshoe crabs, *Tachypleus tridentatus* (*15*). This protein consisted of 175 residues. It contained a total of 16 half-cystines, which are linked with disulfide bridges. Five of them were clustered in the COOH-terminal region. The complete arrangement of these disulfide linkages is now under investigation. *Tachypleus* coagulogen consisted of three polypeptide segments, A-

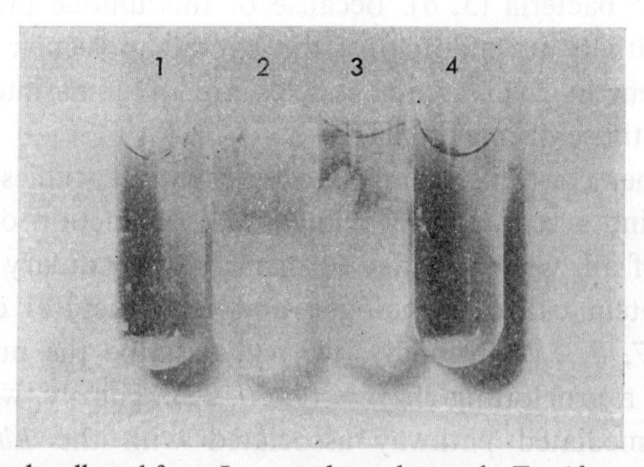

Fig. 1. Hemolymph collected from Japanese horseshoe crab, *T. tridentatus*. Tube 1: 10 min after collection; tube 2: clot formed by addition of bacterial endotoxin (*Escherichia coli* 0111-B4) into the hemolymph; tube 3: a stabilized clot after standing for 1 hr; tube 4: the same as tube 1.

Fig. 2. The complete amino acid sequence of *Tachypleus* coagulogen.
The peptide bonds cleaved by *Tachypleus* clotting enzyme are shown by arrows, and the disulfide bridges so far determined are indicated. A-chain, peptide C, and B-chain correspond, respectively, to the NH_2-terminus to Arg_{18}, from Thr_{19} to Arg_{46}, and from Gly_{47} to the COOH-terminus (*4*, *15*).

chain, peptide C, and B-chain. The clotting enzyme cleaved the Arg_{18}-Thr_{19} linkage and Arg_{46}-Gly_{47} linkage, both located at the NH_2-terminal region. The Arg_{18}-Thr_{19} linkage cleaved by the clotting enzyme was the same type as that cleaved by α-thrombin in the transformation of mammalian fibrinogen to fibrin. Moreover, the COOH-terminal octapeptide sequences of A-chain and peptide C exhibited great homology for each other, and their sequences were very similar to that of *Rhesus* monkey fibrinopeptide B (*8*). The most interesting finding, however, was a partial sequence homology between coagulogen and platelet factor 4, the latter of which is known as antiheparin. When the sequences are aligned for maximum homology, coagulogen was found to contain a number of similar partial sequences with platelet factor 4 (*15*). The first eleven NH_2-terminal residues were especially similar to those of this factor. A comparison of the sequence of *Tachypleus* coagulogen with that of *L. polyphemus* coagulogen, which was recently established in our laboratory is shown in Fig. 3 (*4*). The overall se-

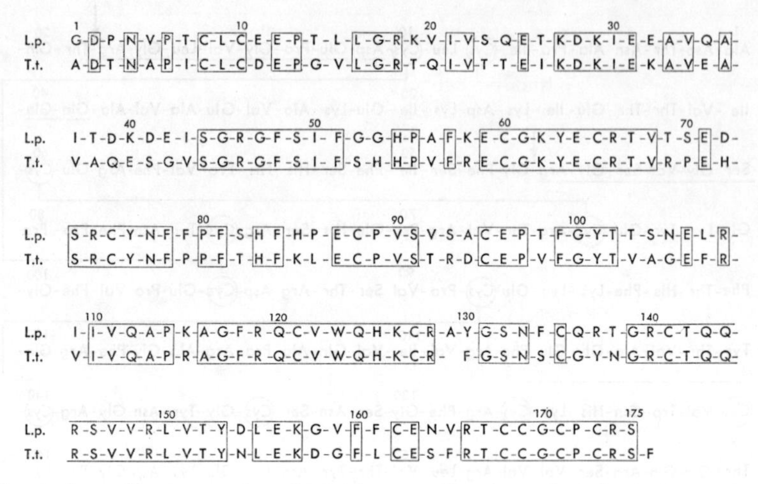

Fig. 3. Amino acid sequence homologies between *Limulus* and *Tachypleus* coagulogens. Identical residues in the sequences are framed (*4*).

quence of *Limulus* coagulogen was very close to that of *Tachypleus* coagulogen, having 70% sequence homology. The 16 half-cystines of these coagulogens were in the same linear position, suggesting a very similar conformation. Moreover, the COOH-terminal tripeptide regions of A-chain and peptide C, Leu-Gly-Arg and Ser-Gly-Arg, were completely conserved in both coagulogens. Therefore, these oligopeptide sequences immediately preceding the bond to be cleaved may be required for the clotting enzyme to split the Arg-Thr and Arg-Gly linkages so as to initiate clot formation.

PURIFICATION AND PROPERTIES OF *TACHYPLEUS* PROCLOTTING ENZYME

Figure 4 summarizes the purification procedures of *Tachypleus* proclotting enzyme from the lysate. The proclotting enzyme was purified by a heparin-Sepharose column followed by chromatographies on DEAE-Sepharose, Sepharose 6B, DE-52, and Sephacryl S-300. Through these procedures, about 2 mg of the purified material was obtained from 1 *l* of hemolymph, and about 100-fold purification was achieved. The purified proclotting enzyme gave a single band on disc-gel electrophoresis at pH 8.3. Furthermore, a single peak with amidase activity due to the clotting enzyme was found at the same position as the protein

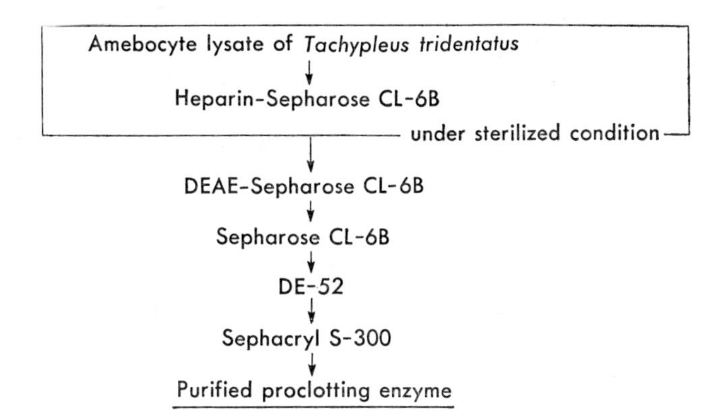

Fig. 4. Procedures for purification of the proclotting enzyme from the *Tachypleus* amebocyte lysate.

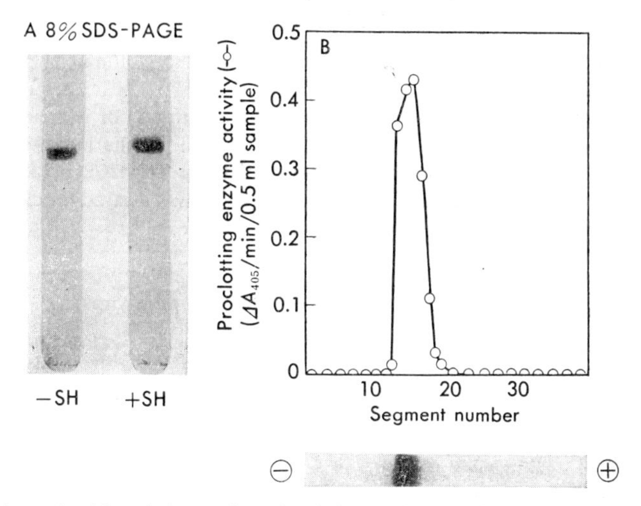

Fig. 5. Polyacrylamide gel electrophoresis of the purified proclotting enzyme.
A: SDS-polyacrylamide gel (8%) electrophoresis (5 μg sample) was performed in the presence (right) or absence (left) of 2-mercaptoethanol. B: polyacrylamide disc gel (8%) electrophoresis was performed at pH 8.3. One of two gels was sliced into 1.5 mm segments and each segment was soaked in 0.5 ml of 0.05 M Tris-HCl buffer, pH 8.0 containing 0.2 M NaCl, and 0.4% bovine serum albumin, and kept overnight at 4°C. The extracts (75 μl) were assayed for proclotting enzyme (○). The other gel was stained with Coomassie brilliant blue R-250 and destained with a mixture of 7.5% acetic acid and 5% methanol.

band stained with Coomassie brilliant blue (Fig. 5B). On SDS-gel electrophoresis with and without reducing agent, the preparation gave a single molecular species, suggesting that the protein consists of a

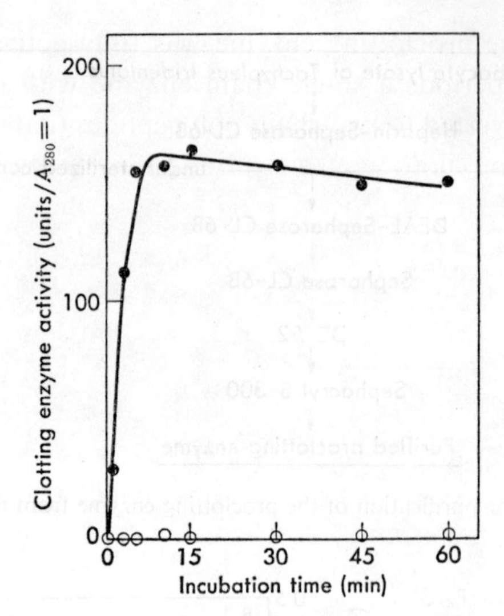

Fig. 6. Time course for activation of proclotting enzyme to clotting enzyme by purified activated factor B.
Proclotting enzyme solutions (99 μg/ml) were incubated with (●) or without (○) purified activated factor B (21.8 μg/ml), in a total volume of 1.0 ml. At the indicated time, 10 μl aliquots were withdrawn and diluted with 190 μl of 0.1 M Tris-HCl, pH 8.0, containing 0.053% bovine serum albumin. The diluted mixture (20 μl) was assayed for clotting enzyme activity by the method described previously (*12*).

TABLE I. Properties of *Tachypleus* and *Limulus* Proclotting Enzymes

Properties	Proclotting enzyme from *T. tridentatus*[a]	Proclotting enzyme from *L. polyphemus*[b]
Molecular weight (by SDS-polyacrylamide gel electrophoresis)	54,000	150,000
γ-Carboxyglutamic acid content	Not detectable	43/1,344 residues
Activation of proclotting enzyme		
LPS-dependency	−	+
Ca^{2+}-dependency	−	+
Limited proteolysis	+	Not examined

[a] This work. [b] Ref. *11*.

single polypeptide chain (Fig. 5A). Figure 6 shows the activation of proclotting enzyme by factor B, which was recently found in the amebocyte lysate (*12*). This factor seems to be one of the serine proteases associated with the *Limulus* coagulation system. In the presence of active factor B, the proclotting enzyme was fully activated within 1 hr, and during these periods a single chain enzyme with molecular weight of 54,000 was converted to two chain polypeptides (data not shown). The properties of the proclotting enzyme are summarized in Table I in comparison with those of the proclotting enzyme purified from *L. polyphemus* by Tai *et al.* (*14*). *Tachypleus* enzyme had a minimum molecular weight of 54,000, and contained 6% hexosamines. This protein appeared not to be vitamin K-dependent like the vertebrate prothrombin family, since its γ-carboxyglutamic acid content was very low. Moreover, no direct activation of the purified proclotting enzyme by endotoxins was observed. These properties differ greatly from those reported on the *Limulus* proclotting enzyme (*14*). These results suggest that the endotoxin-mediated coagulation system consists of a multi-enzyme system. Bacterial endotoxin mediates the activation of factor B or the unknown component(s) existing in the factor B fraction, and its active form converts the proclotting enzyme to an active enzyme, which then catalyzes the gel formation. Thus, this cascade system may provide an extremely high sensitivity of the lysate to endotoxins (Fig. 8).

A NEW (1→3)-β-D-GLUCAN-MEDIATED PATHWAY FOUND IN *LIMULUS AMEBOCYTES*

In the course of these studies, A. Kakinuma of the Central Research Division of Takeda Chemical Industries, informed us that, in addition to endotoxin, a water-soluble antitumor carboxymethylated β-D-glucan (CMPS) activates the *Limulus* coagulation system and forms a clot. This result prompted us to examine which components associated with the coagulation system are activated by β-glucan. Figure 7 shows separation of the amebocyte lysate on a heparin-Sepharose column under a sterilized condition. When each fraction was incubated with the isolated factor B in the presence of endotoxin, a strong amidase activity appeared in both fractions A and G, and the latter contained coagulogen. On the other hand, the factor B mentioned earlier was eluted in fraction B in an inactive form and its activity was detected after incubation with fraction A in the presence of endotoxin. Fractions A, G, and B were separately collected and their amidase activities and clot-forming abilities were measured after treatments of the pooled fractions with β-

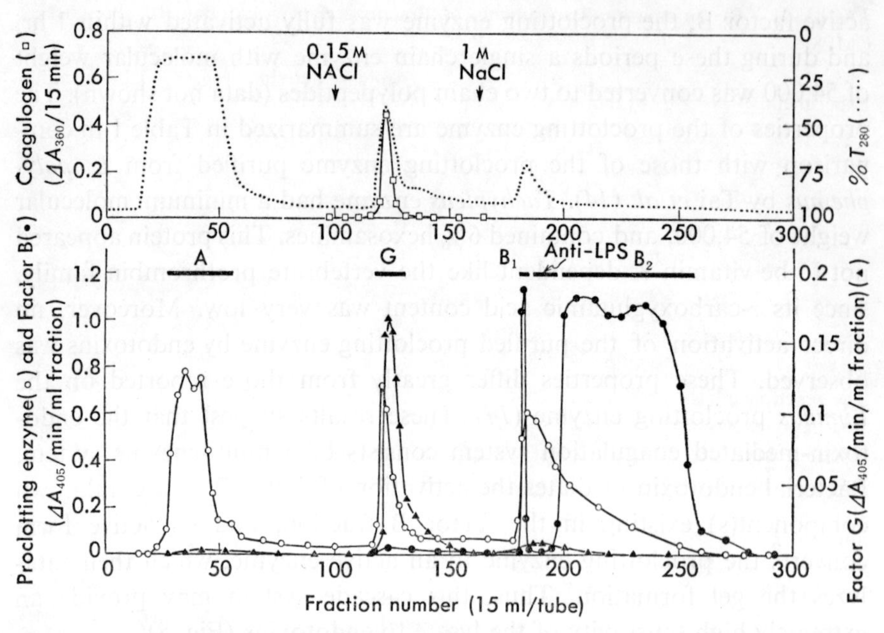

Fig. 7. Heparin-Sepharose CL-6B column chromatography of *T. tridentatus* amebocyte lysate.

The lysate (340 ml) was applied to a column (5×17 cm), pre-equilibrated with 0.05 M Tris-HCl buffer, pH 7.2. The stepwise elution was performed at 4°C first with the equilibration buffer, secondly with the buffer containing 0.15 M NaCl and finally with the buffer containing 1.0 M NaCl. Fractions of 15 ml were collected at a flow rate of 83 ml per hr, and the fractions A, G, B_1, anti-LPS, and B_2 indicated by solid bars were collected. The activities of proclotting enzyme (○), factor G (▲), factor B (●), coagulogen (□), and transmittance at 280 nm (······) were measured, and the detailed procedures were described in the previous paper (*16*).

glucan, instead of endotoxin. As shown in Table II, amidase activity and clot-forming ability were found in fraction G, indicating that this fraction contains a glucan-sensitive factor, tentatively named factor G (Exp. 2). Moreover, a mixture of fraction G with fraction A (Exp. 4) or fraction B (Exp. 6) in the presence of β-glucan showed a stronger amidase activity than fraction G alone. However, there was neither amidase activity nor clot-forming ability in the combination of fractions A and B (Exp. 5). The maximum amidase activity was obtained in the mixture of fractions A, G, and B (Exp. 7). These results suggest that a β-glucan sensitive factor must be contained in fraction G and that the factor induces the activation of the known proclotting

TABLE II. Recombination Experiments of Fractions A, G, and B in the Presence of CMPS

Experiment	Fraction	Amidase activity (μmol pNA released/20 min/ml)	Clot-forming ability
		(1×10^{-4})	
1	Fraction A	0	−
2	Fraction G	135	+
3	Fraction B	0	−
4	Fractions A+G	400	+
5	Fractions A+B	0	−
6	Fractions G+B	278	+
7	Fractions A+G+B	600	+

The reaction mixture containing 50 μl each of a fraction, CMPS (final 24 ng/ml), 0.4 mM Boc-Leu-Gly-Arg-pNA substrate, 80 mM Tris-HCl buffer, pH 8.0, and 5.2 mM $MgCl_2$, in a total volume of 250 μl, was incubated at 37°C for 30 min. Then, 0.8 ml of 0.6 M acetic acid was added to terminate the reaction and the absorbance at 405 nm was measured. In recombination experiments, 50 μl each of fractions A, G or B was mixed together and the amidase activity was measured under the same conditions as above. The clot-forming ability of each fraction was tested using a highly purified coagulogen as follows. The reaction mixture contained 50 μl each of fractions A, G, and B or their combined mixture, 30 μl CMPS (400 ng/ml), 100 μl *Tachypleus* coagulogen (2 mg/ml), and 50 μl 0.4 M Tris-HCl buffer, pH 8.0, containing 26 mM $MgCl_2$, in a total volume of 230 μl. When a clot appeared within 1 hr at 37°C, it was judged that the sample showed a positive reaction.

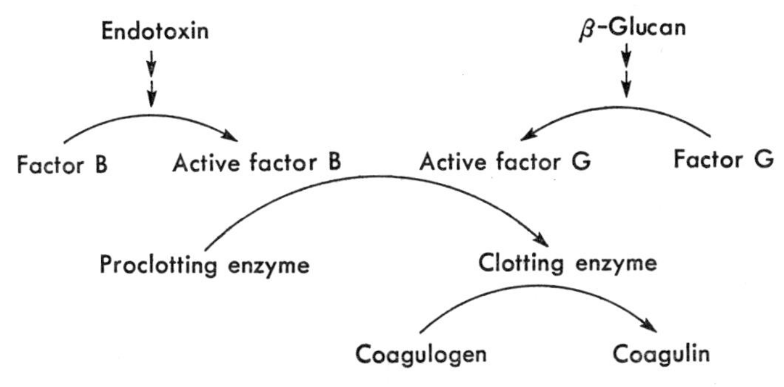

Fig. 8. Tentative mechanism for the coagulation cascade found in *Limulus* amebocyte lysate.
This cascade sequence will need modification as new factors and cofactors are discovered, and further studies will be required for the identification of each compoment.

enzyme eluted in fraction A. Based on these results, we would suggest a new glucan-mediated coagulation pathway in *Limulus* amebocytes, as shown in Fig. 8. Thus, β-D-glucan activates factor G and the active factor G converts the known proclotting enzyme to the clotting enzyme, which then catalyzes the transformation of coagulogen to coagulin gel. Therefore, *Limulus* amebocytes seem to contain two independent coagulation pathways, endotoxin-mediated and glucan-mediated, both of which result in clot-formation (*5*, *12*). The latter pathway seems to correspond to an alternative pathway found in the mammalian complement system.

CHARACTERIZATION OF ANTI-LPS FACTOR FOUND IN *LIMULUS AMEBOCYTES*

As shown in Fig. 7, factor B sensitive to lipopolysaccharide (LPS) was eluted in two separate fractions, B_1 and B_2, suggesting that an inhibitor which disturbs the LPS-mediated activation of factor B may be contained in the fractions between fractions B_1 and B_2. We tentatively called this principle the anti-LPS factor (*16*). The anti-LPS fraction (Fig. 7) was purified further by gel-filtration on a Sephadex G-50 column followed by chromatography on a CM-Sepharose 6B column. Figure 9 shows the elution profile of the anti-LPS factor on a CM-

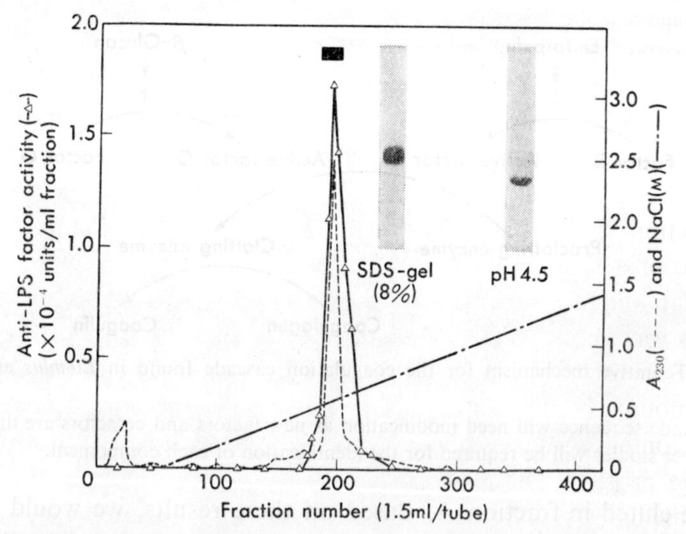

Fig. 9. Chromatography of anti-LPS factor on a CM-Sepharose 6B column. Elution was performed with a linear salt gradient to 0.3 M NaCl at a flow rate of 10 ml per hr. The fractions indicated by a solid bar were collected. The activity of anti-LPS factor was assayed by the previous method (*16*).

TABLE III. Amino Acid Composition of Anti-LPS Factor

Component	Anti-LPS factor Residues per molecule
Asp	5.3(5)
Thr	9.6(10)
Ser	13.1(13)
Glu	14.9(15)
Pro	3.4(3)
Gly	9.8(10)
Ala	8.7(9)
Cys/2	2.3(2)
Val	6.7(7)
Ile	5.5(6)
Leu	10.1(10)
Tyr	3.8(4)
Phe	6.3(6)
Lys	12.3(12)
His	3.5(3)
Trp	6.2(6)
Arg	6.8 (7)
Total	128
Glucosamine	< 0.1
Galactosamine	< 0.1

Amino acid and hexosamine compositions were calculated from extrapolated or average values estimated on samples of 24, 48, and 72 hr hydrolyzates.

Sepharose 6B column used in the final step. The anti-LPS factor appeared as a single peak and the inhibitory activity coincided with a peak detected by the absorbance at 230 nm. Moreover, the protein in this fraction showed a single band on SDS-gel electrophoresis in the presence of a reducing agent, and also on a disc-gel at pH 4.0. The molecular weight of the anti-LPS factor estimated by SDS-gel electrophoresis was about 15,000. The amino acid composition of the purified anti-LPS factor is shown in Table III. The material contained relatively large amounts of threonine, serine, glycine, leucine, and lysine. Hexosamine content was less than 0.1%, suggesting that it consists of a simple basic protein. The effect of the anti-LPS factor on endotoxin- and glucan-mediated activation of the *Limulus* coagulation system is shown in Table IV. In these experiments, a recombination system with proclotting enzyme and factor B or a system with proclotting enzyme and factor G was used, in addition to *Tachypleus* amebocyte lysate. As

TABLE IV. Effect of Anti-LPS Factor on LPS- and CMPS-induced Activation of *Limulus* Coagulation System

Exp.	Sample	Trigger	Amidase activity: Without anti-LSP	Amidase activity: With anti-LPS	Inhibition (%)
			μmol pNA released/30 min/ml ($\times 10^{-4}$)		
1	Amebocyte lysate	LPS	682	115	83.1
2	Fr A+B	LPS	335	30	91.0
			μmol pNA released/17 min/ml ($\times 10^{-4}$)		
3	Amebocyte lysate	CMPS	288	315	0
4	Fr A+G	CMPS	295	316	0
			μmol pNA released/18 min/ml ($\times 10^{-4}$)		
5	Active clotting enzyme		165	170	0
6	Activated factor B+proclotting enzyme (Fraction A)		210	220	0

The reaction mixture containing 20 μl *Tachypleus* amebocyte lysate or the mixtures containing 50 μl each of fractions in Fig. 7 and LPS (final, 24 ng/ml) were incubated at 37°C for 30 min in the presence of 30 μl of the anti-LPS fraction in Fig. 9. After incubation, the LPS-induced amidase activity was measured using chromogenic substrate. The same experiments as above were made using CMPS (final, 24 ng/ml) instead of LPS.

shown in Exp. 1, the appearance of LPS-induced amidase activity due to the clotting enzyme in the lysate was strongly inhibited in the presence of the anti-LPS factor. The same inhibitory effect of the anti-LPS factor on the recombination system with fractions A and B was observed (Exp. 2). However, there was no inhibitory effect of this factor on the glucan-induced activation (Exp. 3), nor did the factor affect the activities of the clotting enzyme and factor B in their activated forms (Exp. 5). These results indicate that the anti-LPS factor inhibits only the activation of the *Limulus* coagulation system mediated with LPS but not with β-glucan. Although we have not reconstituted the LPS-mediated *Limulus* coagulation system using highly purified components, the following possible mode of action of the anti-LPS factor could be considered: 1) its neutralizing effect on bacterial endotoxins, 2) its competitive binding effect on an LPS-interaction site with factor B, and 3) its degradation effect on LPS. Moreover, a bactericidal action of the factor in biological defense of *Limulus* against invading microorganisms can be considered. To resolve these problems, further experimentation will be required.

REFERENCES

1. Dumount, J. H., Anderson, E., and Winner, G. *J. Morphol.*, **119**, 181 (1966).
2. Harada-Suzuki, T., Morita, T., Iwanaga, S., Nakamura, S., and Niwa, M. *J. Biochem.*, **92**, 793 (1982).
3. Levin, J. and Bang, F.B. *Bull. Johns Hopkins Hosp.*, **115**, 265 (1975).
4. Miyata, T., Hiranaga, M., Umetsu, M., and Iwanaga, S. *Ann. N.Y. Acad. Sci.*, in press.
5. Morita, T., Tanaka, S., Nakamura, T., and Iwanaga, S. *FEBS Lett.*, **129**, 318 (1981).
6. Mürer, E. H., Levin, J., and Holme, R. *J. Cell Physiol.*, **86**, 533 (1975).
7. Nakamura, S., Iwanaga, S., Harada, T., and Niwa, M. *J. Biochem.*, **80**, 1011 (1976).
8. Nakamura, S., Takagi, T., Iwanaga, S., Niwa, M., and Takahashi, K. *Biochem. Biophys. Res. Commun.*, **72**, 902 (1976).
9. Nakamura, S., Morita, T., Harada-Suzuki, T., Iwanaga, S., Takahashi, K., and Niwa M. *J. Biochem.*, **92**, 781 (1982).
10. Nakamura, T., Ohtsubo, S., Tanaka, S., Morita, T., and Iwanaga, S. *Seikagaku*, **54**, 822 (1982).
11. Niwa, M. and Waguri, O. *Seikagaku*, **47**, 1 (1975) (in Japanese).
12. Ohki, M., Nakamura, T., Morita, T., and Iwanaga, S. *FEBS Lett.*, **120**, 217 (1980).
13. Ornberg, R. L. and Reese, T. S. *Prog. Clin. Biol. Res.*, **29**, 125 (1979).
14. Tai, J. Y., Seid, R. C., Jr., Huhn, R. D., and Liu, T. Y. *J. Biol. Chem.*, **252**, 4773 (1977).
15. Takagi, T., Hokama, Y., Morita, T., Iwanaga, S., Nakamura, S., and Niwa, M. *Prog. Clin. Biol. Res.*, **29**, 169 (1979).
16. Tanaka, S., Nakamura, T., Morita, T., and Iwanaga, S. *Biochem. Biophys. Res. Commun.*, **105**, 717 (1982).

Proteinase Inhibitors: Medical and Biological Aspects, eds. N. Katunuma et al., pp. 243–250 (1983).
Japan Sci. Soc. Press, Tokyo/Springer-Verlag, Berlin

Rat Plasma Proteinase Inhibitors: The Biosynthesis of α_1-Proteinase Inhibitor and α_2-Macroglobulin

Tilo ANDUS, Volker GROSS, Wolfgang NORTHEMANN, Thuy-Anh TRAN-THI, and Peter C. HEINRICH

*Biochemisches Institut der Universität Freiburg**

There exists a wide spectrum of control mechanisms of extra- and intracellular proteolysis. The subject has recently been reviewed (*6*). One way of regulating proteolysis is by proteinase inhibitors. The major plasma proteinase inhibitors in man are α_1-proteinase inhibitor (α_1-antitrypsin) (α_1-PI) and α_2-macroglobulin (α_2-M). Whereas α_1-PI inhibits only serine proteinases such as elastase, trypsin, and chymotrypsin (*7*), α_2-M forms complexes with nearly all endoproteinases (*5*, *11*). Unlike α_1-PI, α_2-M can hardly be detected in certain species (rat, rabbit) under normal conditions. In the rat α_2-M is strongly induced during inflammation (*9*, *12*) and is thus an acute phase reactant. Both α_1-PI and α_2-M are glycoproteins which are synthesized in the liver and secreted into the blood (*8*, *10*). As an approach to understanding the mechanism(s) of the synthesis of α_1-PI and α_2-M during inflammation, we have studied the biosynthesis of both proteinase inhibitors in a cell-free translation system and in hepatocytes (*1–4*). As a result of this work two forms differing in their apparent molecular weights have been found for α_1-PI as well as for α_2-M, one being the predominant intracellular form, the other representing the form found in the medium.

* Hermann-Herder-Straß 7, D-7800 Freiburg, FRG.

MESSENGER RNA LEVELS OF α_1-PI AND α_2-M DURING INFLAMMATION

It has been found by other investigators that α_1-PI and α_2-M increase in the blood during inflammation. We have shown (Fig. 1) that the levels of translatable mRNA for both inhibitors increase during local inflammation generated by injection of turpentine. α_2-M shows a 66-fold and α_1-PI a 2-fold increase in translatable mRNA 18 hr after injection of turpentine. Concomitantly, a 64% decrease of translatable albumin-mRNA is observed. In livers of untreated rats α_2-M-mRNA is present in very low amounts. Although α_2-M-mRNA increases much more than α_1-PI-mRNA during inflammation, the absolute amount of α_2-M-mRNA is lower by a factor of 8 compared to that of α_1-PI.

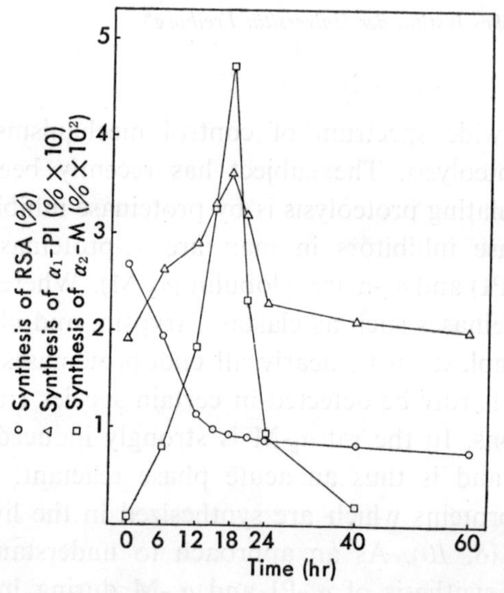

Fig. 1. Levels of translatable mRNA for α_1-PI, α_2-M, and albumin.
At various times after intramuscular injection of turpentine (0.5 ml/100 g body weight) poly(A)-rich RNA was isolated from liver and *in vitro* translated. The relative amounts of radioactivity incorporated into α_1-PI, α_2-M, and albumin are given.

SIZE DETERMINATION OF mRNAs FOR α_1-PI AND α_2-M

Poly(A)-rich RNA isolated from rat liver polysomes was separated on a 2.5% agarose/urea slab gel. The agarose gel was cut into slices, homogenized, and phenol-extracted. The poly(A)-rich RNA recovered from the various slices was translated in a cell-free system from wheat germ and immunoprecipitated with anti-α_1-PI. The antigen-antibody complexes were subjected to SDS-polyacrylamide gel electrophoresis (SDS-PAGE) and fluorography. From a plot of the relative mobilities *versus* the logarithm of the number of nucleotides—calibrated with the ribosomal RNA standards of 5S (121 nucleotides), 5.8S (158 nucleotides), 18S (2,120 nucleotides), and 28S (5,450 nucleotides)—a length of $1{,}930 \pm 170$ nucleotides is estimated for the mRNA of α_1-PI. The size of α_2-M-mRNA was determined by sucrose gradient centrifugation of poly(A)-rich RNA obtained from total liver RNA after extraction with guanidinium HCl and subsequent oligo(dT)-cellulose chromatography. From a sedimentation constant of about 27S a nucleotide number of 5,200 was estimated for α_2-M-mRNA.

IN VITRO SYNTHESIS OF α_1-PI AND α_2-M

Poly(A)-rich RNA from either phenol-extracted polysomes or total guanidinium HCl extracted frozen rat liver was translated into proteins in a cell-free system from wheat germs or reticulocytes. The poly(A)-rich RNA stimulated the incorporation of [^{35}S]methionine 50–100-fold. The radioactively labeled translation products were treated with an antiserum against α_1-PI or α_2-M, respectively. The fluorogram (Fig. 2, A and B, lane 5) revealed one main radioactivity band with an apparent molecular weight of 43,000 for α_1-PI and 162,000 for α_2-M, respectively. When unlabeled α_1-PI or α_2-M was added in increasing concentrations to the translation mixture, a competition was observed (not shown).

Since both proteinase inhibitors are secretory proteins and since with a few exceptions, secretory proteins are synthesized as larger molecular weight precursors with amino-terminal extensions (signal peptides), it seems likely that α_1-PI and α_2-M synthesized *in vitro* as larger molecular weight precursors. The proteolytic processing by the signal peptidase associated with microsomal membranes should result in the cleavage of the signal peptide, and thus lead to polypeptides of lower molecular weights than the *Mr* 43,000 α_1-PI and the *Mr* 162,000 α_2-M precursors. Since both α_1-PI and α_2-M are glycoproteins,

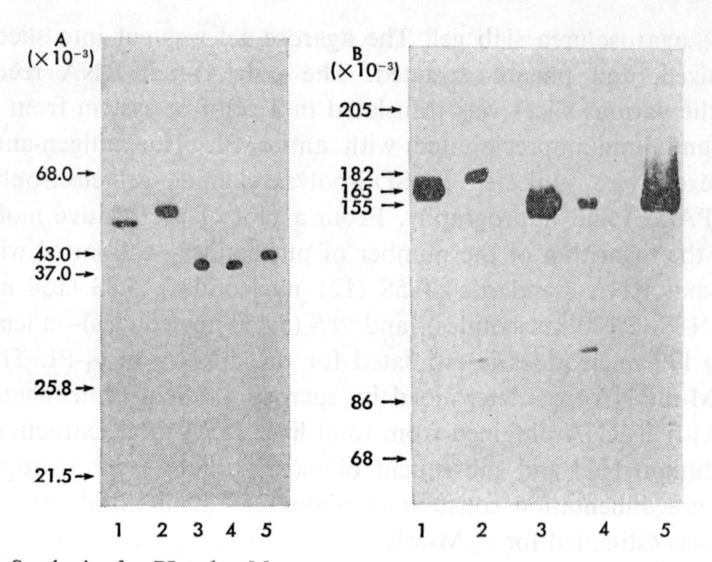

Fig. 2. Synthesis of α_1-PI and α_2-M.
After labeling of hepatocytes with [^{35}S]methionine α_1-PI (A) and α_2-M (B) were immunoprecipitated from the cells (lane 1) and medium (lane 2) of control cultures and from cells (lane 3) and medium (lane 4) of tunicamycin treated cultures. Lane 5 shows α_1-PI and α_2-M synthesized in a cell-free translation system.

the glycosylation concomitant to the proteolytic processing of α_1-PI and α_2-M precursors leads to an increase in molecular weight and thus obscures the proteolytic removal of the signal sequence. Glycosylation had therefore to be prevented under conditions where the polypeptide chain was still inserted into the membrane and the signal sequence cleaved. A commonly used method to prevent glycosylation consists in the administration of tunicamycin to cell cultures.

SYNTHESIS OF α_1-PI AND α_2-M BY RAT HEPATOCYTE PRIMARY CULTURES

Hepatocytes were labeled with [^{35}S]methionine for 2–3 hr. The cells were separated from the medium, disrupted, homogenized, and centrifuged. α_1-PI and α_2-M were immunoprecipitated either from the high-speed supernatant obtained from the cell homogenate or from the medium.

Figure 2, A and B shows that proteins with different electrophoretic mobilities were immunoprecipitated from cells (lane 1) and from medium (lane 2). Apparent molecular weights of 49,000 and 54,000 were determined for α_1-PI in hepatocytes and their medium; the respective values for α_2-M were 176,000 and 182,000. When hepatocytes were preincubated with tunicamycin, which prevents glycosylation of proteins but does not affect their insertion into the membrane, unglycosylated α_1-PI and α_2-M with apparent molecular weights of 41,000 and 162,000 were found in the cells (lane 3) as well as in the medium (lane 4). The molecular weight of the unglycosylated α_1-PI is about 2,000 less than the protein obtained in the cell-free system (lane 5). In the case of α_2-M the unglycosylated form and the *in vitro* synthesized polypeptide exhibited identical electrophoretic mobilities. Due to the high molecular weight of α_2-M, it is evident that our separation system is not suited to resolve such small molecular weight differences of 2,000–3,000, which could be expected for unglycosylated α_2-M and the *in vitro* synthesized α_2-M precursor.

PRECURSOR—PRODUCT RELATIONSHIP BETWEEN CELLULAR AND MEDIUM FORMS OF α_1-PI AND α_2-M

Hepatocytes were pulse-labeled with [^{35}S]methionine for 10 min followed by a chase with unlabeled methionine. When α_1-PI and α_2-M were immunoprecipitated from the cells at different times ranging from 10 to 180 min thereafter, a continuous decrease of radioactivity incorporated into the intracellular forms was found for both proteinase inhibitors (Fig. 3, A and B, lanes 1–6). It can be seen from this figure, A and B, lanes 7–12 that secreted α_1-PI and α_2-M appear in the medium after 60 min.

CHARACTERIZATION OF INTRA- AND EXTRACELLULAR FORMS OF α_1-PI AND α_2-M

Different molecular weight forms of intra- and extracellular α_1-PI and α_2-M could only be detected when glycosylation was not inhibited (see Fig. 2), suggesting that the different molecular weight forms could be due to differences in glycosylation.

Useful tools for structural studies of glycoconjugates are endo- and exoglycosidases. Asparagine-linked oligosaccharides of membrane and secretory glycoproteins are initially synthesized as mannose-rich chains

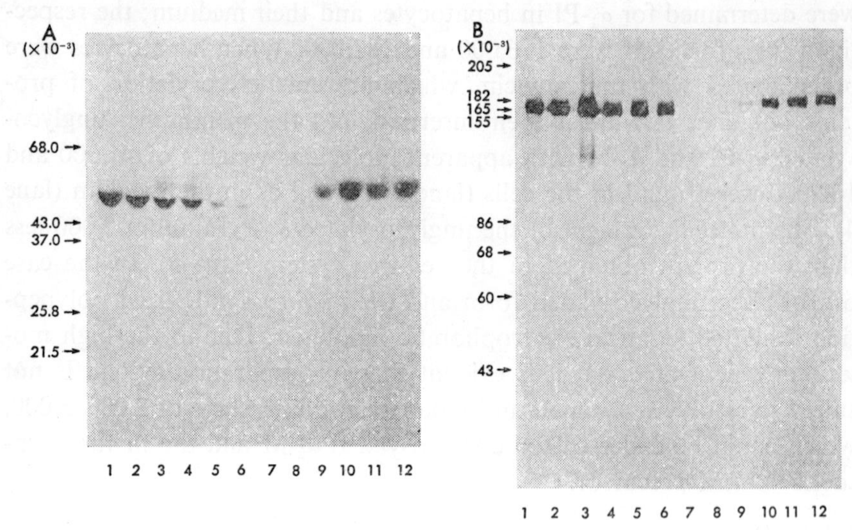

Fig. 3. Pulse-chase kinetics of α_1-PI and α_2-M.
Rat hepatocyte primary cultures were labeled with [^{35}S]methionine for 10 min followed by a chase with unlabeled methionine. α_1-PI (A) and α_2-M (B) were immunoprecipitated from the cells (lanes 1–6) and from the medium (lanes 7–12), 10 min (lanes 1 and 7), 30 min (lanes 2 and 8), 60 min (lanes 3 and 9), 90 min (lanes 4 and 10), 120 min (lanes 5 and 11), and 180 min (lanes 6 and 12) after the chase.

in the rough endoplasmic reticulum. Their transformation to oligosaccharides of the complex type appears later in the Golgi apparatus. We therefore tested the susceptibility of α_1-PI and α_2-M immunoprecipitated either from the hepatocytes or from the hepatocyte medium to endoglucosaminidase H as well as to sialidase. In Fig. 4, it can be seen that the *Mr* 49,000 α_1-PI and the *Mr* 176,000 α_2-M, the predominant forms in the cells, are susceptible to the action of endoglucosaminidase H (lanes 1 and 2). The *Mr* 54,000 α_1-PI and the *Mr* 182,000 α_2-M found in the medium (lanes 3 and 4) and in small amounts also in the cells (lanes 1 and 2) are resistant to endoglucosaminidase H. α_1-PI and α_2-M obtained after endoglucosaminidase H treatment of the intracellular form had nearly the same electrophoretic mobility as α_1-PI and α_2-M found in hepatocytes after tunicamycin treatment. The small difference between the two forms might be due to the fact that endoglucosaminidase H leaves the first *N*-acetylglucosamine molecule of each oligosaccharide chain attached to asparagine. Thus, not all sugars are removed

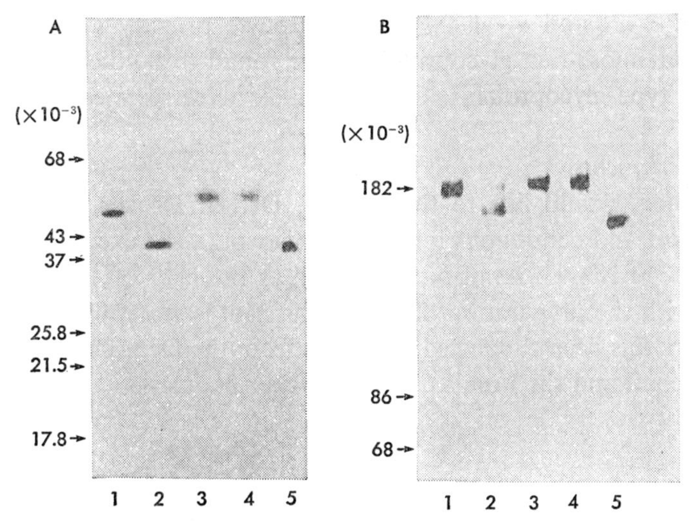

Fig. 4. Treatment of α_1-PI and α_2-M with endoglucosaminidase H.
α_1-PI (A) and α_2-M (B) were immunoprecipitated from [^{35}S]methionine labeled hepatocytes (lanes 1 and 2) and their medium (lanes 3 and 4), incubated without (lanes 1 and 3) or with (lanes 2 and 4) 5 mU of endoglucosaminidase H. Lane 5 shows α_1-PI and α_2-M immunoprecipitated from tunicamycin-treated hepatocytes.

from the protein. The extracellular *Mr* 54,000 α_1-PI and the *Mr* 182,000 α_2-M resistant to endoglucosaminidase H were susceptible to sialidase indicating that sialic acid is the terminal sugar of their oligosaccharide chains (not shown).

To get further insight into the different glycosylation of the main intracellular and medium forms of α_1-PI and α_2-M, the incorporation of radioactively labeled monosaccharides was studied. [^{3}H]mannose was predominantly incorporated into the intracellular forms of both α_1-PI and α_2-M. [^{3}H]fucose and [^{3}H]galactose, on the other hand, were incorporated nearly exclusively into the medium forms of both proteinase inhibitors. When hepatocytes were incubated in a medium containing tunicamycin and the radioactively labeled sugars were added 1 hr thereafter, no radioactivity was found in that region of the gel, where the *Mr* 41,000 α_1-PI and the *Mr* 162,000 α_2-M found after tunicamycin treatment and labeling with [^{35}S]methionine could be expected (Fig. 2, A and B, lanes 3 and 4), indicating that all sugars are bound *N*-glycosidically.

In conclusion, we have shown that two differently glycosylated forms of α_1-PI and α_2-M exist in hepatocyte primary cultures. Intracellular mannose-rich glycoproteins were found to be the precursors of complex type glycoproteins. Only the latter were secreted by the cells.

Acknowledgments

The authors would like to thank Prof. Dr. K. Decker, Freiburg, for his support and continuous interest in this work. The excellent technical assistance of Ms. C. Armbruster as well as the help of Mrs. H. Gottschalk with the preparation of this manuscript is gratefully appreciated. This work has been supported by grants from the Deutsche Forschungsgemeinschaft and the Fonds der Chemischen Industrie.

REFERENCES

1. Andus, T., Gross, V., Tran-Thi, T.-A., and Heinrich, P. C. *FEBS Lett.*, **151**, 10 (1983).
2. Geiger, T., Northemann, W., Schmelzer, E., Gross, V., Gauthier, F., and Heinrich, P. C. *Eur. J. Biochem.*, **126**, 189 (1982).
3. Gross, V., Geiger, T., Tran-Thi, T.-A., Gauthier, F., and Heinrich, P. C. *Eur. J. Biochem.*, **129**, 317 (1982).
4. Gross, V., Kaiser, C., Tran-Thi, T.-A., Schmelzer, E., Witt, I., Plummer, T. H., Jr., and Heinrich, P. C. *FEBS Lett.*, **151**, 10 (1983).
5. Harpel, P. *Methods Enzymol.*, **45**, 639 (1976).
6. Holzer, H. and Heinrich, P. C. *Annu. Rev. Biochem.*, **49**, 63 (1980).
7. Jeppson, J.-O. and Laurell, C.-B. *In* "Proteases and Biological Control," eds. E. Reich, D. B. Rifkin, and E. Shaw, p. 405 (1976). Cold Spring Harbor Laboratories of Cell Proliferation, New York.
8. Koj, A., Regoeczi, E., Toews, C. J., Leveille, R., and Gauldie, J. *Biochim. Biophys. Acta*, **539**, 496 (1978).
9. Metcalfe, J. and Tavill, A. S. *Br. J. Exp. Pathol.*, **56**, 570 (1975).
10. Sarcione, E. J. and Bodgen, A. E. *Science*, **153**, 547 (1966).
11. Starkey, P. M. and Barrett, A. J. *In* "Proteinases in Mammalian Cells and Tissues," ed. A. J. Barrett, p. 662 (1977). Elsevier/North-Holland Biomedical Press, Amsterdam.
12. Van Gool, J., Boers, W., and De Nie, I. *Exp. Mol. Pathol.*, **29**, 228 (1978).

Proteinase Inhibitors: Medical and Biological Aspects, eds. N. Katunuma et al., pp. 251–259 (1983).
Japan Sci. Soc. Press, Tokyo/Springer-Verlag, Berlin

Renin, Prorenin, and Renin Inhibitor

Tadashi INAGAMI, Kunio S. MISONO, Taisuke INAGAKI, and Yukio TAKII

*Department of Biochemistry, Vanderbilt University School of Medicine**

At the time of the discovery of renin in 1898 this substance in renal extract was considered a pressor substance. It was not until 1940 that renin was identified as an enzyme mediating the first step of reactions leading to the production of the pressor peptide angiotensin II from its prohormone angiotensinogen. The decapeptide angiotensin I is formed by renin. It is then converted to the octapeptide angiotensin II by angiotensin converting enzyme. Renin is a peptidase with an exceedingly limited substrate specificity; to date, we know of no substrate for it other than angiotensinogen. Moreover, even in this 70,000-dalton prohormone molecule only Leu_{10}-Leu_{11} peptide bond in horse angiotensinogen or Leu_{10}-Val_{11} in human angiotensinogen is cleaved by renin.

The exceedingly low concentration of this enzyme in the kidney, its instability, and the presence of proteases with non-specific renin-like activity have presented major obstacles to the studies on renin. However, successful purification of this enzyme from mouse submandibular gland in 1972 (*5*) changed this situation and since then steady progress has been made. Particularly significant has been the availability of the affinity matrix utilizing pepstatin (*39*) as ligand for the purification of

* Nashville, Tennessee 37232, U.S.A.

this enzyme and the development of *in vitro* radioimmunoassay of angiotensin I (*11*) for the determination of renin activity.

STRUCTURE OF RENIN

Renin has long defied attempts to identify its active site characteristics with type specific inactivators or inhibitors of proteases until it was found to be sensitive to inhibition by diazo-acetyl-D,L-norleucine methyl ester (DAN) in the presence of Cu^{2+} (*15*). This reagent was known to be a specific inhibitor for acid protease (*31*). Availability of pure renin in larger quantities permitted characterization of its active site in a quantitative manner using various group specific inhibitors which include DAN, 1,2-epoxy-3-*p*-nitrophenoxypropane (EPNP) (*37*), tetranitromethane, acetylimidazole, and phenylglyoxal. Studies revealed that two carboxyl groups, two tyrosyl residues, and an arginine residue are involved in the active site of mouse submandibullar gland renin (*20*). Of the two carboxyl groups, one was specifically modified by DAN and the other by EPNP in a mutually independent manner like pepsin (*21*). These observations indicated that the structure of the active site of renin is closely related to that of acid proteases although renin is functional at neutral pH rather than in acidic pH (*16*), has no general protease activity (*20*) and is inhibited by pepstatin in a manner different from other acid proteases (*1*).

The determination of the complete amino acid sequence of mouse submandibular gland renin (*22*), shown in Fig. 1, became feasible by the development of a rapid and large-scale purification of the submandibular gland enzyme (*20*).

The active form of mouse submandibular renin was found to consist of two separate polypeptide chains (H- and L-chain) connected by by a disulfide bridge (Cys_{250}-Cys_{283}). The amino acid sequences of both H- and L-chain showed extensive sequence identified with acid proteases such as porcine pepsin (44% identical) (*34*), bovine chymosin (36% identical) (*10*), and even with the fungal enzyme penicillopepsin (21% identical) (*12*) as shown in Fig. 1. The two catalytically essential carboxyl groups were identified as Asp_{32} (EPNP specific) and Asp_{215} (DAN specific). Tyr_{75} was also identical with other related enzymes. An im-

```
                       1         10         A B      20          30         40
MOUSE RENIN      S S L T D L I S P V V L T N Y L N S Q Y Y - - G E I G I G T P P Q T F K V I F D T G S A N L W V
PIG PEPSIN                   I G D E P L E N Y L D T E Y F - - G T I G I G T P A Q D F T V I F D T G S S N L W V
BOVINE CHYMOSIN          G E V A S V P L T N Y L D S Q Y F - - G K I Y L G T P P Q E F T V I F D T G S S D F W V
PENICILLOPEPSIN      A A S G V A T N T P T A N - D E E Y I T P V T I G - G T - T - - L N L N F D T G S A D L W V

                 41     46 A B    50          60          70          80 A
MOUSE RENIN      P S T K C S R L Y L A C G I H S L Y E S S D S S S Y M E N G D D F T I H Y G S G R V -
PIG PEPSIN       P S V Y C S - - S L A C S D H N Q F N P D D S S T F E A T S Q E L S I T Y G T G S M -
BOVINE CHYMOSIN  P S I Y C K - - S N A C K N H Q R F D P R K S S T F Q N L G K P L S I H Y G T G S M -
PENICILLOPEPSIN  F S T E L P A - S - Q Q S G H S V Y N P S A T G K - E A S G Y T W S I S Y G D G S S A

                 81        90          100 102A     110         120
MOUSE RENIN      K G F L S Q D S V T V G G I T V T - Q T F G E V T E L - P L I P F M L A Q F D G V
PIG PEPSIN       T G I L G Y D T V Q V G G I S D T N Q I F G - L S E T E P G S F L Y Y A P F D G I
BOVINE CHYMOSIN  Q G I L G Y D T V T V S N I V D I Q Q T V G - L S T Q E P G D V F T Y A E F D G I
PENICILLIPEPSIN  S G N V F T D S V T V G G V T A H G Q A V Q - A A Q Q I S A Q F Q Q D T N N D G L

                 121       130         140         150         160A B
MOUSE RENIN      L G M G F P A Q A V G G V T P V F D H I L S Q G V L K E K V F S V Y Y N R G P H L L
PIG PEPSIN       L G L A Y P S I S A S G A T P V F D N L W D Q G L V S Q D L F S V Y L S S N D D - -
BOVINE CHYMOSIN  L G M A Y P S L A S E Y S I P V F D N M M N R H L V A Q D L F S V Y M D R D G Q - -
PENICILLOPEPSIN  L G L A F S S I N - T V Q P Q S Q T T F F D T V K S S L A Q P L F A V A L K H Q - -

                 161       170         180  184A    190         200
MOUSE RENIN      G G E V V - L G G S D P E H Y Q G D F H Y V S L - S K T D S W Q I T M K G V S V G
PIG PEPSIN       S G S V V L L G G I D S S Y Y T G S L N W V P V - S V E G Y W Q I T L D S I T M D
BOVINE CHYMOSIN  E - S M L T L G A I D P S Y Y T G S L H W V P V - T V Q Q Y W Q F T V D S V T I S
PENICILLOPEPSIN  Q P G V Y D F G F I D S S K Y T G S L T Y T G V D N S Q G F W S F N V D S Y T A G

                 201       210         220         230         238A 240
MOUSE RENIN      S S T L L C E E G C E V V V D T G S S F I S A P T S S L K L - I M Q A L G A - K E
PIG PEPSIN       G E T I A C S G G C Q A I V D T G T S L L T G P T S A I A I N I Q S D I G A - S E
BOVINE CHYMOSIN  G V V V A C E G G C Q A I L D T G T S K L V G P S S D I L - N I Q Q A I G A - T Q
PENICILLOPEPSIN  S Q S G D G F - - - S G I A D T G T T L L L L B D S V V S Q Y Y S Q V S G A Q Q D

                 241       250         260         270         278A B279|
MOUSE RENIN      K R L H E Y V V C C S Q V P T L P D I S F N L G G R A Y T L S S T D Y V L Q Y P N |
                                             S
PIG PEPSIN       N S D G E M V I S C S S I D   L P D I V F T I N G V Q Y P L S P S A Y I L Q - - D
                                             Q
                       D
BOVINE CHYMOSIN  N Q Y   E F D I D C D N L S Y M P T V V F E I N G K M Y P L T P S A Y T S Q - - D
                       G
PENICILLOPEPSIN  S N A G G Y V F X C S B V T B L P V S I S G Y - T A T V P G S L I N Y G P S - - G
                                                                    <--- H-chain ---|
                |280 282A     290         300         310         320       327
MOUSE RENIN     | D K L - C T V A L H A M D I P P P T G P V W V L G A T F I R K F Y T E F D R H N N R I G F A L A R
PIG PEPSIN        D D S - C T S G F E G M D V P T S S G E L W I L G D V F I R Q Y Y T V F D R A N N K V G L A P V A
BOVINE CHYMOSIN   Q G F - C T S G F Q S E N - - - H S - Q K W I L G D V F I R E Y Y S V F D R A N N L V G L A K A I
PENICILLOPEPSIN   N G S T C L G G I Q S N - - - - S G I G F L I F G D I F L K S Q Y V V F D S D G P Q L G F A P Q A
                |--- L-chain --->
```

Fig. 1. Amino acid sequences of mouse submaxillary gland renin are compared with primary structures of pig pepsin (*34*), calf chymosin (*10*), and penicillopepsin (*12*). Numbers designating positions of amino acid residues are those of pig pepsin (*34*). Single letter abbreviations for amino acid residues are; A, alanine; C, cysteine or 1/2 cystine; D, aspartic acid; E, glutamic acid; F, phenylalanine; G, glycine; H, histidine; I, isoleucine; K, lysine; L, leucine; M, methionine; N, asparagine; P, proline; Q, glutamine; R, arginine; S, serine; T, threonine; V, valine; W, tryptophan; Y, tyrosine.

portant structural difference was the existence of free sulfhydryl groups (*21*), perhaps two of them.

The cloning of cDNA for the mRNA of the mouse enzyme permitted the determination of the nucleotide sequence of renin cDNA (*27*). Although results deduced from this method contained several erroneous sequences, it shed light on the structures of renin precursors.

INACTIVE RENIN, RENIN ZYMOGEN

The discovery of an activatable form of renin in human amniotic fluid led to the proposal that a prorenin or renin zymogen might exist (*19*). A series of papers appeared reporting activatable renin in human plasma (*6*), hog, and rabbit kidney (*3*, *17*). Activation was demonstrated by treatment with proteases such as trypsin or pepsin (*7*, *23*), acid treatment (*3*, *6*, *17*, *19*) or cold treatment (*26*, *33*). The latter two modes of activation seem to be due to removal of protease inhibitors in plasma and tissues. Thus, activation of activatable renin is mediated by proteolysis. The protease dependent mode of activation indicated that the activatable renin was a zymogen. An earlier report of a molecular size of activatable renin as large as 60,000 (*6*, *7*, *17*), as opposed to 36,000–40,000 of active renin, also supports this notion. However, failure to observe completely inactive renin made it difficult to establish the identity of this activatable form. Whether the activatable renin was a partially active regulatable enzyme or a mixture of active enzyme and a completely inactive proenzyme was not clear.

Affinity columns to separate active and inactive renin were devised to determine the validity of these hypotheses. Pepstatin-Sepharose retains active renin preferentially but not inactive renin (*40*). Affii-Gel Blue (*40*) and octyl-Sepharose (*4*) were found to bind inactive renin but not the active renin. Using a combination of Affi-Gel Blue and pepstatin-Sepharose, Yokosawa *et al.* demonstrated that human plasma contains a completely inactive renin with a gel-filtration molecular weight of 53,000 and active renin without potential for further activation with a molecular weight of 50,000 (*40*). Completely inactive renin was also found in human kidney (*2*, *4*) and hog kidney (*24*, *36*) with molecular weights of 50,000 and 48,000, respectively. The reason for the lower molecular weight of inactive renin in human kidney compared

TABLE I. Purification of Inactive Rein from Hog Kidney

Step	Total protein (mg)	Specific[b] activity	Purification	Yield (%)
Crude extract[a]	1,600,000	0.0015[c]	1	100
DEAE batch	125,000	0.016[c]	10	82
Pepstatin-Sepharose	60,000	0.03	20	75
DEAE-Sepharose	18,000	0.08	53	63
Octyl-Sepharose	1,000	1.43	950	59
Sephadex G-100	208	5.29	3,530	46
Affi-Gel Blue	16.0	62.9	41,900	42
Con A-Sepharose	0.750	1,000	666,700	31
Sephadex G-100	0.170	2,400	1,600,000	17
DEAE-Sephacel	0.030	5,000	3,333,000	6.3

[a] Prepared from 20 kg of fresh hog kidney. [b] Activity of inactive renin as defined by, inactive renin = total renin activity after activation − renin activity before activation; expressed as μg angiotensin I/mg of protein/hr. [c] Determined after removal of contaminating renin by using affinity chromatography; see ref. *35* for details.

with human plasma inactive renin was not clear. Furthermore, inactive renin was purified from hog kidney by a series of affinity chromatographic steps to an electrophoretically homogeneous form (Table I) and was shown to have a single polypeptide chain of a molecular weight of 48,000 (*35*). The fact that it was not activated by high salt concentration, chaotropes, detergents or denaturants indicated that this form of inactive renin was not an inhibitor-enzyme complex but an inactive zymogen (*35*, *36*).

Supporting but not complete proof of this proposal was obtained from the *in vitro* translation studies of mRNA from mouse submaxillary gland and kidney; the renin-immunoreactive substances were detected after SDS-gel electrophoresis at positions corresponding to the molecular weight of 45,000–50,000 (*8*, *28–30*). Based on the nucleotide sequence of the coding region of mouse renin mRNA the molecular weight of this nascent translation product has been established as 48,000 (*27*).

RENIN-INHIBITOR COMPLEX

An alternative explanation for the inactive form of renin is a renin-inhibitor complex. Indeed, such a possibility has been discussed in the past. Boyd (*3*) and Leckie *et al.* (*17*) reported a 60,000-dalton activatable renin in hog and rabbit kidney, respectively. Although these authors claimed that it was a renin-inhibitor complex, it was not clear whether it was a completely inactive form or a partially active enzyme. Leckie proposed it might represent an intermediate in the pathway of activation of renin zymogen to active renin (*18*). However, no other investigators have been able to reproduce their results to date.

Recently, Ueno *et al.* reported that the renin binding protein (*9*) showed inhibitory activity toward crude renin preparation only when it was used at very high concentrations (*38*). However, this inhibition is now understood to be an artifact due to the destruction of angiotensin I in the assay mixture by strong angiotensin destroying enzyme(s) (angiotensinases). Destruction of the product of renin reaction made it appear that renin inhibitor was present in their preparations extracted from the kidney or mouse submaxillary gland.

While we have been examining renin in cloned mouse neuroblastoma cells (*25*) and rat juxtaglomerular cells (*32*), we observed that renin activity was progressively lost during repeated passages of cells and eventually was lost completely. Search for a possible inactive enzyme revealed the presence of completely inactive renin (*13, 14*). Contrary to renin zymogen, however this inactive renin was not activated by limited proteolysis by a protease such as trypsin. On the other hand, the cell extract developed significant renin activity upon treatment with dithiothreitol (2–10 mM).

In order to definitively demonstrate the presence of renin inhibitor, it was separated from the neuroblastoma cell extract by affinity chromatography on a renin-Sepharose column. The inhibitory component which adhered to the gel was released with 0.5 M NaCl. The released substance was shown to recombine with renin and cause progressive loss of its enzyme activity as the amount of the inhibitor was increased.

The nature of the inhibitor and the inhibitor-renin complex has not yet been clarified completely. Some specific features known to date are enumerated below.

1) The complex remains completely inactive in crude cell extracts which are very dilute in renin concentration. Renin remains inactive during gel-filtration chromatography.

2) Its activation requires brief treatment with sulfhydryl compounds

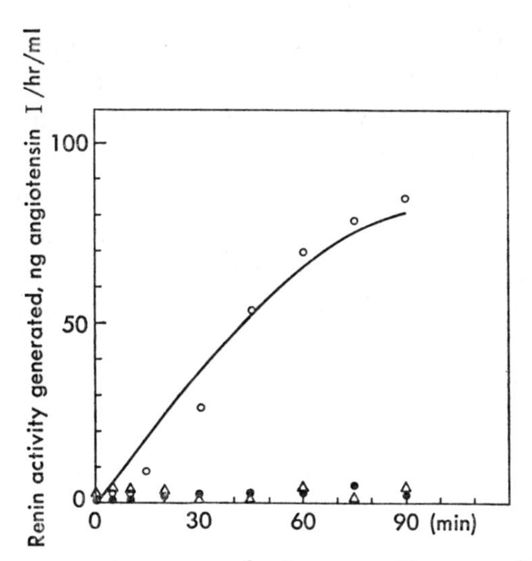

Fig. 2. Activation of renin-inhibitor complex from neuroblastoma cells by dithiothreitol (○).
Trypsin failed to activate the complex (●). Without activation little activity was detectable (△). Abscissa represents the period of incubation for activation at 37°C and pH 7.4.

such as dithiothreitol. The activation and inhibition are reversible (Fig. 2).
3) The inactive complex has a rather broad molecular weight distribution as examined by gel filtration with a peak fraction showing a molecular weight between 50,000 and 45,000. The gel filtration pattern shows the presence of higher molecular weight species of 90,000 and 60,000. This heterogeneous distribution is not due to dynamic association-dissociation equilibrium since rechromatography of the 46,000-dalton peak gives a sharp peak in the same position. Since the molecular weight of renin is 35,000–38,000, the molecular weight of renin inhibitor involved in the 46,000-dalton complex should be in the range of 8,000 to 10,000.

These properties distinguish this newly discovered renin inhibitor from the renin binding protein which has a molecular weight of 60,000 and does not form a stable complex with an expected molecular weight of 100,000 (*9*). It is different from the activatable renin of Boyd (*3*) or Leckie *et al.* (*17*), since these have a greater molecular weight and are dissociable with high salt concentration, detergent or urea. Thus, the activatable renin reported by Boyd and Leckie may well be an intermediate in the pathway of the activation of renin zymogen (prorenin) as suggested by Leckie *et al.* (*18*).

The inactive renin in cultured cells described above is clearly different from the inactive renin zymogen or prorenin in spite of similar molecular weights, since the former is activated only by sulfhydryl compounds but not by proteolysis whereas prorenin or renin zymogen is activatable only by proteolysis (Fig. 2).

These summarized data indicate that a hitherto unknown renin-inhibitor complex has been discovered. It is a stable renin-inhibitor complex with a molecular weight of 46,000 and is activated only by sulfhydryl compounds. Although the demonstration of its presence has been limited to cultured cells such as mouse neuroblastoma cells and rat kidney juxtaglomerular cells, it is likely to be present in other types of tissues.

Physiological significance of the inhibitor is not clear. Since we have not been able to identify a similar complex in plasma, it is likely that the function of the inhibitor is limited to the regulation of intracellular function of renin in the intracellular production of angiotensin as has been demonstrated in the cultured neuroblastoma cells (*25*), and kidney juxtablomerular cells (*32*) and perhaps in many tissues where renin functions intracellularly.

Thus, the system involving inactive renins discussed above may be summarized by the following scheme.

$$\text{Renin zymogen} \xrightarrow{\text{Proteolysis}} \text{Active renin} \underset{\text{-S.S-}}{\overset{\text{-SH}}{\rightleftharpoons}} \text{Renin-inhibitor complex}$$

Acknowledgments
Preparation of this manuscript was supported by research grants from NIH HL-14192, HL-22288, and HL-24112, a Grant-in-Aid from the American Heart Association and an Investigatorship Award to K.S.M. from the Tennessee Heart Association.

REFERENCES

1. Aoyagi, T., Morishima, H., Nishizawa, R., Kunimoto, S., Takeuchi, T., Umezawa, H., and Ikezawa, H. *J. Antibot.*, **25**, 689 (1972).

2. Atlas, S. A., Sealey, J. E., Dharmgrongartama, B., Hesson, T. E., and Laragh, J. H. *Hypertension*, **3** (Suppl. I), 30 (1981).
3. Boyd, G. W. *Circ. Res.*, **35**, 426 (1974).
4. Chang, J. J., Kisaragi, M., Okamoto, H., and Inagami, T. *Hypertension*, **3**, 509 (1981).
5. Cohen, S., Taylor, J. M., Murakami, K., Michelakis, A. M., and Inagami, T. *Biochemistry*, **11**, 4286 (1972).
6. Day, R. P. and Luetscher, J. A. *J. Clin. Endocrinol. Metab.*, **38**, 923 (1974).
7. Day, R. P. and Luetscher, J. A. *J. Clin. Endocrinol. Metab.*, **40**, 1085 (1975).
8. Dykes, C., Bhat, K., Taylor, J. M., and Inagami, T. *Biomed. Res.*, **1**, 565 (1980).
9. Funakawa, S., Funae, Y., and Yamamoto, K. *Biochem. J.*, **176**, 977 (1978).
10. Foltman, B. and Pedersen, V. B. *In* "Acid Protease, Structure, Function and Biology," ed. J. Tang, p. 3 (1977). Plenum Press, New York.
11. Haber, E., Koerner, T., Page, L. B., Kliman, B., and Purnode, A. *J. Clin. Endocrinol. Metab.*, **29**, 1349 (1969).
12. Hsu, I. N., Delbare, L. T. J., James, M., and Hofman, T. *Nature*, **266**, 140 (1977).
13. Inagaki, T., Okamura, T., and Inagami, T. *Brain Res.*, **250**, 373 (1982).
14. Inagaki, T., Inagami, T., Rightsel, W. A., and Muirhead, E. E. *Circulation*, **66** (Suppl. II), 362 (1982).
15. Inagami, T., Misono, K., and Michelakis, A. M. *Biochem. Biophys. Res. Commun.*, **56**, 503 (1974).
16. Inagami, T. *In* "Biochemistry of Blood Pressure Regulation," ed. Soffer, p. 39 (1981). Wiley Interscience, New York.
17. Leckie, B. *Clin. Sci.*, **44**, 301 (1973).
18. Leckie, B., McConnell, A., and Jordan, J. *In* "Acid Protease, Structure, Function and Biology," ed. J. Tang, p. 249 (1977). Plenum Press, New York.
19. Lumbers, E. R. *Enzymologia*, **40**, 329 (1971).
20. Misono, K. S., Holladay, L. A., Murakami, K., Kuromizu, K., and Inagami, T. *Arch. Biochem. Biophys.*, **217**, 574 (1982).
21. Misono, K. S. and Inagami, T. *Biochemistry*, **19**, 2616 (1980).
22. Misono, K. S., Chang, J. J., and Inagami, T. *Proc. Natl. Acad. Sci. U.S.*, **79**, 4358 (1982).
23. Morris, B. J. and Lumbers, E. R. *Biochim. Biophys. Acta*, **289**, 385 (1972).
24. Murakami, K., Takahashi, S., Hirose, S., Ohsawa, T., and Inagami, T. *In* "Heterogeneity of Renin and Renin Substrate," ed. H. P. Sambhi, p. 45 (1981). American Elsevier, New York.
25. Okamura, T., Clemens, D. L., and Inagami, T. *Proc. Natl. Acad. Sci. U.S.*, **78**, 6940 (1981).
26. Osmond, D. H., Ross, L. J., and Scaiff, K. D. *Can. J. Physiol.*, **51**, 705 (1973).
27. Panthier, J. J., Foote, S., Chambraud, B., Strosberg, A. D., Corvol, P., and Rougeon, F. *Nature*, **298**, 90 (1982).
28. Poulsen, K., Vuust, J., and Lund, T. *Clin. Sci.*, **59**, 297 (1980).
29. Poulsen, K., Vuust, J., Lykkegard, S., Nielsen, A. H., and Lund, T. *FEBS Lett.*, **56**, 105 (1979).
30. Pratt, R. E., Dzau, V. J., and Ouelette, A. J. *Nucleic Acid Res.*, **9**, 3433 (1981).
31. Rajagopalan, T. C., Stein, W. H., and Moore, S. *J. Biol. Chem.*, **241**, 4295 (1966).
32. Rightsel, W. A., Okamura, T., Inagami, T., Pitcock, J. A., Brooks, B., Brown, P., and Muirhead, E. Z. *Circ. Res.*, **50**, 822 (1982).
33. Sealey, J. E. and Laragh, J. H. *Circ. Res.*, **36/37** (Suppl. I), 10 (1975).

34. Sepulveda, P., Marciniszin, J., Jr., Liu, D., and Tang, J. *J. Biol. Chem.*, **250**, 5082 (1975).
35. Takii, Y. and Inagami, T. *Biochem. Biophys. Res. Commun.*, **12**, 200 (1982).
36. Takii, Y. and Inagami, T. *Biochem. Biophys. Res. Commun.*, **94**, 182 (1980).
37. Tang, J. *J. Biol. Chem.*, **246**, 4510 (1971).
38. Ueno, N., Miyazaki, H., Hirose, S., and Murakami, K. *J. Biol. Chem.*, **256**, 12023 (1981).
39. Umezawa, H., Aoyagi, T., Morishima, H., Matsuzaki, M., Hamada, M., and Takeuchi, T. *J. Antibiot.*, **22**, 259 (1970).
40. Yokosawa, N., Takahashi, N., Inagami, T., and Page, D. L. *Biochim. Biophys. Acta*, **569**, 211 (1979).

REGULATION OF INTRACELLULAR PROTEINASES

Proteinase Inhibitors: Medical and Biological Aspects, eds. N. Katunuma et al., pp. 263–273 (1983).
Japan Sci. Soc. Press, Tokyo/Springer-Verlag, Berlin

Bifunctional Activities and Possible Modes of Regulation of Some Lysosomal Cysteinyl Proteases

George KALNITSKY, Ranjit CHATTERJEE, Hari SINGH, Mark LONES, and Andrzej PASZKOWSKI

*Department of Biochemistry, University of Iowa College of Medicine**

We have isolated five lysosomal proteases in apparently homogeneous form from rabbit lung: cathepsins B, B2, L, C, and benzoyl-arginine-β-naphthylamide (BANA) hydrolase (Table I). Each of these enzymes is a cysteinyl protease and requires a reducing or -SH compound for activation. The enzymes range in molecular weight from relatively small (21,000–29,000) to medium (~52,000) to large (~140,000). The optimum pH for each enzyme is generally in the slightly acid range except that two of the enzymes, BANA hydrolase (as an aminopeptidase) and cathepsin C (in transpeptidation reactions), function at neutral or slightly alkaline pH values. Typical peptide substrates are listed for each enzymes, plus an effective inhibitor where this is known.

We have characterized these apparently homogeneous enzymes with respect to some of their physical and chemical properties and their specificities. In this necessarily limited review, we plan to discuss some of these findings under the following headings: (A) A comparison of the properties of two similar enzymes, BANA hydrolase and cathepsin H. As a result, we propose the name cathepsin I for BANA hydrolase. (B) The dual or bifunctional activities of four of the five lung proteases we

* Iowa City, Iowa 52242, U.S.A.

TABLE I. Some Properties of Lung Lysosomal Cathepsins

Cathepsin	Substrate	Molecular weight	Opt. pH	Inhibitor
B1	Z-Ala-Arg-Arg-βNA	~29,000	6.2	Leupeptin
	Bz-Arg-βNA (BANA)			
BH	Bz-Arg-βNA	26,000	6.5	Cytosolic inhib.,
	Leu-βNA		7.2	Leu-CH_2Cl
L	Azocasein	21,000	5.0	Leupeptin
B2	Bz-Arg-NH_2	52,000	5.7	Leupeptin
	Bz-Gly-Arg		5.0	
C	Gly-Arg-βNA	140,000		
	hydrolysis		6.2–6.4	
	transpeptidation		7.5–8.5	

BH, BANA hydrolase or cathepsin I (see next section); B2, lysosomal carboxypeptidase B; C, dipeptidylaminopeptidase I, or DAP I.

have isolated. It becomes obvious that such dual specificities may result in important intracellular functions for these enzymes. (C) Finally, in an attempt to answer the question how these cysteine proteases are regulated inside the cell, we present evidence relating to two possible mechanisms of regulation of specific lysosomal proteases.

PROPERTIES OF BANA HYDROLASE (OR CATHEPSIN I) *VS.* CATHEPSIN H

In the course of isolating cathepsin B (or B1), from rabbit lung we isolated another enzyme which was even more effective than cathepsin B in hydrolyzing the substrate, BANA. This new enzyme was called BANA hydrolase. In accordance with the accepted practice of enzyme nomenclature for lysosomal proteases, we propose that the enzyme BANA hydrolase (*22*, *23*) be referred to as cathepsin I (for Iowa). In this section, we want to outline the similarities and the differences between cathepsins H (*12*, *13*) and I (*22*, *23*) in a rather large number of properties.

Both of these enzymes are lysosomal, both are sulfhydryl enzymes and they have similar molecular weights. In addition, both enzymes hydrolyze the substrates BANA and Leu-βNA. Beyond these similarities, there appear to be a large number of differences in the properties of these two cathepsins.

There are differences in their activities on several substrates. Cathepsins H and I have significantly different activities with BAA (or

Bz-Arg-NH_2) as substrate (481 *vs.* 4,500 nmol $NH^+{}_3$ liberated/min/mg protein). With L-amino acid-β-naphthylamides (1 mM), the order of decreasing rates for the two enzymes is different. Furthermore, a striking difference is observed with collagen as substrate. Cathepsin H, whether obtained from rat liver (*12*, *13*) or human liver (*21*), shows no detectable collagenolytic activity. In contrast, both lung cathepsin I and lung cathepsin B demonstrate strong collagenolytic activity against type I skin collagen in solution (*24*).

The extent of inhibition by a number of inhibitors is different. Fifty percent inhibition of cathepsins H and I is achieved by 1×10^{-6} M and 1×1^{-8} M Leu-CH_2Cl, respectively (after 5 min incubation at 37°C). Again, this 100-fold difference does not appear to be due to species or tissue differences, since the concentration of Leu-CH_2Cl needed to achieve 50% inhibition of cathepsin B is 1×10^{-5} M whether this enzyme is obtained from rat liver or rabbit lung. With Leu-βNA as substrate, 1 mM puromycin inhibited cathepsins H and I 20 and 85%, respectively. With BANA or Bz-Arg-βNA as substrate, the inhibitions achieved with 0.01 mM Tos-Lys-CH_2Cl were 35 and 98%, respectively. With cathepsin H, 50% inhibition is obtained at a concentration of 5×10^{-6} M leupeptin (*2*), whereas with cathepsin I, even at a leupeptin concentration 12-times higher (6×10^{-5} M) the inhibition is only 19% (*24*).

Cathepsins H and I also show definite differences in a number of their physical properties, such as optimum pH (6.0 *vs.* 6.5–7.0), the number of isozymes (2 *vs.* 6), and their iso-electric points (7.1 *vs.* 5.9–6.5), respectively. There are also significant differences between the stabilities of the two enzymes. Alkali (pH 8.5, 4 hr, 20°C) inactivated cathepsins H and I 80–90% and 18%, respectively, and treatment at higher temperature (60°C, 30 min) resulted in inactivations of 10–20% and ~85%, respectively. These differences in a large number of properties indicate that these two enzymes are different, although they may be related. The production of specific antisera and the determination of their amino acid sequences will be helpful in answering this point.

BIFUNCTIONAL ACTIVITIES OF LYSOSOMAL PROTEASES

As more lysosomal proteases are obtained in apparently homogeneous form and more information is obtained about their specificities, it is becoming apparent that several of these enzymes have bifunctional activities.

Cathepsin B is known to digest a number of proteins (see review by Barrett, *1*). Bovine cathepsin B hydrolyzes a number of peptide bonds in the oxidized insulin B chain, clearly demonstrating endopeptidase activity (*11*, *19*). With collagen as substrate, cathepsin B exhibits collagenolytic activity, cleaving the non-helical, N-terminal telopeptide region of native collagen in solution, specifically the Gly-Ile and Ser-Val bonds (*8*). With aldolase as substrate, cathepsin B hydrolyzes peptides bonds and inactivates the enzyme. However, it does so by sequentially liberating dipeptides from the carboxy-terminal end of the aldolase molecule (*3*). Therefore, cathepsin B appears to exhibit both endopeptidase and dipeptidyl-carboxypeptidase activities.

Cathepsin B2 or lysosomal carboxypeptidase B, obtained from rabbit lung in apparently homogenous form (*14*), clearly shows carboxypeptidase activity with several peptides as substrates. For example, with the tetrapeptide, Thr-Pro-Arg-Lys, Lys is very rapidly liberated and digestion stops there. With angiotensin I (Asp-Arg-Val-Tyr-Ile-His-Pro-Phe-His-Leu) (100 nmol, 571 : 1, substrate : enzyme ratio), within 30 min, 100% of the leucine and only 7% of the histidine are liberated into the medium. No other amino acids are released. Substance P, which contains eleven amino acid residues plus a C-terminal amide group, is virtually inactive as a substrate. However, with oxidized insulin B as substrate, this enzyme clearly exhibits both carboxypeptidase and endopeptidase activities (Fig. 1). The results can be accounted for as follows: (a) carboxypeptidase action liberates the C-terminal Ala; (no Lys, Pro, Thr, Tyr, Phe, Arg, or Cys are liberated); (b) there is an endopeptidase cleavage of the peptide bond between Leu_{15}-Tyr_{16}, followed by (c) carboxypeptidase activity liberating Leu, Ala, Glu, Val, Leu, His, Ser, and Gly. Thus two Ala and two Leu residues are liberated, plus one Glu and one Val, plus smaller amounts of His, Ser, and Gly. In addition, two peptides were isolated at the end of the digestion period. One was a tetradecapeptide containing the amino acids Tyr_{16} through Lys_{29}. The other contained Phe_1 through Lys_{29} without the C-terminal Ala. The isolastion of these peptides supports the proposed.mechanism of action of lyosomal carboxypeptidase B (or cathepsin B2) on oxidized insulin B chain (*14*).

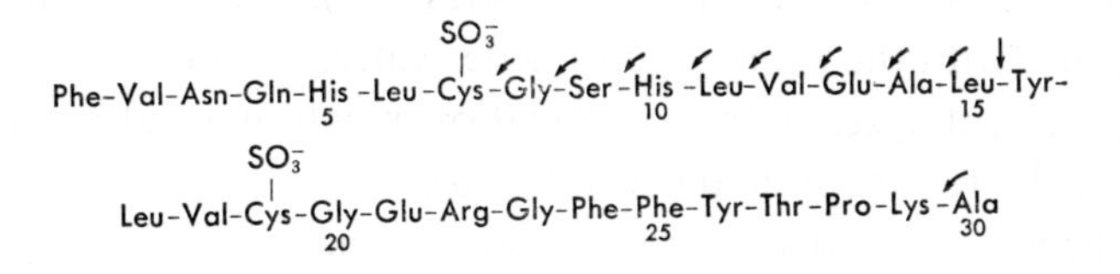

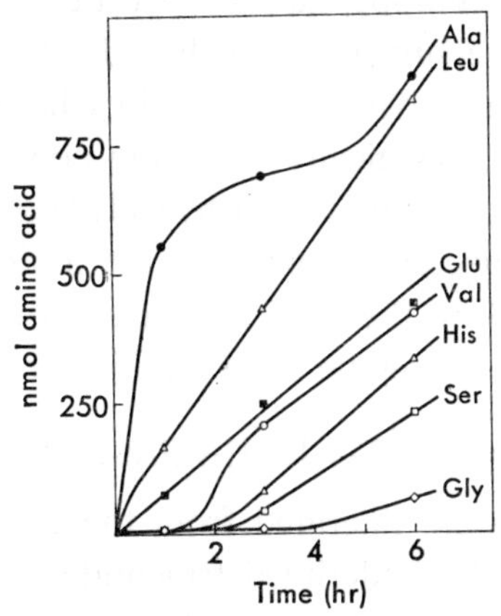

Fig. 1. Digestion of oxidized insulin B chain by cathepsin B2.
Substrate, 800 nmol; substrate : enzyme ratio=229 : 1; pH 6.0, 10 mM dithioerythritol, 1 mM EDTA, 37°C. Samples were removed from digestion mixture at the indicated times and subjected to amino acid analysis. ● Ala; △ Leu; ■ Glu; ○ Val; △ His; □ Ser; ◇ Gly.

Cathepsin C (or dipeptidylpeptidase I or dipeptidylaminopeptidase I (DAP I)) has been previously obtained from spleen and characterized by Metrione (*17*). Our preparation from rabbit lung contained no carboxypeptidase, aminopeptidase, endopeptidase, or dipeptidase activities, nor was this enzyme active on substrates for DAP II, III, or IV. This enzyme demonstrated strong hydrolytic activity on Gly-Arg-β-naphthylamide (Gly-Arg-βNA), yielding the dipeptide+βNA. The pH optimum for this reaction was ~6.2–6.4 (*20*).

However, at pH 7–8, this enzyme has strong transpeptidase activity, converting the dipeptide derivative Gly-Arg-4-methyoxy-βNA to the tetrapeptide Gly-Arg-Gly-Arg-4-methoxy-βNA. This confirms the much earlier work of Fruton and his colleagues carried out with impure preparations of this enzyme (*9*, *10*). Thus, at a slightly acid pH, this enzyme exhibits dipeptidyl-aminopeptidase activity, readily liberating dipeptides from dipeptide-βNA derivatives and from the N-terminal portions of large polypeptide hormones (*15*). However, at a slightly alkaline pH, with the same dipeptide derivative as substrate, the enzyme very rapidly catalyzes a transpeptidation reaction resulting in the formation of the tetrapeptide derivative (*20*).

Cathepsins H and I each show strong enzymatic activity in hydrolyzing typical synthetic aminopeptidase substrates (such as amino acid-βNAs) and endopeptidase substrates (such as Bz-Arg-βNA) (*12, 13, 22, 23*). The aminopeptidase activity of cathepsin I has been demonstrated with a large number of biologically occurring peptides (*5*). With these peptides as substrates, this enzyme will hydrolyze all amino acids from the amino terminal group, except proline; the aminopeptidase activity is halted one residue before the proline residue in the peptide. A good example is the digestion of angiotensin III inhibitor,

Arg-Val-Tyr-Ile-His-Pro-Ile,

where the amino acids were liberated from the N-terminal end in the indicated sequence and the underlined tripeptide, His-Pro-Ile, was isolated at the end of the digestion period. Aminopeptidase activity is completely blocked by the presence of an *N*-acetyl group (as in *N*-acetyl renin substrate), by a pyroglutamyl group at the N-terminal (as in bradykinin potentiator B) or by a penultimate prolyl residue, as in the tetrapeptide Thr-Pro-Arg-Lys, substance P and bradykinin, where the Pro immediately follows the N-terminal residue.

Cathepsin I also clearly exhibits endopeptidase activity with a number of biologically-occurring peptides as substrates (*5*). For example, acetyl renin substrate contains 14 amino acids, with its N-terminal end blocked with an acetyl group. Despite this lack of a free α-NH_2 group, acetyl renin substrate is readily digested by cathepsin I (*23*). Analysis of the digestion products (*5*) demonstrated endopeptidase cleavage of the peptide bonds between Arg-Val, Tyr-Ile, His-Leu, and Leu-Val. The resulting peptides are then digested by the aminopeptidase activity of cathepsin I, when the presence of Pro will not interfere.

CH_3-CO-Asp-Arg-Val-Tyr-Ile-His-Pro-Phe-His-Leu-Leu-Val-Tyr-Ser-

The underlined peptides were isolated at the end of the digestion period (substrate, 128 nmol; substrate : enzyme=174 : 1; 60 min).

POSSIBLE REGULATORY MECHANISMS OF SOME LYSOSOMAL PROTEASES

Cathepsin C (or DAP I) obtained from rabbit lung (*20*) has an absolute requirement for Cl^-, and its activity is markedly enhanced by the presence of sulfhydryl compounds. This enzyme is composed of subunits with a molecular weight of 20,000–25,000 (as determined by four different methods). In an attempt to obtain and confirm a molecular weight for the native enzyme, we utilized the velocity enzyme sedimentation technique (at pH 4.6, 0.5 mg enzyme/ml, 20°C) and, to our surprise, under these conditions the enzyme did not sediment. Increasing the enzyme concentration and raising the pH were not effective. When we added the substrate, the enzyme sedimented nicely (at lower or higher pH), indicating that the presence of the substrate favored association of the subunits. At a lower temperature (7°C) at least two molecular weight species were observed: ~90,000 and ~137,000. This probably indicates a reversible equilibrium between the subunits under these conditions.

The question arises: which is the active enzyme—the lower molecular weight monomers (20,000–25,000), the intermediate molecular weight enzyme or the larger molecular weight enzyme? The question was answered using the reacting enzyme sedimentation technique (*25*, *26*). The substrate in these experiments, Gly-Arg-fluorocoumarin, does not absorb light, whereas the product of the enzymatic hydrolysis, fluorocoumarin, does absorb light at 380 nm. With 0.2 μg of enzyme in the small reservoir of the cell, during the acceleration of the rotor the enzyme is layered over the assay mixture. As the enzyme band sediments through the assay mixture, the substrate is hydrolyzed and the A_{380} is recorded along the length of the cell with the photoelectric scanner at 4 min intervals. The increase in product between two successive scans is the measure of enzyme activity and is used to follow the progress of the reacting enzyme band (Fig. 2). Plotting the log χ (log of distance from center of rotation to peak of maximum A_{380}) *vs.* time yields a straight line (Fig. 3). From the slope of the line, the corrected $S_{20,w}$ is calculated.

From these experiments we conclude the following. There is only one active enzyme band. There is no appearance of active subunits, nor

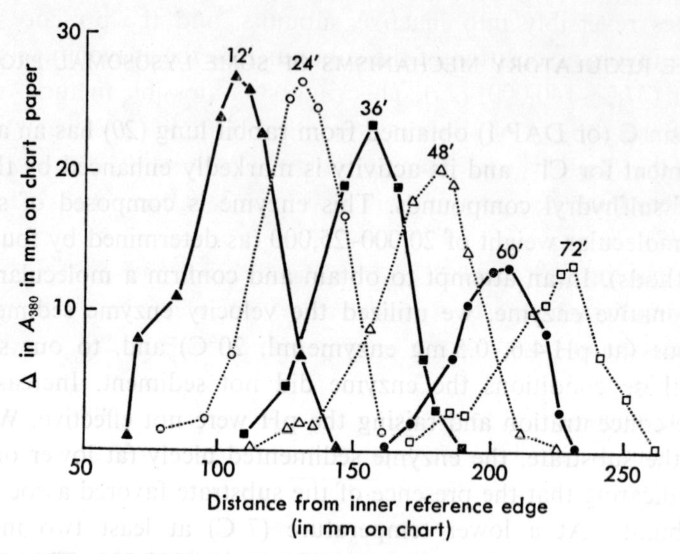

Fig. 2. Reacting enzyme sedimentation of cathepsin C.
Medium contained 0.2 μg enzyme, 0.2 mM Gly-Arg-fluorocoumarin, 10 mM cacodylate buffer pH 6.0, 10 mM β-mercaptoethanol-HCl, 50% D_2O in a total volume of 0.4 ml at 23°C.

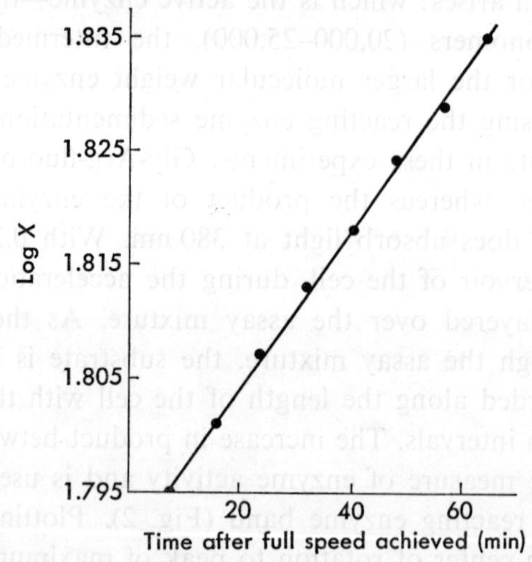

Fig. 3. Reacting enzyme sedimentation of cathepsin C. Plot of log χ *vs.* time. $S_{obs}=6.9$ S $S_{corr}=7.87$ S (corr. for density and viscosity of solvent).

do we have a slowly sedimenting enzyme. Only one active form of the enzyme sediments and that is the higher molecular weight. The enzyme dissociates reversibly into inactive subunits, and the presence of the substrate favors association of the subunits into the enzymatically active oligomer (*Mr* ~140,000) (*20*). This suggests a possible method of regulation of the activity of this enzyme *in vivo* (*20*). Of passing interest is the fact that cathepsin C from rabbit lung (*Mr* ~140,000) appears to be smaller than the same enzyme (*Mr* ~200,000) obtained from beef spleen (*17*).

All cysteinyl proteinases require the presence of a thiol compound for maximum activity. However, various enzymes react differently to different reducing compounds. For example, at 0.11 mM cysteine, cathepsin I is activated ~95% of its maximum (obtained with 2.0 mM cysteine), whereas cathepsin B is only activated to ~50% of its maximum. Again, with 0.125 mM β-mercaptoethanol, cathepsin I is activated to ~95% of its maximum, whereas cathepsin B is activated to only ~30%.

Captopril, a synthetic -SH compound, (D-3-mercapto-2-methylpropanoyl-L-proline) inhibits angiotensin-converting enzyme (*7*) and is being used clinically in an attempt to control hypertension (*27*). However, we find that low concentrations of captopril (0.06–0.11 mM) have markedly different abilities to activate different lysosomal cysteinyl proteases (*6*). This ability ranges from no effect with cathepsin B2, to a slight effect with cathepsin B, to a marked and complete activation of cathepsins I and L (Table II). With each enzyme, 100% activation is achieved with 2.0 mM cysteine. Higher concentrations of captopril have no further effects on cathepsins B or B2. However, increasing the concentration of captopril *decreased* the activities of cathepsins I and L to ~35% of their maximal activities. Thus, a synthetic compound like captopril, depending on its concentration, has the capacity for both increasing and decreasing the activity of cathepsins I and L (*6*).

We next tried to determine if similar effects could be obtained with two naturally-occurring thiol-containing peptides, glutathione (GSH) and cysteinyl-glycine (Cys-Gly). GSH, up to 2.0 mM, was not a very effective reducing agent for either cathepsin B or I (*6, 22*). Cys-Gly, whose structure is in some respects similar to that of captopril, did exhibit some similar properties.

The Cys-Gly was synthesized (*18*) by W. Phares, Lawrence Livermore Laboratories, Livermore, California. The dipeptide migrated as a

TABLE II. Effects of Captopril and Cys-Gly on Several Lung Lysosomal Cysteinyl Proteases

Enzyme	Captopril		Cys-Gly	
Cathepsin	Concentration (mM)	Activation (%)	Concentration (mM)	Activation (%)
B	0.11	25	0.11	60
	1.25	25	0.25	65
			1.25	60
I	0.11	85	0.06	100
	1.25	35	1.25	35
L	0.06	145		
	2.0	65		
	4.0	32		
B2	0.1	0		
	1.25	0		

The enzymes were pre-incubated with the indicated concentrations of captopril or Cys-Gly before the substrates were added and the enzymes assayed for activity. Cathepsin B, 1.5 μg, Z-Ala-Arg-Arg-4MeOBNA (pH 6.0); cathepsin I, 1.5 μg, Bz-Arg-βNA (pH 6.0) or Leu-βNA (pH 7.0); cathepsin L, 0.1 μg, azocasein (pH 5.0); cathepsin B2, ~5 μg, Bz-Arg-NH_2 (pH 5.7).

single spot on paper chromatography in butanol-acetic acid-water (12 : 13 : 5) with an R_f of 0.4. Amino acid analysis (after oxidation of the cysteine to cysteic acid with performic acid) gave a ratio of cysteic acid: Gly of 1.08 : 1.00.

With 0.25 mM Cys-Gly, cathepsin B was activated to ~65% of its maximum. This activity did not decrease significantly with higher concentrations of the dipeptide. However, 0.06 mM concentration of Cys-Gly completely activated cathepsin I with Bz-Arg-βNA as substrate. (This activation was equivalent to that achieved with 1.0 mM dithioerythritol or 2.0 mM cysteine). With increasing concentrations of this dipeptide, the activity decreased to ~35%. Similar results were obtained with cathepsin I and Leu-βNA as substrate. Thus, this naturally-occurring dipeptide exhibits the potential for both strongly increasing and decreasing the activity of cathepsin I *in vivo* (*6*).

We do not know if these effects occur *in vivo*. GSH, or γ-Glu-Cys-Gly, is present in all types of cells which have been studied thus far (*4*) and occurs in concentrations up to 10 mM (*28*). In the γ-glutamyl cycle, GSH is converted to the dipeptide Cys-Gly (*28*). Therefore, at the concentrations of Cys-Gly we have employed to activate (0.01–0.06 mM) and inhibit (1.25 mM) cathepsin I, a regulatory role of this dipeptide is theoretically possible with this enzyme.

REFERENCES

1. Barrett, A. M. *In* "Proteinases in Mammalian Cells and Tissues," ed. A. J. Barrett, p. 181 (1977). North-Holland Publ. Co., Amsterdam.
2. Barrett, A. J. and McDonald, J. D. eds. "Mammalian Proteinases: A Glossary and Bibliography." Vol. 1, Endopeptidases, p. 276 (1982). Academic Press, New York.
3. Bond, J. and Barrett, A. J. *Biochem. J.*, **189**, 17 (1980).
4. Boyland, E. and Chasseaud, L. F. *Adv. Enzymol.*, **32**, 173 (1969).
5. Chatterjee, R. and Kalnitsky, G. *Fed. Proc.*, **39**, 1685 (1980).
6. Chatterjee, R. and Kalnitsky, G. *Fed. Proc.*, **41**, 1143 (1982).
7. Cushman, D. W., Cheung, H. S., and Ondetti, M. A. *Biochemistry*, **16**, 5484 (1977).
8. Etherington, D. J. *Biochem. J.*, **137**, 547 (1974).
9. Fruton, J. S., Hearn, W. R., Ingraham, V. M., Wiggans, D. S., and Winitz, M. *J. Biol. Chem.*, **204**, 891 (1953).
10. Jones, M. E., Hearn, W. R., Fried, M., and Fruton, J.S. *J. Biol. Chem.*, **195**, 645 (1952).
11. Keilova, H. *In* "Tissue Proteinases," eds. A. J. Barrett and J. T. Dingle, p. 45 (1971). North-Holland Publ. Co., Amsterdam.
12. Kirschke, H., Langner, J., Wiederanders, B., Ansorge, S., Bohley, P., and Broghammer, U. *Acta Biol. Med. Ger.*, **35**, 285 (1976).
13. Kirschke, H., Langer, J., Wiederanders, B., Ansorge, S., Bohley, P., and Hanson, H. *Acta Biol. Med. Ger.*, **36**, 185 (1977).
14. Lones, M., Chatterjee, R., Singh, H., and Kalnitsky, G. *Arch. Biochem. Biophys.*, **221**, 64 (1983).
15. McDonald, J. K. and Schwabe, C. *In* "Proteinases in Mammalian Cells and Tissues," ed. A. J. Barrett, p. 314 (1977). North-Holland Publ. Co., Amsterdam.
16. Meister, A. *Science*, **180**, 33 (1973).
17. Metrione, R. M., Okuda, Y., and Fairclough, G. F. *Biochemistry*, **9**, 2427 (1970).
18. Olson, C. K. and Binkley, F. *J. Biol. Chem.*, **186**, 731 (1950).
19. Otto, K. *In* "Tissue Proteinases," eds. A. J. Barrett and J. T. Dingle, p. 1 (1971). North-Holland Publ. Co., Amsterdam.
20. Paszkowski, A., Singh, H., and Kalnitsky, G. Abstr. of XI Int. Congress of Biochemistry, No. 03-7-S14, p. 231 (1979). Toronto.
21. Schwartz, W. N. and Barrett, A. J. *Biochem. J.*, **191**, 487 (1980).
22. Singh, H. and Kalnitsky, G. *J. Biol. Chem.*, **253**, 4319 (1978).
23. Singh, H. and Kalnitsky, G. *J. Biol. Chem.*, **255**, 369 (1980).
24. Singh, H., Kuo, T. and Kalnitsky, G. *In* "Protein Turnover and Lysosomal Function," eds. H. Segal and D. Doyle, p. 315 (1978). Academic Press, New York.
25. Taylor, B. L., Barden, R. E., and Utter, M. F. *J. Biol. Chem.*, **247**, 7383 (1972).
26. Taylor, B. L., Frey, W. H., Barden, R. E., Scrutton, M. C., and Utter, M. F. *J. Biol. Chem.*, **253**, 3062 (1978).
27. Vidt, D. G., Bravo, E. L., and Fouad, F. M. *N. Engl. J. Med.*, **306**, 214 (1982).
28. Vina, J., Hems, R., and Krebs, H. A. *Biochem. J.*, **170**, 627 (1978).

Proteinase Inhibitors: Medical and Biological Aspects, eds. N. Katunuma et al., pp. 275–281 (1983).
Japan Sci. Soc. Press, Tokyo/Springer-Verlag, Berlin

Role of Medullasin in the Defense Mechanism against Cancer Development

Yosuke Aoki

*Department of Biochemistry, Jichi Medical School**

Medullasin is the name of a protease found in human bone marrow cells (from the Latin, *medulla ossium*) (*1*, *4*, *5*). This protease is considered to play an important role in the defense mechanism of human beings. In this paper one aspect of the biological significances of medullasin, *i.e.*, its role in preventing cancer development, will be presented. Mammals resist cancer development in their body with several factors, one being certain blood cells. Macrophages exert their antineoplasmic effect by both directly damaging cancer cells and by activating lymphocytes to attack them. B lymphocytes produce antibodies against cancers, killer T-cells sensitized with cancer cells attack them, and K-cells injure cancer cells antibody-dependently. In addition to these lymphocytes and macrophages, the natural killer (NK)-cell which kills cancer cells without prior sensitization has attracted the attention of many oncologists and immunologists (*8*, *12*). Nude mice which lack T-cell but have a high level of NK-cell activity develop cancers less frequently than other kinds of mice (*11*), and in beige mice which lack NK-cells but have a normal number of T-cells transplanted tumor cells grow faster than in other kinds of mice (*13*). For these the NK-cell is

* Minamikawachi, Kawachi, Tochigi 329-04, Japan.

considered to play an important role in the defense mechanism against cancer development. The NK-cell activity was stimulated by interferon (*15*) and its inducers (*14*), and also by interleukin II (*7*). Although proteases such as trypsin, papain, and pronase reduced NK-cell activity (*9*, *10*), medullasin markedly enhanced it (*4*). Therefore, the effect of medullasin on human NK-cell activity was studied and is reported in this paper.

PROPERTIES OF MEDULLASIN

Medullasin was purified from bone marrow cells by the method described (*1*, *5*). Its molecular weight is 31,800, and optimum pH 8.5. Histidine, serine, and one carboxyl group were proved to be essential to the protease activity. Medullasin is similar to elastase in that it hydrolyzes ester substrates for elastase and is inhibited by elastatinal, an inhibitor of elastase. However, it is not identical to elastase, because it possesses no elastolytic activity (as shown in Fig. 1), and no ability to hydrolyze amide substrates for elastase. No isozymes of medullasin were demonstrated in crude extracts of human bone marrow cells in contrast

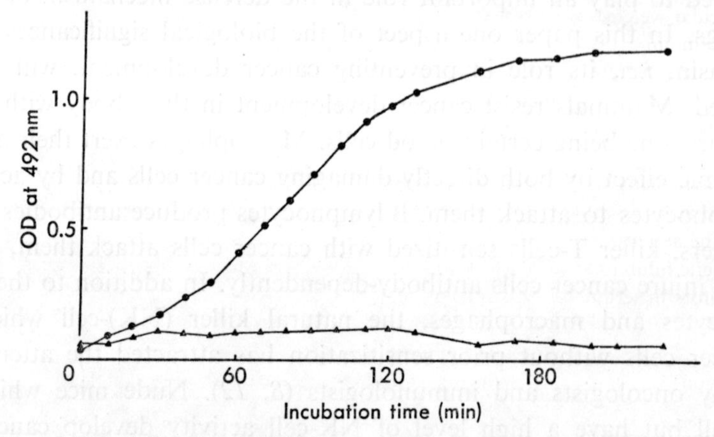

Fig. 1. Elastolytic activity of medullasin.
Tubes containing Congo red-elastin (1 mg/ml of 0.02 M sodium borate buffer, pH 8.8) and medullasin (50 μg) or porcine pancreas elastase (30 μg, Boehringer) in a final volume of 3.25 ml were incubated at 25°C. After various incubation periods tubes were centrifuged and optical density in the supernatant at 492 nm was measured to determine elastolytic activity. As shown in this figure, medullasin showed no elastolytic activity in contrast to pancreas elastase. ▲ medullasin; ● pancreas elastase.

with granulocyte elastase (*6*). Medullasin activity was inhibited by plasma protease inhibitors such as α_1-antitrypsin, α_2-macroglobulin, and α_2-plasmin inhibitor. Both granulocytes and erythroblasts contain medullasin, but it was not detected in lymphocytes, monocytes, thrombocytes or mature erythrocytes. This protease was shown to be located in the inner membrane of mitochondria in erythroblasts, however, in granulocytes it seems to be mainly located in lysosome granules since it is easily released from granulocytes by various stimuli.

PHYSIOLOGICAL ROLES OF MEDULLASIN IN HUMAN BEINGS

Medullasin activity in bone marrow cells is higher in human beings than in other animals (*1*). The activity was elevated in mature granulocytes obtained from patients with Behçet's disease, uveitis of unknown etiology, central retinal vein occlusion of young adults, and rheumatoid arthritis in the active phase, and decreased to a normal level when

TABLE I. Properties of Medullasin

Enzymological properties		Biological properties
Molecular weight	31,800	1. Induces inflammation by causing degeneration of endothelial cells in vessels and retaining macrophages in the lesion
pH optimum	8.5	2. Activates lymphocyte functions
Synthetic substrates	Ester substrates for elastase	3. Potentiates the effect of mitogens on lymphocyte functions
Enzyme substrates	Apo-pyridoxal enzymes	4. Stimulates NK-cell activity
Elastolytic activity	No	5. Potentiates superoxide production by macrophages
Amino acids essential to the activity	Serine, histidine one carboxyl group	6. Stimulates granulocyte chemotaxis
Synthetic inhibitors	DFP, PMSF	7. Activity in granulocytes increases in chronic inflammatory diseases
Inhibitors in serum	α_1-Antitrypsin, α_2-macroglobulin, α_2-plasmin inhibitor	8. Regulates δ-aminolevulinic acid synthetase levels in erythroblasts
Location	Erythroblasts and granulocytes	9. Plays a significant role in the development of anemia of chronic disorders
Location in erythroblasts	Inside of inner mitochondrial membrane	
Differences in activity among various animals	Marked (highest in human beings)	

* DFP, diisopropylfluorophosphate; PMSF, phenylmethylsulfonylfluoride.

patients became in remission. An injection of a small amount of medullasin (20 μg) into guinea pig skin caused an inflammation characterized by the infiltration of a large number of macrophages (*2*). Chemotactic activity of macrophages was inhibited by treatment with a small amount of medullasin (*2*). These results suggest that medullasin in mature granulocytes plays an important role in the development and maintenance of inflammation, especially in humans. Besides its phlogistic activity, medullasin modified lymphocyte functions. Both DNA and RNA synthesis increased by treatment with medullasin. Furthermore, this protease potentiated the effect of mitogens on DNA synthesis of human lymphocytes (*3*). Both the enzymological and biological properties of medullasin are summarized in Table I.

ROLE OF MEDULLASIN IN THE DEFENCE MECHANISM AGAINST CANCER DEVELOPMENT

Besides chronic inflammatory diseases, medullasin activity in mature granulocytes was found to be elevated in patients with cancer such as uterine cancer, ovarian cancer, malignant lymphoma, and pancreatic cancer. The protease activity in mature granulocytes also increased during the carcinogenetic process of rats fed on acetylaminofluorene or 3′-methyl-dimethylaminoazobenzene. In contrast to cancer patients, the activity of rats increased prior to the development of cancer and decreased below the normal level when cancers appeared. These results suggest that medullasin in granulocytes plays certain roles in the defense mechanism against cancer development, and therefore, the effect of medullasin on human NK-cell activity was examined. As shown in Fig. 2, treatment of human lymphocytes with a small amount of medullasin (20 μg/ml, 37°C, 60 min) resulted in a marked increase in NK-cell activity. This increase was observed whether K-562 cells or Molt-4 cells were employed as target cells. The increment of NK-cell activity was small when the cytotoxicity was measured immediately after the medullasin treatment, but gradually increased when lymphocytes treated with medullasin were incubated at 37°C in a 5% CO_2 atmosphere, reaching a peak at 24 hr after the treatment. The optimum amount of medullasin employed for the treatment of lymphocytes was 20 μg/ml when lymphocytes were treated at 37°C for 60 min. This amount is

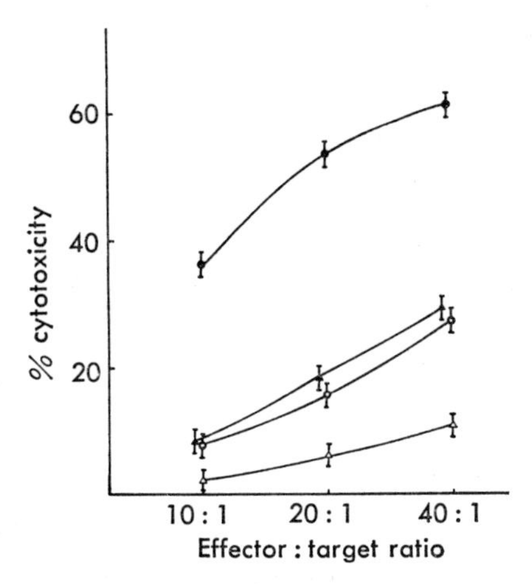

Fig. 2. Enhancement of NK-cell activity by medullasin (*4*).
Human lymphocytes (2×10^6/ml) suspended in RPMI 1640 were treated with medullasin (20 μg/ml) at 37°C for 60 min in a humidified 5% CO_2 incubator. After the addition of 10% fetal calf serum lymphocyte suspensions were further incubated at 37°C for 24 hr before cytotoxicity assay. Each NK-cell activity was assayed using the following cells: ○ lymphocytes treated in the absence of medullasin (target cells: K-562); ● lymphocytes treated with medullasin (target cells: K-562); △ lymphocytes treated in the absence of medullasin (target cells: Molt-4); ▲ lymphocytes treated with medullasin (target cells: Molt-4). Bars, standard deviation of triplicate determinations.

considered to be physiological since human peripheral blood contains 5–15 μg of medullasin per ml. The effect of medullasin on the enhancement of NK-cell activity depends upon its proteolytic activity, since the treatment of lymphocytes with medullasin in the presence of phenylmethylsulfonyl fluoride (0.1 mM) or elastatinal (20 μg/ml) abolished the enhancement of the activity. Human lymphocytes were separated into those bearing Fc receptor for immunoglobulin G (IgG) and those without Fc receptor by a method employing ox erythrocytes bound with rabbit IgG to see whether or not the NK-cell stimulated by medullasin treatment is of a classical type. NK-cell activity of the lymphocyte fraction bearing the Fc receptor was stimulated by medullasin, but that without the receptor was not. Therefore, NK-cells stimulated by medullasin treatment are considered to be those of a classical type.

Interferon is known to increase NK-cell activity both *in vitro* and *in vivo*. Also, various agents to stimulate NK-cell activity were reported to do so through the induction of interferon production as described

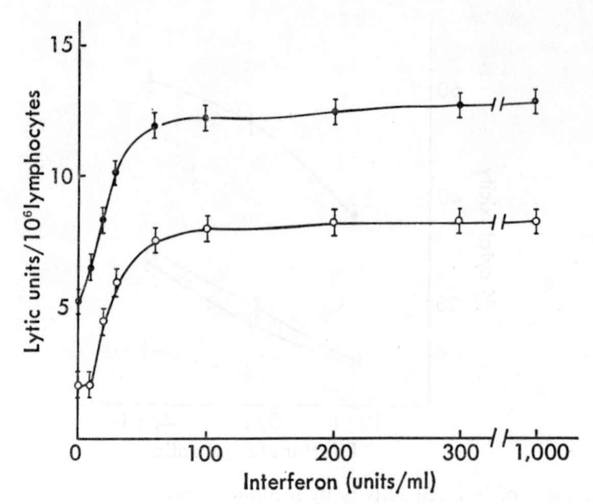

Fig. 3. Effect of medullasin on the NK-cell activity stimulated with interferon (*4*). Tubes containing human lymphocytes (2×10^6/ml) in RPMI 1640 were treated with 20 μg/ml medullasin at 37°C for 60 min, followed by the addition of 10% fetal calf serum and various amounts of interferon-α. Control tubes were those treated in the absence of medullasin. Tubes were then incubated at 37°C for 24 hr in a humidified 5% CO_2 atmosphere to be assayed for natural cytotoxicity. Each tube contained the following lymphocytes: ○ lymphocytes stimulated with interferon; ● lymphocytes treated with medullasin and further stimulated with interferon. Bars, standard deviation of triplicate determinations.

in the introduction. Therefore, the effect of medullasin on NK-cell activity stimulated by interferon was examined. As shown in Fig. 3 medullasin further enhanced the activity stimulated by interferon. Furthermore, no interferon activity was detected in the incubation medium containing lymphocytes stimulated by medullasin. From these results the stimulatory effect of this protease on NK-cell activity is considered not to be mediated through interferon production.

In conclusion, medullasin in granulocytes is thought to play an important role in the defense mechanism against cancer development by stimulating NK-cell activity, although the possibility remains that it plays also certain roles in the defense mechanism by altering the functions of other kinds of lymphocytes and macrophages. The effect of medullasin on NK-cell activity was proved not to be mediated through interferon production.

REFERENCES

1. Aoki, Y. *J. Biol. Chem.*, **253**, 2026 (1978).
2. Aoki, Y. and Machinami, R. *Arth. Rheumat.*, in press.
3. Aoki, Y. and Machinami, R. *Asian Med. J.*, **25**, 873 (1982); **26**, 15 (1983); **26**, 73 (1983).
4. Aoki, Y., Sumiya, M., and Oshimi, K. *J. Clin. Invest.*, **69**, 1223 (1982).
5. Aoki, Y., Urata, G., Takaku, F., and Katunuma, N. *Biochem. Biophys. Res. Commun.*, **65**, 567 (1975).
6. Feinstein, G. and Janoff, A. *Biochim. Biophys. Acta*, **403**, 493 (1975).
7. Henney, C. S., Kuribayashi, K., Kern, D. E., and Gillis, S. *Nature*, **291**, 335 (1981).
8. Herberman, R. B., Nunn, M. E., Larvin, D. H., and Asofsky, R. *J. Natl. Cancer Inst.*, **51**, 1509 (1973).
9. Koide, Y. and Takasugi, M. *J. Natl. Cancer Inst.*, **59**, 1099 (1977).
10. Peter, H. H., Pavie-Fisher, J., Fridman, W. H., Aubert, C., Cesarini, J. P., Roubin, R., and Kourilsky, F. M. *J. Immunol.*, **115**, 539 (1975).
11. Rygaad, J. and Povlson, C. O. *Transp. Rev.*, **28**, 43 (1976).
12. Takasugi, M., Mickey, M. R., and Terasaki, P. I. *Cancer Res.*, **33**, 2898 (1973).
13. Talmadge, J. E., Meyers, K. M., Prieur, D. J., and Starkey, J. R. *Nature*, **284**, 622 (1980).
14. Trinchieri, G., Santoli, D., Dee, R. R., and Knowles, B. B. *J. Exp. Med.*, **147**, 1229 (1978).
15. Zarling, J. M., Eskra, L., Borden, E. C., Horoszewicks, J., and Carter, W. A. *J. Immunol.*, **123**, 63 (1979).

Proteinase Inhibitors: Medical and Biological Aspects, eds. N. Katunuma et al., pp. 283–292 (1983).
Japan Sci. Soc. Press, Tokyo/Springer-Verlag, Berlin

Intracellular Protein Degradation: Studies on Transplanted Mitochondrial and Microinjected Cytosol Proteins

Peter J. EVANS, Susan M. RUSSELL, Fergus DOHERTY, and R. John MAYER

*Department of Biochemistry, University of Nottingham Medical School, Queen's Medical Centre**

BACKGROUND

The mechanism(s) and regulation of intracellular proteinolysis are not understood. Most "classical" studies have involved the use of radiolabelled amino acids or amino acid precursors which naturally label all intracellular proteins. Subsequently the destructive fate of specific proteins or completely or partially resolved protein mixtures have been measured (*9*, *10*). This methodology is thwart with difficulties of precursor choice, experimental design and isotopic reutilization which render some of the data of little or no value for the elucidation of mechanisms of intracellular proteinolysis.

When great care is taken with the choice of radiolabelled protein precursors, experimental design, subcellular fractionation procedures, choice of protein analytical method and statistical analysis of the degradation rates obtained, then accurate evaluation of intracellular protein degradation rates can be carried out (*19*, *23*). These analyses suggest that populations (groups, families or sets) of degradation rates may be

* Nottingham NG7 2UH, U.K.

associated with proteins in defined cytomorphological locations in the cell (*19*). The information for protein destructive rates in each cytomorphological site must be very precise, residing both in the cytomorphologically defined specific proteins, much as topogenic sequences in organelle biogenesis or protein secretion, and in currently unknown cellular site-specific recognition markers akin to surface or intracellular ligand-specific receptors.

The resolution of these problems can be achieved by introducing specific proteins or defined protein mixtures into target cells. The introduction of "hot" (radiolabelled) proteins into "cold" cells offers for the first time the chance to study defined proteins unhampered by radiolabelling all "background" proteins so that the mechanism(s) of intracellular proteinolysis and their regulation can be unravelled. Introduction of proteins into target cells is a relatively "young art" and therefore a brief description of the aims and objectives of such studies will be given, followed by a detailed consideration of the fate of such proteins in hepatocytes, hepatomas and 3T3-L1 cells.

INTRODUCING MACROMOLECULES INTO CELLS

The wide range of molecular and cell biological phenomena studied after microinjection of biological macromolecules into target cells is shown in Table I. The techniques used require either needle microinjection or red cell mediated microinjection whereby red cells or red cell ghosts loaded with the proteins of interest are fused by means of polyethylene glycol (PEG) or Sendai virus to the target cells. Clearly examples of the whole range of cellular activities from gene expression, mRNA translation, assembly and distribution of organelle specific macromolecules and toxin action, to the destruction of proteins is covered. Within the limitations of the new technology is offered a new dimension of cell biological research whereby specific macromolecular events can be observed in the presence of the numerous other continuous cellular activities.

INTRODUCING HETEROLOGOUS SOLUBLE PROTEINS INTO CELLS

Proteins have been introduced into target cells by fusion with eryth-

TABLE I. Purposes of Microinjection

Purpose	Method	Reference
1. Introduction of mRNA into target cells, *e.g.*, globin mRNA	Needle microinjection	*3*
2. Introduction of gene(s) into target cells, *e.g.*, "cloned" globin genes	Needle microinjection	*22*
3. Studies on ribosomal assembly, *e.g.*, ribosomal protein S6	Needle microinjection	*13*
4. Studies on fate of small nuclear RNA molecules, *e.g.*, U_1-U/, tRNA 5S RNA	Needle microinjection	*6*
5. Studies on cytomorphological location of proteins, *e.g.*, actin	Needle microinjection	*14*
6. Studies on cytotoxins in target cells, *e.g.*, diphtheria toxin	Red cell mediated microinjection	*27*
7. Studies on growth factors in target cells, *e.g.*, macrophage growth factors	Red cell mediated microinjection	*25*
8. Studies on diffusional rate in cytoplasm, *e.g.*, fluorescent heterologous protein	Red cell mediated microinjection	*26*
9. Studies on protein degradation, *e.g.*, of *heterologous* and *homologous* proteins	Red cell mediated microinjection	*2, 7, 11, 16, 28*

rocytes (*8, 15, 20*) or erythrocyte ghosts (*12*). These studies have all involved the use of soluble proteins, mainly heterologous proteins, *e.g.*, bovine serum albumin, which were radiolabelled before use by radioiodination. Proteins microinjected heterologously into HeLa cells are degraded by an apparent first-order process with half-lives ranging from 3 hr (non-histone chromosomal protein) to 75 hr (lactate dehydrogenase) (*28*). The injected proteins are distributed throughout the cytoplasm. No proteolytic fragments derived from microinjected proteins could be observed and some half-lives of proteins appeared to correlate with molecular weight. This conclusion was also drawn for rat liver ^{125}I-cytosol proteins heterologously microinjected into fibroblasts (*16*). These authors also observed that the catabolism of microinjected heterologous rat liver cytosol proteins is increased in media lacking serum, insulin, fibroblast growth factor, and dexamethasone. The authors also made a cautionary observation that the rate of degradation of microinjected ^{125}I-bovine serum albumin varied with different preparations of the modified protein. Fluorescently labelled proteins (fluorescein conjugated) may be degraded more rapidly than their iodinated (^{125}I) counterparts (*28*). Previously it had been observed (*21*) that fluorescein-conjugated protein assumes a so-called peri-nuclear distribution similar to intracellular organelles (*e.g.*, lysosomes (*21*), Golgi apparatus (*17*), and mitochondria (*5*)) during its destruction.

A key question in intracellular proteinolysis is the quantitative role of the intracellular vesicular destructive apparatus (including the structures identified as lysosomes) in the destruction of proteins from defined cytomorphological sites. Bigelow *et al.* (*2*) have exploited the fact that sucrose is indigestible in this vacuolar system to study the distribution of sucrose-conjugated protein "fragments" and therefore estimate the autophagic contribution to specific protein degradation rate. Microinjected [^{14}C]-sucrose-bovine serum albumin is degraded in L929 fibroblasts at a similar rate to ^{125}I-bovine serum albumin injected into other cell lines ($t_{1/2}$, 20 hr). [^{14}C]-sucrose-conjugated peptides derived from [^{14}C]-sucrose-bovine serum albumin or [^{14}C]-sucrose-bovine serum-albumin further conjugated with poly-L-lysine *both* entering the cells by endocytosis are *both* present exclusively within the intracellular vacuolar lysosomal aparatus. However, one half-life after microinjection of [^{14}C]-sucrose conjugated bovine serum albumin or [^{14}C]-sucrose-conjugated pyruvate kinase, 62% and 48% respectively of the [^{14}C]-sucrose conjugated degradation products are found in the post lysosomal supernatant fraction. Since controls showed little leakage of [^{14}C]-sucrose conjugated degradation products from the lysosomal fraction during subcellular fractionation, the authors concluded that at least half of the observed degradation of proteins injected into the cytosol of the target cells was extra-lysosomal (*cf. 1*). Hendil (*11*), using microinjected [^{3}H]-dextran as indigestible material further concluded that in fibroblasts there is a slow accumulation of [^{3}H]-dextran in the lysosomal vacuolar apparatus and that the minimally estimated rate of uptake (8.7%/24 hr) can account for only a minor fraction of the degradation of microinjected soluble proteins in growing cells.

INTRODUCING MEMBRANE PROTEINS INTO CELLS

1. *Homologous Membrane Protein Transplantation in Hepatocytes*

Transplantation of homologous membrane proteins has been achieved by organelle or organelle-vesicle cell fusion mediated by polyethylene glycol. In this way deliberate miscompartmentalization of mitochondrial or outer mitochondrial membrane proteins has been carried out in order to study their subsequent intracellular distribution and destruction. This approach allows evaluation of cytomorphology in determining proteolytic rates. Liver mitochondrial preparations or outer mitochondrial membrane vesicle preparations are first labelled by reductive methylation [^{3}H] before fusion by PEG with target hepatocytes in suspension (*7*). Reductive methylation derivatizes some lysine residues in the membrane proteins to yield methylated [^{3}H]-lysine residues. This amino acid derivative is an end product of degradation and accumulates in the culture medium.

Conditions were initially established where hepatocyte monolayers could be cultured for up to 140 hr with the cells showing *in vivo* catabolic rates so that *in vivo* rates of transplanted mitochondrial destruction might be observed if such rates indeed occur after fusion (*7*).

The results in Fig. 1A show that after mitochondrial-hepatocyte fusion, the transplanted material (putative mitoplast?) is initially not destroyed in a quasi first order manner, but subsequently such degradation does take place. The average half-life of the transplanted proteins is 69.2±5 hr (*19*) which is of the same order as that measured for mitochondrial proteins in rat liver *in vivo* (*18*, *19*). Such an obser-

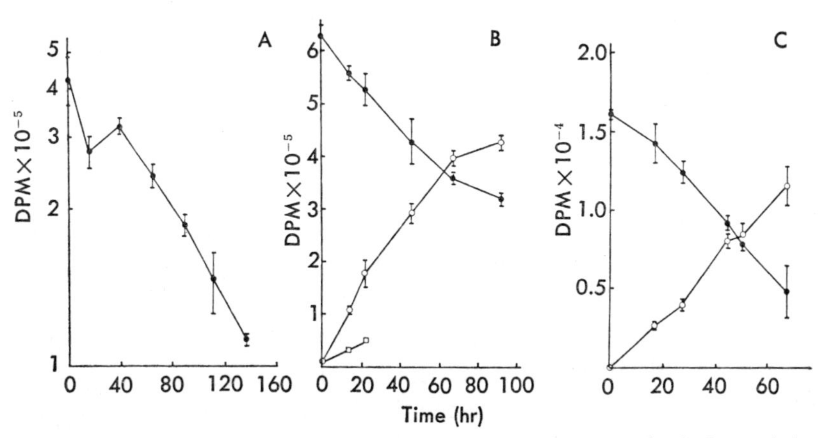

Fig. 1. Degradation in hepatocyte monolayers of transplanted reductively methylated [^{3}H] mitochondrial proteins (A), transplanted reductively methylated [^{3}H] outer mitochondrial membrane vesicular proteins (B), [^{3}H]-pargyline-labelled monoamine oxidase in outer mitochondrial membrane vesicles (C).

● TCA insoluble radioactivity; ○ TCA soluble radioactivity; □ chloroquine inhibition. $t_{1/2}$=A: 69.2±5 hr; B: 70.5±9 hr; C: 55 hr.

vation could be coincidental, therefore reductively methylated [^{3}H] outer mitochondrial membrane vesicles were fused with hepatocytes and degradation studied. The average half-life of transplanted outer mitochondrial membrane proteins (70.5±9 hr) (Fig. 1B) is remarkably similar to that measured (*18*, *19*) in liver *in vivo* ($t_{1/2}$, 60–70 hr). Furthermore, the destruction of outer mitochondrial membrane proteins is partially (57%) inhibited by chloroquine which demonstrates that the material is intracellular and is present in a chloroquine sensitive lysosomal compartment. Interestingly, the concept of two populations of degradation rates in the outer mitochondrial membrane in liver *in vivo* (*19*) reconciles conveniently with incomplete inhibition by a lysosomotropic agent. Two populations of degradation rates are further supported by a 40–50% inhibition of the degradation of transplanted reductively methylated [^{3}H] outer mitochondrial membrane proteins by 0.5 mM-leupeptin (results not shown) when assessed over a 40 hr culture period. Finally, although the average half-lives of mitoplast (?) and outer mitochondrial membrane vesicular proteins may be coincidentally very similar *in vivo* and when transplanted into hepatocyte monolayers, it is not likely that coincidence would explain the fact that the rate of destruction of a single enzyme would be the same *in vivo* and *in vitro*. The results in Fig. 1C show that the outer mitochondrial membrane enzyme monoamine oxidase, derivatized by the suicide-inhibitor [^{3}H]-pargyline is destroyed in hepatocyte monolayers with a half-life (55 hr) which is *exactly* the same as that determined in liver *in vivo* (*18*). This data on the degradation of transplanted mitochondrial proteins shows for the first time that some feature(s) of the proteins in sub-organelle structures or some unknown factors present in the membranes themselves can survive the insults of subcellular fractionation, reductive methylation, and PEG mediated fusion to ensure that the transplanted proteins are destroyed at *in vivo* like rates in hepatocyte monolayers.

Mitochondrial or outer mitochondrial membrane vesicular preparations, derivatized with fluorescein isothiocyanate, can be conveniently observed after fusion with hepatocytes by fluorescence microscopy (results not shown). Fluorescent material enters the cells and undergoes "translocation/redistribution" to appear in vesicular structures having an apparent peri-nuclear distribution similar to lysosomes (*21*), endogenous mitochondria (*5*) or Golgi apparatus (*17*) in tissue culture cells.

It is very apparent that the time needed for such "redistribution" corresponds to time when degradation of transplanted mitochondrial proteins (Fig. 1A), outer mitochondrial membrane proteins (results not shown) or monoamine oxidase (Fig. 1C) does not follow quasi linear destruction rates. It is therefore tempting to propose that the transplanted mitochondrial membrane material must assume an intracellular distribution within the cytoplasm analogous to endogenous intracellular organelles before *in vivo* like protein destruction rates can occur. This proposal again emphasizes that information in the transplanted structures is sufficient to ensure *in vivo* like degradation rates in contrast to proteins entering cells by pinocytosis (*24*) or receptor mediated endocytosis (*17*) which are destroyed extremely quickly ($t_{1/2}$'s, approx. 30–60 min). Whatever the mechanism of internalization of material transplanted into the hepatocyte plasma membrane (*i.e.*, endocytic mechanism) the intracellular fate of the transplanted proteins is not like the fate of material internalized by receptor mediated endocytosis or pinocytosis (*17*, *24*). Material transplanted into the hepatocyte plasma membrane may be internalized by some continuous surface-surveillance mechanism or by a surveillance mechanism invoked by the fusion insult. Such alternative mechanisms have been recruited to explain the lymphocytic capping phenomenon (*4*). Surveillance mechanisms may normally preclude protein miscompartmentation *e.g.*, mistaken dispatch of proteins during organelle biogeneis. Proteinase inhibitor data suggests a partial lysosomal role in destroying transplanted mitochondrial proteins in hepatocyte monolayers at *in vivo* rates. The fluorescence microscopy also points to the vesicular nature of the internalized material. Together these observations may mean that the rate limiting step in mitochondrial protein degradation may be vesicle-lysosome fusion or vesicle-lysosome protein exchange.

2. Heterologous Membrane Protein Transplantation into Morris 7288c Hepatoma Cells

Heterologous membrane protein transplantation is achieved by fusing reductively methylated [^{3}H] outer membrane vesicles with hepatoma cells. The results indicate that heterologous transplantation results in destruction of liver outer mitochondrial membrane proteins at a much faster average rate, $t_{1/2}$, 24.9±3.6 hr (*27*) than homologously trans-

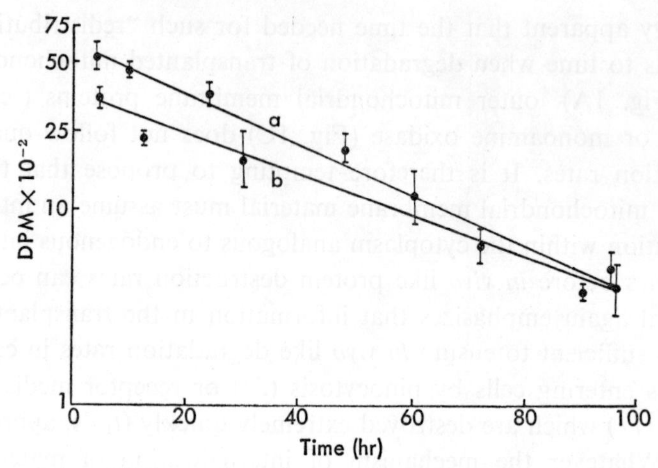

Fig. 2. Degradation of [³H]-pargyline-labelled monoamine oxidase (a) and [³H]-pargyline-fluorescein labelled outer mitochondrial membrane transplanted into Morris hepatoma 7288c cells (b).

planted material (Fig. 2). Furthermore, [³H]-pargyline derivatized monoamine oxidase is destroyed at a similar rate, $t_{1/2}$, 23.6±5.4 hr (*23*) which is not significantly altered by conjugating outer mitochondrial membrane vesicles with fluorescein (Fig. 2). Observation of such derivatized vesicles by fluorescence microscopy (results not shown) reveals a very rapid (within 10 min) patching and subsequent capping of the transplanted material resulting in the internalization of the material into intracellular vesicular structures. This internalization and vesicularization process contrasts completely with the hepatocyte mechanism. Although the fate of transplanted outer mitochondrial membranes and membrane proteins is very different in hepatomas and hepatocytes, again a surface surveillance mechanism must operate to remove miscompartmented proteins.

INTRODUCING HOMOLOGOUS CYTOSOL PROTEINS INTO CELLS

Previously red cell mediated microinjection has been used for the transfer of heterologous radioiodinated proteins or protein mixtures into target cells. Reductively methylated [³H] 3T3-L1 fibroblast cytosol proteins are destroyed after homologous microinjection into 3T3-L1 cells as shown in Fig. 3. Interestingly, microinjected cytosol proteins are

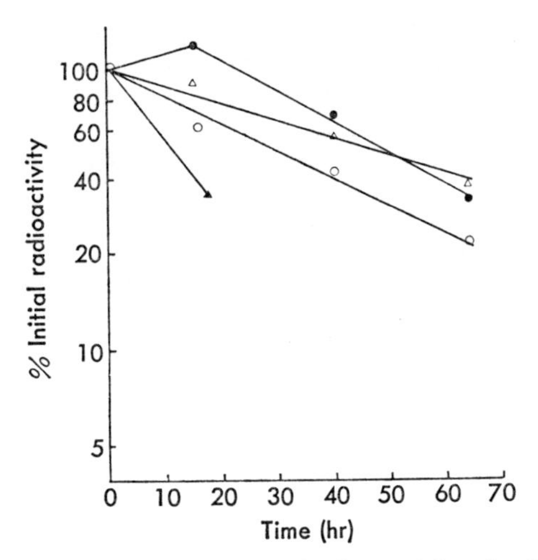

Fig. 3. Degradation of endogenous (△) proteins in non-injected cells, microinjected (○) and endogenous (●) proteins in injected cells, and endocytosed homologous cytosol (▲) proteins by 3T3-L1 preadipocytes.

degraded at similar rates to endogenous [^{14}C]-leucine labelled cytosol proteins in the same cells and to endogenous cytosol proteins in non-injected cells ($t_{1/2}$, 40–50 hr). However, endocytosed poly-L-lysine conjugated [^{3}H]-cytosol proteins are destroyed considerably faster ($t_{1/2}$, approx. 20 hr). Reductive methylation, although extensively modifying the proteins (up to 40% of cytosol protein lysine residues), does not considerably alter the average *in vitro* susceptibility of derivatized proteins to proteolytic enzymes (results not shown) or significantly affect the average intracellular degradation rates after homologous microinjection. Furthermore, the microinjected proteins are found in the 3T3-L1 cell cytosol (at least 40–50%) in contrast to endocytosed poly-L-lysine conjugated cytosol proteins which are found (96%) in the intracellular lysosomal/vacuolar fraction (results not shown). This data on the distribution and degradation of homologously microinjected cytosol proteins again indicates (*cf.* heterologous transfer, *2*, *11*) that a mechanism independent of the intracellular vacuolar lysosomal apparatus may be involved in the destruction of cytosol proteins in 3T3-L1 cells.

REFERENCES

1. Ballard, F. J. *In* "Essays in Biochemistry," Vol. 13, eds. P. N. Campbell and W. N. Aldridge, p. 1 (1977). Academic Press, London.

2. Bigelow, S., Hough, R., and Rechsteiner, M. *Cell*, **25**, 83 (1981).
3. Bravo, R. and Celis, J. E. *Exp. Cell Res.*, **126**, 481 (1980).
4. Corps, A. N., Metcalfe, J. C., and Pozzan, T. *Biochem. J.*, **204**, 229 (1982).
5. Couchman, J. R. and Rees, D. A. *Eur. J. Cell Biol.*, **27**, 47 (1982).
6. De Robertis, E., Lenhard, S., and Pariot, R. S. *Nature*, **295**, 572 (1982).
7. Evans, P. J. and Mayer, R. J. *Biochem. Biophys. Res. Commun.*, **107**, 51 (1982).
8. Furusawa, M., Yamaizumi, M., Nishimura, T., Uchida, T., and Okada, Y. *In* "Methods in Cell Biology," Vol. 14, ed. D. M. Prescott, p. 73 (1976). Academic Press, London.
9. Goldberg, A. L. and Dice, J. F. *Annu. Rev. Biochem.*, **43**, 835 (1974).
10. Goldberg, A. L. and St. John, A. C. *Annu. Rev. Biochem.*, **45**, 747 (1976).
11. Hendil, K. B. *Exp. Cell Res.*, **135**, 157 (1981).
12. Kaltoft, K. and Celis, J. E. *Exp. Cell Res.*, **115**, 423 (1978).
13. Kaltoft, H. and Richter, D. *Biochemistry*, **21**, 741 (1982).
14. Kreis, T. E., Winterhalter, K. H., and Birchmeier, W. *Proc. Natl. Acad. Sci. U.S.*, **76**, 3814 (1979).
15. Loyter, A., Zakai, N., and Kulka, R. G. *J. Cell Biol.*, **66**, 292 (1975).
16. Neff, N. T., Bourret, L., Miao, P., and Dice, J. F. *J. Cell Biol.*, **91**, 184 (1981).
17. Pastan, I. H. and Willingham, M. C. *Science*, **214**, 504 (1981).
18. Russell, S. M., Burgess, R. J., and Mayer, R. J. *Biochem. J.*, **192**, 321 (1980).
19. Russell, S. M., Burgess, R. J., and Mayer, R. J. *Biochim. Biophys. Acta*, **714**, 34 (1982).
20. Schlegel, R. A. and Rechsteiner, M. C. *Cell*, **5**, 371 (1975).
21. Stacey, D. W. and Allfrey, V. G. *J. Cell Biol.*, **75**, 807 (1977).
22. Wagner, T. E., Hoppe, P. C., Jollick, J. D., Scholl, D. R., Hodinka, R. L., and Gault, J. B. *Proc. Natl. Acad. Sci. U.S.*, **78**, 6376 (1981).
23. Wilde, C. J., Saxton, J., and Mayer, R. J. *Biochim. Biophys. Acta*, **714**, 46 (1982).
24. Willingham, M. C., Maxfield, F. R., and Pastan, I. H. *J. Cell Biol.*, **82**, 614 (1979).
25. Wille, W. and Willecke, K. *Exp. Cell Res.*, **130**, 95 (1980).
26. Wojcieszyu, J. W., Schlegel, R. A., Wu, E., and Jacobson, K.A. *Proc. Natl. Acad. Sci. U.S.*, **78**, 4407 (1981).
27. Yamaizumi, M., Uchida, T., Takamatsu, K., and Okada, Y. *Proc. Natl. Acad. Sci. U.S.*, **79**, 461 (1982).
28. Zavortink, M. and Rechsteiner, M. *Cell*, **25**, 83 (1981).

Proteinase Inhibitors: Medical and Biological Aspects, eds. N. Katunuma et al., pp. 293–300 (1983).
Japan Sci. Soc. Press, Tokyo/Springer-Verlag, Berlin

Accumulation of Autolysosomes in Hepatocytes

Koji FURUNO, Toyoko ISHIKAWA, and Keitaro KATO

*Faculty of Pharmaceutical Sciences, Kyushu University**

Autophagy is a mechanism whereby cellular components are sequestered into vacuoles and transferred into the lysosomal compartment to be digested. It is well known that autophagic vacuoles appear prominently in cells where the rate of proteolysis is accelerated by amino acid deprivation or glucagon (*1*, *11*, *12*, *14*). Such phenomenon is interpreted that enhanced autophagy contributes to the increased degradation of intracellular proteins. In contrast, autophagic vacuoles accumulate when late steps in the autophagic/lysosomal pathway are blocked by mechanisms such as impairment of fusion between autophagosomes and lysosomes, or inhibition of lysosomal proteolitic activity. Marzella and Glaumann (*9*) have found that vinblastine causes an accumulation of autophagic vacuoles in rat hepatocytes, and have suggested that the drug stimulates lysosomal protein degradation *in vivo*. However, Kovács *et al.* (*7*) demonstrated that vinblastine inhibited protein degradation in isolated hepatocytes, and suggested that the accumulation of autophagic vacuoles could probably be explained by the action of the microtubule poison on autophagosome-lysosome interactions. We have recently provided evidence (*2*, *5*) for the accumulation of autolysosomes in rat

* Maedashi 3-1-1, Higashi-ku, Fukuoka 812, Japan.

hepatocytes exposed to leupeptin which effectively inhibits the lysosomal degradation of intracellular proteins (*6*, *8*, *13*). The hepatocytes treated with leupeptin displayed autolysosomes characterized by their regression phase, suggesting that digestion of segregated cell organelles in the autolysosomes proceeds very slowly (*5*). The autolysosomes were also found to show a remarkable increase in sedimentation coefficient and equilibrium density, thus making their isolation possible (*3*). Lipid analysis of the isolated autolysosomes revealed that poor lipid content was responsible for the enhanced autolysosomal density.

MORPHOLOGICAL ANALYSIS OF HEPATOCYTES

The characteristic features of hepatocytes stained for acid phosphatase activity in rats after the administration of leupeptin are shown in Fig. 1. Numerous autophagic vacuoles, autophagosomes and autolysosomes developed in the vicinity of bile canaliculi and also near the Golgi apparatus in hepatocytes within 30 min after the injection. It could be seen that these autophagic vacuoles contain various cell components, among which mitochondria, rough-surfaced endoplasmic reticulum and glycogen were easily recognizable (Fig. 2). At 1 hr, sequestered organelles in the autolysosomes were embedded in the dense matrix and were in a poor state of morphological preservation, as compared with organelles within the autophagosomes, thus indicating that degradation had to some extent already occurred. Development of the autolysosomes was accompanied by the reciprocal disappearance of preexisting secondary lysosomes, indicating the secondary lysosomes are a predominant source of lysosomal enzymes. From 1 to 8 hr, the autolysosomes varied to a great extent in both size and shape as a result of coalescence with each other. With the enlargement of the autolysosomes, the sequestered organelles were quite altered in structure and degraded into electron-lucent unidentifiable debris (Fig. 2b). Later, the dense cores which were cytochemically negative for acid phosphatase activity arose within the autolysosomes. They were budded off into the cytoplasm in the form of residual bodies and were occasionally discharged into the space of Disse (*5*). From 9 to 12 hr, small and dense lysosomes of irregular outline appeared, probably in consequence of condensation of autolysosomal contents. About 12 hr after leupeptin treatment, the

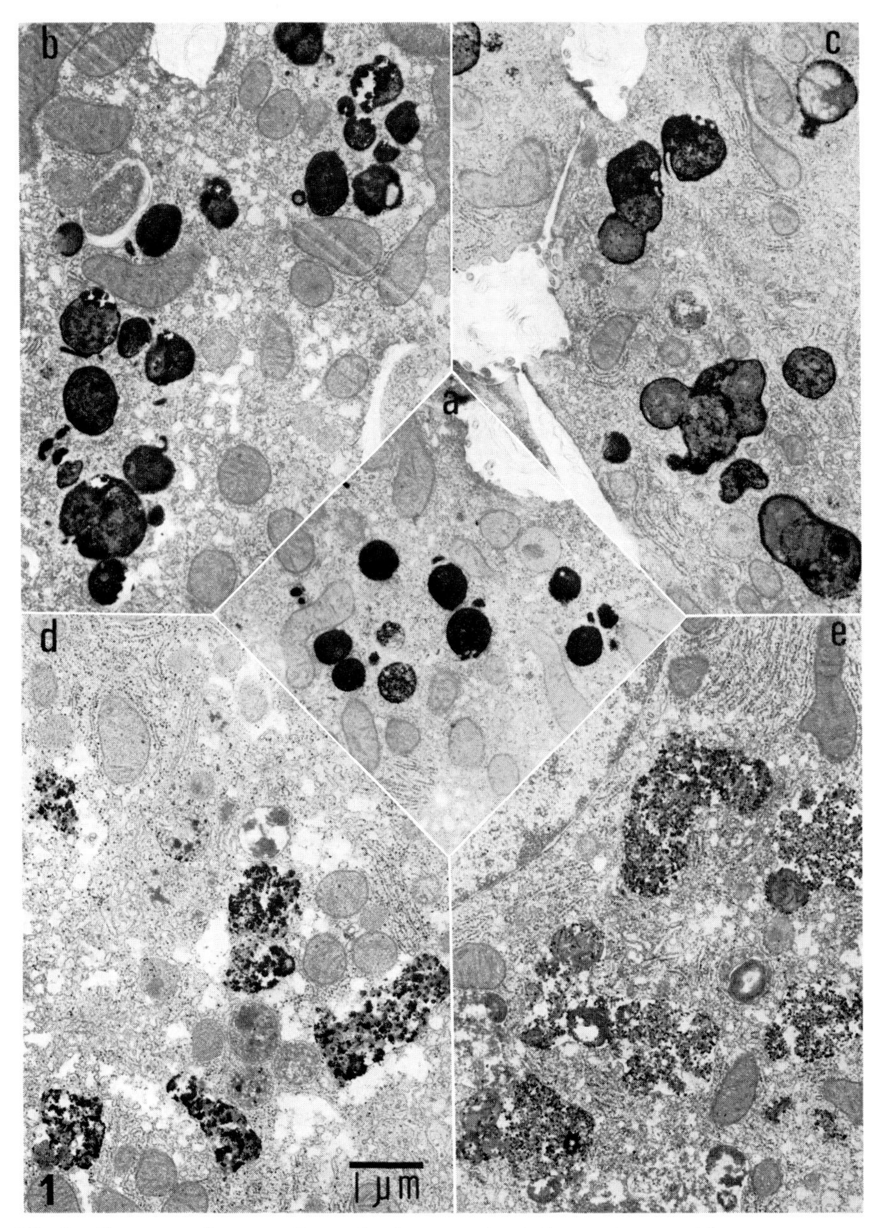

Fig. 1. Sequence of events of autophagic processing in leupeptin-treated hepatocytes. Liver sections from control (a) and leupeptin-treated rats killed at 30 min (b), 1 hr (c), 4 hr (d), and 8 hr (e) after injection were stained for acid phosphatase activity. Autolysosomes were numerous by 30 min after leupeptin treatment and subsequently coalesced to form large irregular-shaped clusters. Lead reaction products precipitated on the body and along the limiting membranes at an early stage, and appeared as small dots with a tendency to decrease in size at the later stage. ×12,000.

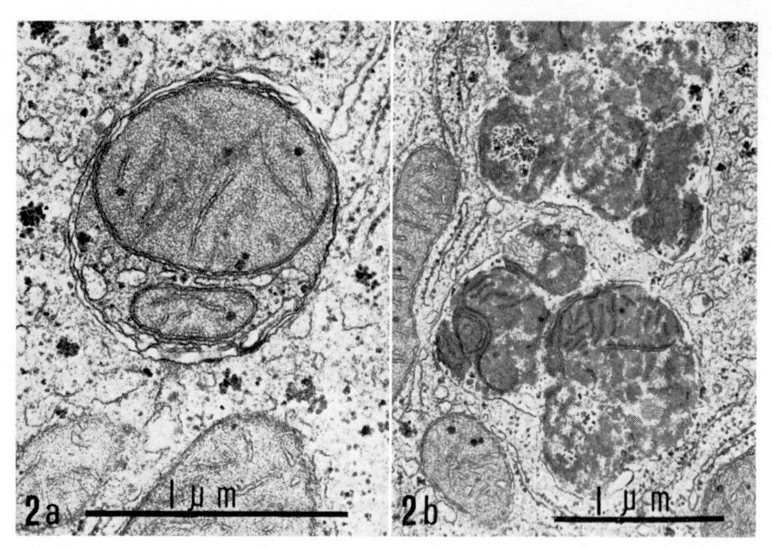

Fig. 2. Degradation of sequestered cellular organelles within autolysosomes.
a: mitochondria in newly formed autophagosomes are well preserved. ×48,000. b: mitochondria in clustered autolysosomes are barely recognizable by their intracrista. ×31,000.

autolysosomes decreased in number and secondary lysosomes of normal shape and size appeared reciprocally.

PHYSICAL PROPERTIES OF AUTOLYSOSOMES

The livers from normal and leupeptin-treated rats were homogenized and crude lysosomal fractions were obtained by sedimenting $600 \times g$ supernatant at $10{,}000 \times g$ for 20 min. The fractions were suspended in iso-osmotic Percoll at a density of 1.11 g/ml and subjected to centrifugation to give self-generating gradients. Figure 3 shows the distribution pattern of acid phosphatase and protein. Acid phosphatase in normal rat livers distributed at a median density of 1.09 g/ml. The main peak of protein with a density of 1.10 g/ml consisted predominantly of mitochondria as determined by cytochrome *c* oxidase activity. In leupeptin-treated rat livers, the lysosomal enzyme activity mostly migrated to the higher density of 1.14 g/ml. Denser particles were identified as autolysosomes by electron micrography (Fig. 5). The migration occurred within 30 min after leupeptin injection and was completed by 60 min. The density of autolysosomes did not increase beyond 1.14 g/ml and

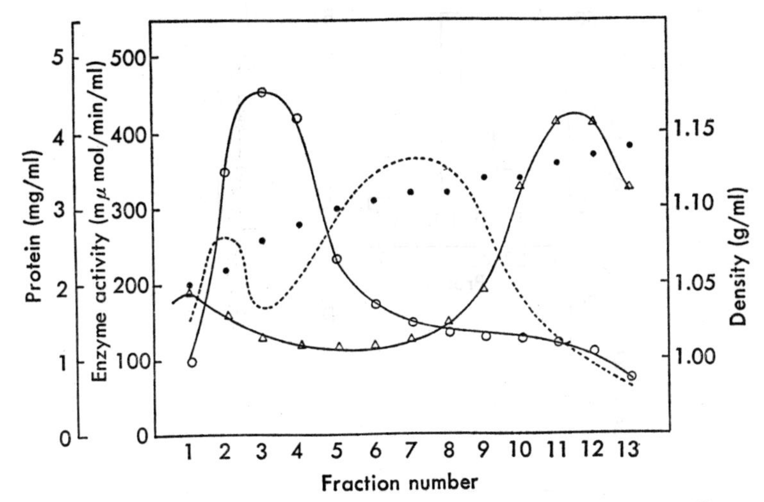

Fig. 3. Density distribution of acid phosphatase activity in iso-osmotic self-generating gradients after centrifugation of crude hepatic lysosomal fractions from normal (○) and leupeptin-treated (△) rats.
Rats were injected intraperitoneally with leupeptin at a dose of 20 mg/kg body weight and killed at 1 hr. Gradients were collected in 13 fractions of 2 ml each. ---- protein; ···· density.

began to return to normal about 12 hr after the injection. The buoyant density of the main peak of protein was not affected by leupeptin treatment at any time examined.

Figure 4 shows the distribution of cathepsin D in the liver subcellular fractions at various times from 30 min to 4 hr after leupeptin injection. The activity of cathepsin D sedimented predominantly with fraction III in untreated rat livers. After injection of leupeptin, the enzyme activity became sedimentable with particles of greater sedimentation rate. This sedimentation shift was maximal at around 4 hr after the injection and returned to normal after about 12 hr. The result probably represents the increase in autolysosomal size as evidenced by morphological examination. Parallel with this sedimentation shift of autolysosomes, an increase in distribution of cathepsin D in the supernatant fraction was observed. It is likely that leakage of lysosomal contents occurred during homogenization because of the fragility of autolysosomes.

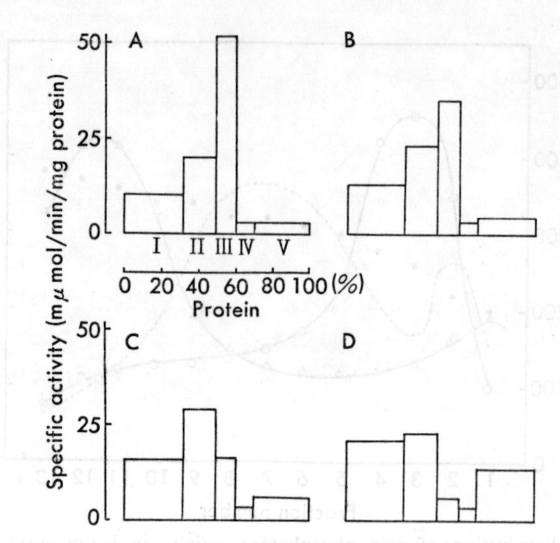

Fig. 4. Distribution of cathepsin D in subcellular fractions of rat livers following injection of leupeptin.
The livers were homogenized at various times after the injection. Centrifugation of the homogenates at $600 \times g$ for 5 min, $1{,}500 \times g$ for 10 min, $10{,}000 \times g$ for 20 min, and $105{,}000 \times g$ for 60 min yielded pellets designated fractions I, II, III, and IV, respectively. The final supernatant was designated fraction V. A: control. B: 0.5 hr. C: 1 hr. D: 4 hr.

CHARACTERIZATION OF ISOLATED AUTOLYSOSOMES

The autolysosomes were isolated from rat livers treated with leupeptin by a combination of differential and Percoll density gradient centrifugation techniques (*3*). Appearance of the purified autolysosome fraction is presented in Fig. 5. The enrichment of the lysosomal enzyme activities in this fraction over the homogenate was 12-, 14-, 22-, and 24-fold for β-glucuronidase, acid phosphatase, β-*N*-acetylglucosaminidase, and cathepsin D, respectively. The lower level of enrichment of the lysosomal enzyme activities in the autolysosome fraction compared to that in the tritosomes (*15*) is probably due to undigested protein materials accumulated in the autolysosomes. While endoplasmic reticulum and mitochondria were frequently found within the autolysosomes, the activities of the reference enzymes of these organelles were negligible in the purified autolysosome fraction. Table I shows the lipid/protein ratios of the autolysosomes, which showed much lower lipid content than other organelles. This finding suggests that lipids are degraded

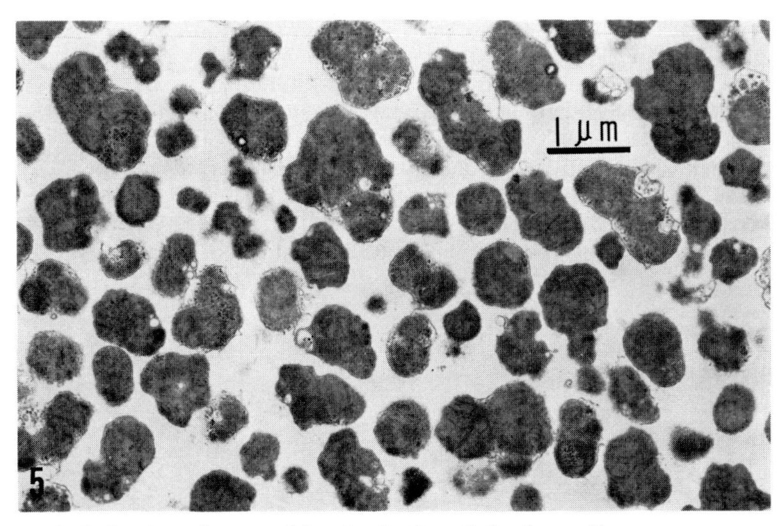

Fig. 5. The isolated autolysosomal fraction by Percoll density gradients. The fraction contains scarcely detectable contaminants. ×15,000.

TALBE I. Lipid Content in the Isolated Autolysosomes and Other Subcellular Organelles

	Phospholipid	Cholesterol
Autolysosomes	48	2.9
Microsomes	434	17.5
Mitochondria	179	3.6

Values represent lipid to protein ratios (μg/mg).

more rapidly than proteins in the autolysosomes in leupeptin-treated cells, and the resulting accumulation of undigested proteins render the density of the autolysosomes heavier. The comparison of the properties of our autolysosomes with those prepared by other investigators (*4*, *10*) will be described elsewhere.

REFERENCES

1. Deter, R. L. *J. Cell Biol.*, **48**, 473 (1971).
2. Furuno, K., Ishikawa, T., and Kato, K. *J. Biochem.*, **91**, 1485 (1981).
3. Furuno, K., Ishikawa, T., and Kato, K. *J. Biochem.*, **91**, 1943 (1981).
4. Gray, R. H., Sokol, M., Brabec, R. K., and Brabec, M. J. *Exp. Mol. Pathol.*, **134**, 72 (1981).
5. Ishikawa, T., Furuno, K., and Kato, K. *Exp. Cell Res.*, **144**, 15 (1983).
6. Knowles, S. E. and Ballard, F. T. *Biochem. J.*, **156**, 609 (1976).

7. Kovács, A. T., Reith, A., and Seglen, P. O. *Exp. Cell Res.*, **137**, 191 (1982).
8. Libby, P. and Goldberg, A. L. *Biochem. J.*, **188**, 213 (1980).
9. Marzella, L. and Glaumann, H. *Lab. Invest.*, **42**, 18 (1980).
10. Marzella, L., Ahlberg, J., and Glaumann, H. *J. Cell Biol.*, **93**, 144 (1982).
11. Schworer, C. M. and Mortimore, G. E. *Proc. Natl. Acad. Sci. U.S.*, **76**, 3169 (1979).
12. Schworer, C. M., Shiffer, K. A., and Mortimore, G. E. *J. Biol. Chem.*, **256**, 7652 (1981).
13. Seglen, P. O., Grinde, B., and Solkeim, A. E. *Eur. J. Biochem.*, **95**, 215 (1979).
14. Woodside, K. H., Ward, W. F., and Mortimore, G. E. *J. Biol. Chem.*, **249**, 5458 (1974).
15. Yamamoto, K., Ikehara, Y., Kawamoto, S., and Kato, K. *J. Biochem.*, **87**, 237 (1980).

Proteinase Inhibitors: Medical and Biological Aspects, eds. N. Katunuma et al., pp. 301–306 (1983).
Japan Sci. Soc. Press, Tokyo/Springer-Verlag, Berlin

Cotranslational Mode of Protein Translocation across Membranes

Matthias MÜLLER, Peter WALTER, Reid GILMORE, and Günter BLOBEL

*Laboratory of Cell Biology, The Rockefeller University**

The collective term topogenesis has been introduced (*3*) to encompass partial or complete protein translocation across membranes as well as subsequent posttranslocational protein traffic. It has been proposed (*3*) that the information for intracellular protein topogenesis resides in discrete topogenic sequences that constitute a permanent or transient part of the polypeptide chain. Four types of topogenic sequences were distinguished (*3*): 1) *Signal sequences* initiate translocation of proteins across specific membranes and are decoded by protein translocators. 2) *Stop-transfer sequences* interrupt the translocation process and yield asymmetric integration of proteins into translocation-competent membranes. 3) *Sorting sequences* act as determinants for posttranslocational traffic. 4) *Insertion sequences* interact with the lipid bilayer directly and thereby anchor a protein to the hydrophobic core of the lipid bilayer.

Translocation is understood here as transport of an entire polypeptide chain across membrane(s), proceeding unidirectionally from the protein biosynthetic compartment. Two modes of translocation have been distinguished: a cotranslational mode and a posttranslational mode only the former being discussed here in detail. Cellular mem-

* New York, New York 10021, U.S.A.

branes that have been shown to be competent for cotranslational translocation are the rough endoplasmic reticulum and the prokaryotic plasma membrane.

The present paper summarizes recent work on the isolation and characterization of two components required for the cotranslational translocation of proteins across and integration into the endoplasmic reticulum. They provide the strongest support to date for the postulate of the signal hypothesis (*4*) that protein translocation is a receptor-mediated process in which specificity is achieved by signal sequences in the proteins to be translocated and by signal-sequence-specific translocation systems of distinct cellular membranes.

SIGNAL RECOGNITION PARTICLE (SRP)

Only after the development of an *in vitro* translocation system (*5*) that was able to reproduce translocation across the rough endoplasmic reticulum (isolated in the form of closed microsomal vesicles) with apparent fidelity did it become possible to assay and to characterize translocation activities of the rough endoplasmic reticulum *in vitro*. One approach was dissecting the translocation activity by salt extraction of microsomes (*18*, *23*) which yielded membrane vesicles that were largely translocation-inactive; translocation activity, however, could be restored by readdition of the salt extract. This finding provided an assay for the purification of the active component which turned out to be an 11S ribonucleoprotein that consisted of six non-identical polypeptide chains (*Mr*=72,000, 68,000, 54,000, 19,000, 14,000, and 9,000 daltons) (*18*) and one molecule of 7S RNA (*20*). The RNA has been identified by partial sequence analysis (*20*) to be the previously described (*24*) and recently sequenced (*10*, *17*) small cytoplasmic 7SL RNA (7S RNA, ScL). Both RNA and protein are required for signal recognition particle (SRP) activity. In dog pancreas at physiological salt concentration (150 mM potassium ions) the bulk of the 11S ribonucleoprotein appears to be about equally distributed between a membrane-bound and a free or ribosome/polysome-associated form (P. Walter and G. Blobel, in preparation).

Studies on the role of the 11S ribonucleoprotein in the translocation process revealed that it is involved in the recognition of the signal sequence and therefore it was termed SRP. When SRP was present in the cell-free translation system in the absence of salt-extracted microsomal membranes, it was found to inhibit selectively the translation only of mRNA for secretory proteins but not of mRNA for cytosolic proteins (*21*). Moreover, SRP was found to bind with a relatively low affinity (apparent $K_d < 5 \times 10^{-5}$ M) to ribosomes but was shown to bind with a 6,000-fold higher affinity (apparent $K_d < 8 \times 10^{-9}$ M) when ribosomes were engaged in the translation of mRNA for secretory proteins (*21*). Most interestingly, this high-affinity binding of SRP caused a site-specific and signal sequence-induced arrest of chain elongation. The elongation-arrested peptide of nascent preprolaction was shown to be ~70 amino acid residues long (*19*). Because the signal sequence of nascent bovine preprolactin comprises 30 amino acid residues and because about 40 amino acid residues of the nascent chain are buried (protected from proteases) in the large ribosomal subunit (ref. in *19*) it was concluded that it is the signal sequence of the nascent chain (fully emerged on the outside of the large ribosomal subunit) that causes high-affinity binding of SRP, which in turn modulates translation and causes arrest in chain elongation.

SRP RECEPTOR

Most strikingly the SRP-induced elongation arrest is released upon binding of the elongation-arrested ribosome to salt-extracted microsomal membranes resulting in chain elongation and translocation into the microsomal vesicles (*6*, *13*, *19*). Using this arrest-releasing activity as an assay a protein of 72,000 daltons (*7*, *14*) has been purified from detergent-solubilized microsomal membranes by SRP-affinity chromatography (*7*). Based upon its affinity for SRP it is referred to as the SRP receptor. It is an integral membrane protein of the endoplasmic reticulum that consists of a large cytoplasmic domain of 60,000 daltons (*12*) that can be severed from the membrane in an intact form by treatment with a variety of proteases and can be added back to the proteolyzed membranes to reconstitute activity (*6*, *11*, *22*).

MECHANISM OF COTRANSLATIONAL TRANSLOCATION

The drawing in Fig. 1 represents a model (taken from ref. *19*) illustrating schematically both facts and speculations about translocation of secretory (*16*, *19*, *21*) and lysosomal (A. Erickson *et al.*, in preparation) proteins across and integration of membrane proteins (*1*) into the rough endoplasmic reticulum. It was proposed (*19*) that an equilibrium exists between a free, soluble form of SRP, SRP bound to ribosomes, and SRP bound to the SRP receptor (Fig. 1A and B). Upon translation of a mRNA coding for a signal sequence (Fig. 1C), there is an enhancement of the apparent affinity of SRP to the translating ribosomes by several orders of magnitude. Concomitantly, and presumably through the ribosome, SRP arrests the elongation of the initiated polypeptide chain, preventing its completion in the cytoplasm (Fig. 1D). Translation arrest is released only upon interaction of the SRP-arrested ribosome with the SRP receptor (Fig. 1E).

We have estimated (*7*, *21*) that one equivalent of dog pancreas microsomal membranes contains approximately 500 fmol of bound ribosomes, approximately 20 fmol of SRP, and about 100 fmol of SRP receptor. Thus, the content of both SRP and SRP receptor is less than that of bound ribosomes. This suggests (*19*) that the ribosome-SRP-SRP receptor interaction might be a transient one, merely targeting the SRP-arrested ribosome to a specific membrane site where per-

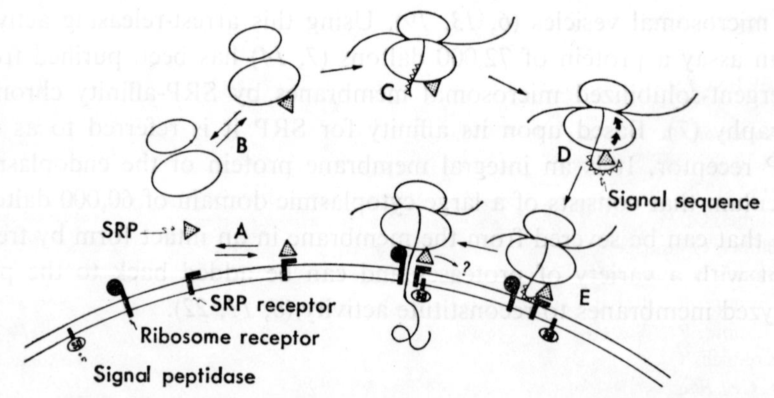

Fig. 1. Model for the function of SRP and SRP receptor in the translocation process.

sistent binding might occur *via* other integral membrane proteins. These could be represented by ribophorins I and II each of which has been reported to be present in microsomal membranes in stoichiometric amounts to those of membrane bound ribosomes (*8*, *9*). It should be noted, however, that Bielinska *et al.* (*2*) have argued against the involvement of ribophorins in chain translocation and that these proposals suggesting an involvement of several integral membrane proteins in the formation of a productive ribosome-membrane junction are at this moment purely speculative.

Remarkably, SRP-dependent translocation across the endoplasmic reticulum membrane as well as SRP-induced translation arrest was recently found to occur also with a bacterial secretory protein (*15*) suggesting a considerable degree of conservation of the mode of cotranslational translocation during evolution.

The ability of SRP to arrest chain elongation and the finding that microsomal membranes release this arrest are of teleologic interest. If this mechanism also operates *in vivo*, it would provide the cell with a means to stop the synthesis of proteins that might be harmful if completed in the cytoplasm (*e.g.*, lysosomal hydrolases), unless sites on the rough endoplasmic reticulum are available so that translocation and segregation into the intracisternal space are ensured. Furthermore, modulation of the arrest releasing activity either by other not yet identified components or by direct modification of SRP and/or the SRP receptor may provide the cell with an on/off switch for translocation-coupled protein synthesis and thereby provide a mechanism for a fast and regulatable response to a variety of physiological stimuli.

REFERENCES

1. Anderson, D. J., Walter, P., and Blobel, G. *J. Cell Biol.*, **93**, 501 (1982).
2. Bielinska, M., Rogers, G., Rucinsky, T., and Boime, I. *Proc. Natl. Acad. Sci. U.S.*, **76**, 6152 (1979).
3. Blobel, G. *Proc. Natl. Acad. Sci. U.S.*, **77**, 1496 (1980).
4. Blobel, G. and Dobberstein, B. *J. Cell Biol.*, **67**, 835 (1975).
5. Blobel, G. and Dobberstein, B. *J. Cell Biol.*, **67**, 852 (1975).
6. Gilmore, R., Blobel, G., and Walter, P. *J. Cell Biol.*, **95**, 463 (1982).
7. Gilmore, R., Walter, P., and Blobel, G. *J. Cell Biol.*, **95**, 470 (1982).
8. Kreibich, G., Freienstein, C. M., Pereyra, B. N., Ulrich, B. L., and Sabatini, D. D. *J. Cell Biol.*, **77**, 488 (1978).
9. Kreibich, G., Ulrich, B. L., and Sabatini, D. D. *J. Cell Biol.*, **77**, 464 (1978).

10. Li, W. Y., Reddy, R., Henning, D., Epstein, P., and Busch, H. *J. Biol. Chem.*, **257**, 5136 (1982).
11. Meyer, D. I. and Dobberstein, B. *J. Cell Biol.*, **87**, 498 (1980).
12. Meyer, D. I. and Dobberstein, B. *J. Cell Biol.*, **87**, 503 (1980).
13. Meyer, D. I., Krause, E., and Dobberstein, B. *Nature*, **297**, 647 (1982).
14. Meyer, D. I., Louvard, D., and Dobberstein, B. *J. Cell Biol.*, **92**, 579 (1982).
15. Müller, M., Ibrahimi, I., Chang, C. N., Walter, P., and Blobel, G. *J. Biol. Chem.*, **257**, 11860 (1982).
16. Stoffel, W., Blobel, G., and Walter, P. *Eur. J. Biochem.*, **120**, 519 (1981).
17. Ullu, E., Murphy, S., and Melli, M. *Cell*, **29**, 195 (1982).
18. Walter, P. and Blobel, G. *Proc. Natl. Acad. Sci. U.S.*, **77**, 7112 (1980).
19. Walter, P. and Blobel, G. *J. Cell Biol.*, **91**, 557 (1981).
20. Walter, P. and Blobel, G. *Nature*, **299**, 691 (1982).
21. Walter, P., Ibrahimi, I., and Blobel, G. *J. Cell Biol.*, **91**, 545 (1981).
22. Walter, P., Jackson, R. C., Marcus, M. M., Lingappa, V. R., and Blobel, G. *Proc. Natl. Acad. Sci. U.S.*, **76**, 1795 (1979).
23. Warren, G. and Dobberstein, B. *Nature*, **273**, 569 (1978).
24. Zieve, G. and Penman, S. *Cell*, **8**, 19 (1976).

Proteinase Inhibitors: Medical and Biological Aspects, eds. N. Katunuma *et al.*, pp. 307–312 (1983).
Japan Sci. Soc. Press, Tokyo/Springer-Verlag, Berlin

Proteolytic Processing of Enzyme Precursors by Liver and Adrenal Cortex Mitochondria

Tsuneo OMURA, Akio ITO, Yoshiie OKADA, Yasuhiro SAGARA, and Hideyu ONO

*Department of Biology, Faculty of Science, Kyushu University**

The biogenesis of mitochondria is dependent on both cytoplasmic and mitochondrial protein-synthesizing systems (*14*). However, except for a small number of inner membrane proteins, which are synthesized inside of the organelle, the majority of mitochondrial proteins are apparently synthesized by cytoplasmic ribosomes, mostly as larger precursor peptides, to be imported post-translationally into the organelle (*9*, *15*). The import of the precursor peptides into mitochondria is accompanied by their proteolytic processing into mature size peptides, which was first shown by Maccecchini *et al.* (*3*) for the subunits of F_1-ATPase. The mechanism of import of precursor peptides into mitochondria has been the subject of recent intensive studies.

We have been studying the biosynthesis and import of several mitochondrial enzymes located in different submitochondrial compartments for some years. They include monoamine oxidase in the outer membrane (*13*), sulfite oxidase in the inter-membrane space (*10–12*), cytochrome P-450(11β) in the inner membrane (*7*), and adrenodoxin in the matrix space (*8*). Monoamine oxidase was apparently synthesized as a mature size peptide in an RNA-dependent *in vitro* translation

* Hakozaki 6-10-1, Higashi-ku, Fukuoka 812, Japan.

system (*13*), but all the other enzymes were synthesized as precursor peptides, which were significantly larger than the peptides of the corresponding mature enzyme proteins (*7, 8, 11, 12*). In this report, we present our recent observations on the proteolytic processing of those precursor peptides by liver and adrenal cortex mitochondria.

BIOSYNTHESIS AND IMPORT OF SULFITE OXIDASE, CYTOCHROME P-450 (11β), AND ADRENODOXIN

Sulfite oxidase was synthesized as a larger precursor when rat liver RNA was translated in a reticulocyte lysate system (*12*) (Table I). Similarly, the *in vitro* translation of bovine adrenal cortex RNA gave larger precursor peptides of cytochrome P-450(11β) and adrenodoxin (*7, 8*) (Table I). On the other hand, monoamine oxidase of outer mitochondrial membrane was synthesized *in vitro* apparently as a mature size peptide (*13*) (Table I). Another outer membrane protein, porin, has also recently been shown to be synthesized as a mature size peptide (*2, 5*). The existence of an extra peptide, which is to be cleaved off at the step of the incorporation of precursor peptides into mitochondria, seems to be a common property of the mitochondrial proteins located in the compartments inside of the outer membrane.

When the *in vitro* translation products were incubated with mitochondria *in vitro*, the precursor peptides of sulfite oxidase, cytochrome P-450(11β) and adrenodoxin, were efficiently incorporated into the inside of the organelle concomitant with their conversion into mature sizes, and the processed peptides could be recovered from correct sub-mitochondrial compartments. Judging from the inhibitory actions of uncouplers, energy-transfer inhibitors, and ionophores, the import process was dependent on the energization of the inner membrane. The import

TABLE I. Precursor Peptides of Mitochondrial Enzymes

Enzyme	Location in mitochondria	Molecular weight	
		Mature form	Precursor
Monoamine oxidase	Outer membrane	59,000	(59,000)
Sulfite oxidase	Inter-membrane space	57,000	60,000
Cytochrome P-450 (11β)	Inner membrane	43,000	48,000
Adrenodoxin	Matrix	12,000	22,000

and processing of the precursor peptides did not show strict species specificity nor tissue specificity. Pre-adrenodoxin translated from bovine adrenal cortex messenger RNA (mRNA) was efficiently imported and processed by rat liver mitochondria.

PROTEOLYTIC PROCESSING OF PRE-SULFITE OXIDASE BY INNER MITOCHONDRIAL MEMBRANE

The import of pre-sulfite oxidase into the inter-membrane space of mitochondria across the outer membrane seems to require the participation of the inner membrane (*12*), and the processing of the precursor to a mature size peptide was effected by incubation with isolated mitoplast as well as with intact mitochondria. None of the outer membrane, inter-membrane space fraction, or matrix fraction could catalyze the conversion of pre-sulfite oxidase to the mature size (Fig. 1). The processing activity was tightly associated with the mitoplast and sensitive to treatment with detergents. Since the processing of pre-sulfite oxidase by mitoplast was energy-dependent, however, the inhibitory action of the detergents on the processing could be due to de-energization of the inner membrane and not to inactivation of the processing protease, which is possibly present on the outer surface of the mitoplast.

The processing of pre-sulfite oxidase by mitochondria or mitoplast

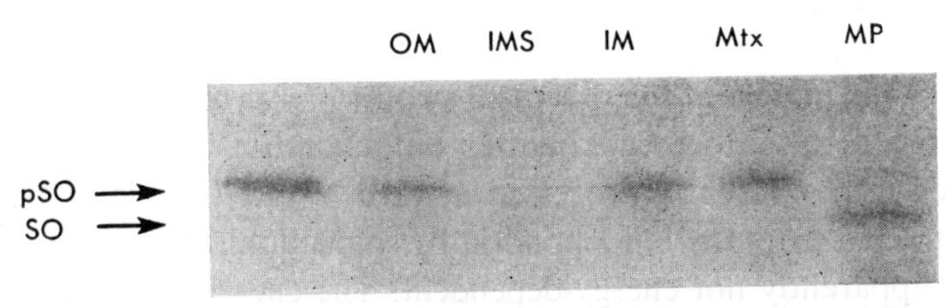

Fig. 1. Sub-mitochondrial distribution of pre-sulfite oxidase-processing activity. *In vitro* translation products of rat liver RNA were incubated for 60 min at 30°C with 100 μg each of the subfractions of rat liver mitochondria as follows. Leftmost lane, no addition; OM, outer membrane fraction; IMS, inter-membrane space fraction; IM, inner membrane; Mtx, matrix fraction; MP, mitoplast. After incubation, the precursor and mature forms of sulfite oxidase were recovered from the incubation mixtures by immunoprecipitation, and the immunoprecipitates were analyzed by SDS-polyacrylamide gel electrophoresis and fluorography. pSO and SO in the figure denote the positions of pre-sulfite oxidase and mature sulfite oxidase, respectively.

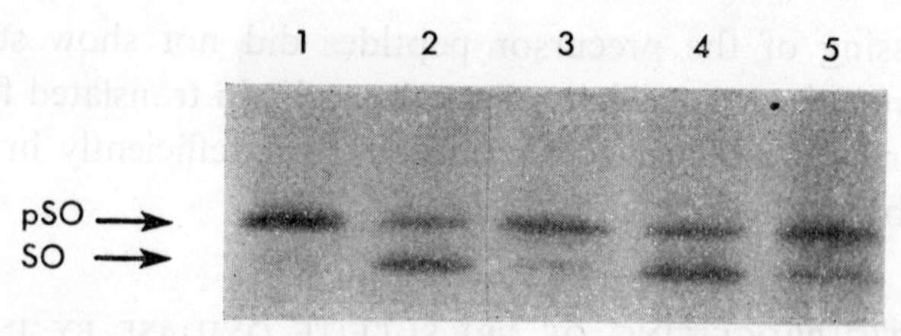

Fig. 2. Effects of metal chelators on the processing of pre-sulfite oxidase by mitoplast. *In vitro* translation products of rat liver RNA were incubated with the mitoplast from rat liver mitochondria in the presence of various metal chelators. 1, no addition; 2, incubated with mitoplast; 3, 2+1 mM GTP; 4, 2+1 mM EGTA; 5, 2+1 mM EDTA. After the incubation, processing of pre-sulfite oxidase was examined as described in the legend to Fig. 1.

was strongly inhibited by some metal chelators including guanosine triphosphate (GTP) and EDTA (Fig. 2). It was not inhibited by phenyl methylsulfonyl fluoride (PMSF), aprotinin, and various microbial proteinase inhibitors including pepstatin, leupeptin, antipain, and chymostatin. The processing enzyme seems to be a metal proteinase, and different from known lysosomal proteinases. The optimum pH of the processing activity was 7.5–8.0. However, solubilization of the processing activity has so far been un successful, and the nature of the membrane-bound enzyme responsible for the processing of pre-sulfite oxidase is still to be elucidated.

PROTEOLYTIC PROCESSING OF CYTOCHROME P-450(11β) AND ADRENODOXIN BY SOLUBLE MATRIX FRACTION

Although the import of the precursor peptides of cytochrome P-450(11β) and adrenodoxin into adrenal cortex mitochondria was energy-dependent, they were efficiently processed *in vitro* by the soluble matrix fraction extracted from the mitochondria by sonication. The processing step itself is apparently not energy-dependent. The energization of the inner membrane is possibly required in the penetration of the precursor peptides across the membrane. In accordance with the observation that rat liver mitochondria can import pre-adrenodoxin, the soluble matrix fraction from the liver mitochondria was also able to process the precursor peptide to the mature size.

The processing of pre-adrenodoxin by the matrix fraction was strongly inhibited by metal chelators, and not inhibited by PMSF or

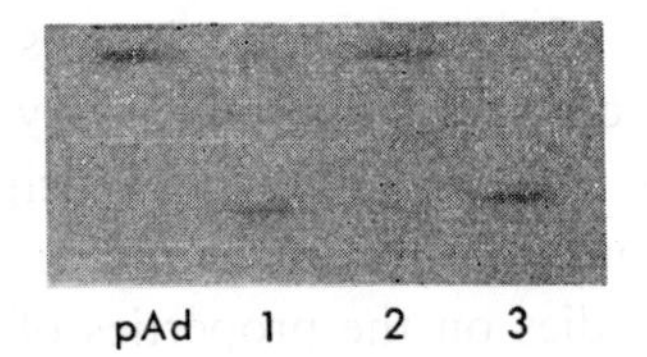

Fig. 3. Inhibition of pre-adrenodoxin-processing activity of mitochondrial matrix by EDTA and its reversion by Mn^{2+}.
In vitro translation products of bovine adrenal cortex RNA were incubated with the matrix fraction from bovine adrenal cortex mitochondria in the presence of EDTA and/or $MnCl_2$. pAd, no addition; 1, incubated with matrix fraction; 2, 1+10 μM EDTA; 3, 1+10 μM EDTA+100 μM $MnCl_2$. After the incubation, processing of pre-adrenodoxin was examined as described in the legend to Fig. 1. pAd denotes pre-adrenodoxin.

various microbial proteinase inhibitors. The metallo-protein nature of the processing enzyme was clearly demonstrated by the reactivation of the processing activity of the EDTA-treated matrix fraction by the addition of certain heavy metals (Fig. 3). Mn^{2+} was most effective in the reactivation. Judging from the filtration of the matrix fraction through a column of Sephadex G-150, the molecular weight of the processing enzyme was about 66,000 daltons.

In contrast with the processing of pre-adrenodoxin, the conversion of the precursor of cytochrome P-450(11β) to the mature form catalyzed by the matrix fraction was not inhibited by metal chelators including EDTA, GTP, and Zincon. It was strongly inhibited by *N*-ethylmaleimide or monoiodoacetamide. It is not certain, however, whether the latter reagents acted on the processing enzyme itself or reacted with the precursor peptides. The processing of pre-adrenodoxin by the matrix fraction was also sensitive to *N*-ethylmaleimide.

PRESENCE OF MULTIPLE PROCESSING ENZYMES IN MITOCHONDRIA

Our observations indicated the presence of multiple processing proteases in mitochondria. The protease catalyzing the processing of pre-sulfite oxidase was bound to the inner membrane, whereas the processings of both pre-cytochrome P-450(11β) and pre-adrenodoxin were effected by the soluble matrix fraction. Moreover, the processings of the latter two precursor peptides were differently affected by some inhibitors: the processing of pre-cytochrome P-450(11β) was not sensitive to metal chelators, whereas that of pre-adrenodoxin was strongly inhibited by metal chelators and reactivated by certain heavy metals. Similar metal proteases have recently been reported to participate in the processing of the precursors of ornithine transcarbamylase (*1*, *6*) and F_1-ATPase subunits (*4*). Further studies on the properties of those processing proteases in mitochondria will provide us with valuable information on the mechanism of import of various mitochondrial enzyme precursors from the cytoplasm into the correct mitochondrial compartments.

REFERENCES

1. Conboy, J. G., Fenton, W. A., and Rosenberg, L. E. *Biochem. Biophys. Res. Commun.*, **105**, 1 (1982).
2. Freitag, H., Janes, M., and Neupert, W. *Eur. J. Biochem.*, **126**, 197 (1982).
3. Maccecchini, M. L., Rudin, Y., Blobel, G., and Schatz, G. *Proc. Natl. Acad. Sci. U.S.*, **76**, 343 (1979).
4. McAda, P. C. and Douglas, M. G. *J. Biol. Chem.*, **257**, 3177 (1982).
5. Mihara, K., Blobel, G., and Sato, R. *Proc. Natl. Acad. Sci. U.S.*, **79**, 7102 (1982).
6. Miura, S., Mori, M., Amaya, Y., and Tatibana, M. *Eur. J. Biochem.*, **122**, 641 (1982).
7. Nabi, N., Kominami, S., Takemori, S., and Omura, T. *Biochem. Biophys. Res. Commun.*, **97**, 687 (1980).
8. Nabi, N. and Omura, T. *Biochem. Biophys. Res. Commun.*, **97**, 680 (1980).
9. Neupert, W. and Schatz, G. *Trends Biochem. Sci.*, **6**, 1 (1981).
10. Ono, H., Ito, A., and Omura, T. *J. Biochem.*, **91**, 107 (1982).
11. Ono, H. and Ito, A. *J. Biochem.*, **91**, 117 (1982)
12. Ono, H. and Ito, A. *Biochem. Biophys. Res. Commun.*, **107**, 258 (1982).
13. Sagara, T. and Ito, A. *Biochem. Biophys. Res. Commun.*, **109**, 1102 (1982).
14. Schatz, G. and Mason, T. L. *Annu. Rev. Biochem.*, **43**, 51 (1974).
15. Schatz, G. *FEBS Lett.*, **103**, 203 (1979).

Subject Index

Acid phosphatase 97
α-Actinin 69
Active site cleft, subsites 203
Actomyosin degradation 198
Additivity of residue contributions 65, 66
Adrenodoxin 308, 310
Aldolase 148
Alkaline proteinases, myometrium 193
Amino acid
  compositions 131, 215
  sequence of inhibitor 92
Aminopeptidases 7, 9–11, 13, 267
Amino terminal sequences, cathepsins B and L 127
Anti-lipopolysaccharide (LPS) factor 229
β_1-Anticollagenase 225
Antithrombin III 90, 115
Antitumor 10
Arphamenines 9, 10
Aspartic proteinases
  inhibitors 203
  pepstatin 5
  sources 201
  peptide substrates 207
Association equilibrium constant, K_{assoc} 56, 61–66
Autolysosomes 294
Autophagy 293
  vacuoles 293

Bestatin 10, 11, 13
Bound leupeptin 20
Bowman-Birk protease inhibitor 46

Ca-ionophore 70
Calcium activated neutral protease (CANP) 167, 173
  m- 174
  μ- 176
Calmodulin 171
Calpain 165, 173
  I and II 167
Calpastatin 165
Cancer 121, 275
Captopril-effects on Cathepsin
  B 271
  B2 271

I 271
L 271
Cardiomyopathic hamsters 32
Cathepsin 25, 74, 125, 135
B inhibitor 125, 135
B, H, and L 125, 135
C 263
D in myometrium 196
H, bovine spleen 126
M 148
Cell culture 97
Cerebrospinal fluid 119
Chymotrypsin 62–65
Circular dichroism spectra 132
Coagulogen 229
Collagen 211
inhibitor 214
-Sepharose 212
Comparison with serum inhibitors 97
Complement 37
Covalent modification 184
Cultured muscle cells 97
Cysteine proteinases
homology 125
inhibitors, leupeptin, E-64, antipain 3, 25, 97
cathepsin B, H, L 125, 135
inhibitors 125, 135
Cysteinylglycine, effects on
BANA hydrolizing protease 263
cathepsins B and L 271
Cystratin 129
Cytochrome P-450 (11β) 308, 310

Delayed-type hypersensitivity 9, 10
Dipeptidylaminopeptidase 264
Disulfide rear-rangement 219
Double headed inhibitor 46
Dystrophy
hereditary, chicken 72
patients 19

E-64 27, 70
and its analogs 25, 70
Ebelactones 7
Eglin 93
Elastase 62, 87
-α_1-proteinase inhibitor complex 88
Emphysema 121
Endogenous inhibitor 157, 177
Endopeptidase 267
Endotoxemia 92
Enzyme-inhibitor complex 135, 218, 225

Factor
G 229
XIII 90
Fluorescence microscopy 288
Fructose-1,6-bisphosphatase 148, 185

Gelatinase 212, 225
inhibitor 225
β-D-Glucan 229
Glucose oxidase 222
Glutathione
cycle 223
oxidized 217
reduced 217
Glycoproteins 215
Granulocyte 277

Haptoglobin 114
Hepatocytes 284
Hepatomas 284
HMP shunt activity 223
Homology 128
Horseshoe crab, *Tachypleus tridentatus* & *Limulus polyphemus* 229
Human endogenous protein inhibitor 129

Immune responses 7, 9, 13
Immunological disease 37
Inflammation 277
Inhibition mechanism, peanut inhibitor 51
Inhibition of lung metastases 132
Inhibitor 3–9, 12–14

renin 256
therapy 92
yeast 181
Inhibitory specificity 59
Initiation of proteolysis 184
Interferon 279
Interleukins 10
Intestinal absorption 18
Intracellular proteases 97

Latent collagenase 212
Legume Protease Inhibitors 45
Leupeptin 3, 4, 6, 11–14, 178, 257, 294
derivatives 81
Leupeptinic acid 81
Leupeptinol 81
Lipopolysaccharide 230
Limited proteolysis 147
Limulus
coagulation system 229
test 229
Lysosome 91, 153
enzymes 85
membrane inhibitor 160
Lysosomotrophic amines 97

α_2-Macroglobulin 115, 243
Mechanism of inhibition 94
Medullasin 275
Membrane bound cathepsin 151
Metabolism, Ep475 29
Metallo enzyme 223
Metastasis 77
Microinjection 284
Mitochondria 286
Modulation of inhibitor activity 139
Monoamine oxidase 308
Muscle cell inhibitor 106
Muscle protease 97
Muscle wasting diseases 32
Muscular dystrophy 11, 13, 25, 100, 121
Duchenne muscular dystrophy 69
Mutagenesis 78, 80
Myoblast fusion 97, 108
Myocardial infraction 34

Natural killer (NK) cell 275
Neutrophils 85
New lysosomal proteinase 148

Ovomucoid third domain 56–58, 60, 62, 63, 67

Peanut inhibitor 46
Pepsin inhibitor peptide
NH_2-terminal peptide 204
synthetic analogues 205
Pepstatins, isovaleryl 204
Phagocytosis 222
Pregnancy dependent protease 198
Processing enzymes, mitochondria 311
Proclotting enzyme 229
Prorenin, activation 254
Protease 3, 6, 13
cysteinyl 271
lysosomal 263
plasma 39
α_1-Proteinase inhibitor 195, 243
Proteolytic processing, enzyme precursors 307

Radioimmunoassay (RIA) 20
Reactive site 55, 56, 63
Redox potential 222
Regulation, cathepsin C
BANA hydrolyzing protease 263
cathepsin C 263
Renin
active site, amino acid sequence 252
-inhibitor complex, reversible activation of 258
zymogen 254
Residual bodies 294
Rheumatoid arthritis 226
Ribonucleoprotein 302
Rough endoplasmic reticulum 302

Saccharomyces cerevisiae 181
Second inhibitor 20
Secretory protein 305
Septicemia 88
Sequence to reactivity algorithm 56, 59
Serine proteinases 55, 59
  inhibitors 55
Signal recognition particle (SRP) 302, 303
Single amino acid replacement 59, 60
Sister chromatid exchange 78
Soybean trypsin inhibitor (Kunitz) 45
Streptomyces griseus proteinase
  A, SGPA 62–65
  B, SGPB 57, 62–65
Substrates
  cathepsins band I 26
  -unspecific proteolysis 87
Subtilisin 62–64
Sulfite oxidase 308, 309
Sulfhydryl group 218, 225

Synovial fluid 119
Synthetic inhibitors 37

Thiol-disulfide interchange 223, 225
Thiol protease inhibitor 25, 91
  α_2- 115
Thiol-Sepharose 214
Tissue specific inhibitor 194
Toxicity of E-64 29
Translocation
  contranslational 302
  proteins 301
Transplantation 286
  arrested 304
Trypsin 37
Tumor promoter 77

Uterine myometrium 191

Vertebrate collagenase 211

Author Index

Adachi, T. 25
Afting, E.-G. 191
Andus, T. 243
Aoki, Y. 275
Aoyagi, T. 3
Ardelt, W. 55

Bird, J.W.C. 97
Blobel, G. 301
Brzin, J. 125

Colella, R. 97
Chatterjee, R. 263

Doherty, F. 283
Dunn, B.M. 201
Duswald, K.-H. 85

Empie, M.W. 55
Evans, P.J. 283

Fekete, E. 97
Fritz, H. 85
Fujii, S. 37
Furuno, K. 293

Gilmore, R. 301
Giraldi, T. 125
Gross, V. 243

Hanada, K. 25
Heinrich, P.C. 243
Hirado, M. 37
Hiranaga, M. 229
Hitomi, Y. 37
Holzer, H. 181
Horecker, B.L. 147

Ikari, N. 37
Ikenaka, T. 45
Imahori, K. 173
Inagaki, T. 251
Inagami, T. 251
Ishikawa, T. 293
Ishiura, S. 69
Ito, A. 307
Iwanaga, S. 229

Jochum, M. 85

Kalnitsky, G. 263
Kashiwagi, K. 25

Kato, I. 55
Kato, K. 293
Katunuma, N. 25, 135
Kawashima, S. 173
Kay, J. 201
Kirschke, H. 97
Kominami, E. 25, 135
Kopitar, M. 125
Kregar, I. 125

Laskowski, M. Jr. 55
Lenney, J.F. 113
Li, Q.-S. 97
Ločnikar, P. 125
Lones, M. 263
Longer, M. 125

Macartney, H.W. 211
Machleidt, W. 125
Mayer, R.J. 283
Melloni, E. 147
Misono, K.S. 251
Miyata, T. 229
Morita, T. 229
Müller, M. 301
Murachi, T. 165

Nakamura, T. 229
Neumann, S. 85
Niinobe, M. 37
Nonaka, I. 69
Norioka, S. 45
Northemann, W. 243

Oguma, K. 25
Ohmura, S. 25
Ohtsubo, S. 229
Okada, Y. 307
Omura, T. 307
Ono, H. 307

Park, S.J. 55
aszkowski, A. 263

Pontremoli, S. 147
Popović, T. 125

Ritonja, A. 125
Roisen, F.J. 97
Russell, S.M. 283

Sagara, Y. 307
Sakai, Y. 37
Sava, G. 125
Seemüller, U. 85
Singh, H. 263
St. John, A.C. 97
Sugita, H. 69
Suzuki, K. 173

Takano, E. 165
Takii, Y. 251
Takio, K. 135
Tamai, M. 25
Tanaka, K. 165
Tanaka, S. 229
Tanaka, W. 17
Tashiro, M. 55
Titani, K. 135
Towatari, T. 25
Tran-Thi, T.-A. 243
Tschesche, H. 211
Turk, V. 125

Umezawa, K. 77
Umezawa, H. 3

Valler, M.J. 201
Vitale, Lj. 125

Wakamatsu, N. 135
Walter, P. 301
Wieczorek, M. 55
Witte, J. 85
Wood, L. 97

Yorke, G. 97